METHODS IN CELL BIOLOGY

VOLUME XVII

Chromatin and Chromosomal Protein Research. II

Methods in Cell Biology

Series Editor: **DAVID M. PRESCOTT**

DEPARTMENT OF MOLECULAR, CELLULAR
AND DEVELOPMENTAL BIOLOGY
UNIVERSITY OF COLORADO
BOULDER, COLORADO

Methods in Cell Biology

VOLUME XVII

Chromatin and Chromosomal
Protein Research. II

Edited by

GARY STEIN and JANET STEIN

DEPARTMENT OF BIOCHEMISTRY AND
MOLECULAR BIOLOGY
UNIVERSITY OF FLORIDA
GAINESVILLE, FLORIDA

LEWIS J. KLEINSMITH

DIVISION OF BIOLOGICAL SCIENCES
UNIVERSITY OF MICHIGAN
ANN ARBOR, MICHIGAN

1978

ACADEMIC PRESS • New York   San Francisco   London
A Subsidiary of Harcourt Brace Jovanovich, Publishers

ACADEMIC PRESS, INC.
111 Fifth Avenue, New York, New York 10003

United Kingdom Edition published by
ACADEMIC PRESS, INC. (LONDON) LTD.
24/28 Oval Road, London NW1

LIBRARY OF CONGRESS CATALOG CARD NUMBER: 64–14220

ISBN 0–12–564117–6

PRINTED IN THE UNITED STATES OF AMERICA

CONTENTS

23. Methods for Assessing the Binding of Steroid Hormones in Nuclei

G. C. Chamness, D. T. Zava, and W. L. McGuire

24. Methods for Assessing the Binding of Steroid Hormones in Nuclei and Chromatin

J. H. Clark, J. N. Anderson, A. J. W. Hsueh, H. Eriksson, J. W. Hardin, and E. J. Peck, Jr.

25. Proteins of Nuclear Ribonucleoprotein Subcomplexes

Peter B. Billings and Terence E. Martin

LIST OF CONTRIBUTORS

Numbers in parentheses indicate the pages on which the authors' contributions begin.

VINCENT G. ALLFREY, The Rockefeller University, New York, New York (253)

J. N. ANDERSON, Department of Cell Biology, Baylor College of Medicine, Houston, Texas (335)

PETER B. BILLINGS, Department of Biology, University of Chicago, Chicago, Illinois (349)

G. D. BIRNIE, Wolfson Laboratory for Molecular Pathology, The Beatson Institute for Cancer Research, Bearsden, Glasgow, Scotland (13)

HARRIS BUSCH, Department of Pharmacology, Baylor College of Medicine, Houston, Texas (163)

ROGER CHALKLEY, Department of Biochemistry, College of Medicine, University of Iowa, Iowa City, Iowa (1, 235)

G. C. CHAMNESS, Department of Medicine, University of Texas Health Science Center, San Antonio, Texas (325)

JEN-FU CHIU, Department of Biochemistry, Vanderbilt University School of Medicine, Nashville, Tennessee (211)

J. H. CLARK, Department of Cell Biology, Baylor College of Medicine, Houston, Texas (335)

ROBERT J. COHEN, Department of Biochemistry and Molecular Biology, University of Florida, Gainesville, Florida (89)

DAVID E. COMINGS, Department of Medical Genetics, City of Hope National Medical Center, Duarte, California (115)

J. DERKSEN, Department of Genetics, Catholic University, Nijmegen, Holland (133)

H. ERIKSSON, Department of Cell Biology, Baylor College of Medicine, Houston, Texas (335)

INGEMAR ERNBERG, Department of Tumor Biology, Karolinska Institutet, Stockholm, Sweden (401)

CARL M. FELDHERR, Department of Anatomy, University of Florida College of Medicine, Gainesville, Florida (75)

FRANCES K. FRANOLICH, Department of Biophysics, The University of Houston, Houston, Texas (101)

WILLIAM T. GARRARD, Department of Biochemistry, University of Texas Health Science Center, Dallas, Texas (27)

SIDNEY R. GRIMES, Department of Biochemistry, Vanderbilt University School of Medicine, Nashville, Tennessee (211)

RONALD HANCOCK, Swiss Institute for Experimental Cancer Research, Lausanne, Switzerland (27)

J. W. HARDIN, Department of Cell Biology, Baylor College of Medicine, Houston, Texas (335)

ROSS HARDISON, Department of Biochemistry, School of Medicine, University of Iowa, Iowa City, Iowa (235)

LUBOMIR S. HNILICA, Department of Biochemistry, Vanderbilt University School of Medicine, Nashville, Tennessee (211)

PETER HOFFMANN,[1] Department of Biochemistry, College of Medicine, University of Iowa, Iowa City, Iowa (1)

[1] *Present address:* Department of Experimental Therapeutics, Roswell Park Memorial Institute, Buffalo, New York.

A. J. W. HSUEH, Department of Cell Biology, Baylor College of Medicine, Houston, Texas (335)

AKIRA INOUE, The Rockefeller University, New York, New York (253)

VALERIE M. KISH, [2] Worcester Foundation for Experimental Biology, Shrewsbury, Massachusetts (377)

LEWIS J. KLEINSMITH, Division of Biological Sciences, The University of Michigan, Ann Arbor, Michigan (285)

MARGARIDA O. KRAUSE, Department of Biology, University of New Brunswick, Fredericton, New Brunswick, Canada (51)

L. Y. LEE, Department of Biophysics, The University of Houston, Houston, Texas (101)

SHIRLEY LINDE, Department of Biomathematics, The University of Texas System Cancer Center, M. D. Anderson Hospital and Tumor Institute, Houston, Texas (101)

N. H. LUBSEN, Department of Genetics, Faculty of Sciences, Catholic University, Nijmegen, The Netherlands (81)

JOSEPH J. MAIO, Department of Cell Biology, Albert Einstein College of Medicine, Bronx, New York (93)

TERENCE E. MARTIN, Department of Biology, University of Chicago, Chicago, Illinois (349)

KEIJI MARUSHIGE, Laboratories for Reproductive Biology and Department of Biochemistry, Division of Health Affairs, University of North Carolina, Chapel Hill, North Carolina (5)

YASUKO MARUSHIGE, Laboratories for Reproductive Biology, Division of Health Affairs, University of North Carolina, Chapel Hill, North Carolina (59)

ALAN MCCLEARY, Division of Biological Sciences, The University of Michigan, Ann Arbor, Michigan (285)

W. L. MCGUIRE, Department of Medicine, University of Texas Health Science Center, San Antonio, Texas (325)

MASAMI MURAMATSU, Department of Biochemistry, Tokushima University School of Medicine, Tokushima, Japan (141)

LARRY NOODEN, Division of Biological Sciences, The University of Michigan, Ann Arbor, Michigan (285)

MARK O. J. OLSON, Department of Pharmacology, Baylor College of Medicine, Houston, Texas (163)

TOSHIO ONISHI, Department of Biochemistry, Tokushima University School of Medicine, Tokushima, Japan (141)

WILLIAM D. PARK, Department of Biochemistry and Molecular Biology, University of Florida, Gainesville, Florida (293)

E. J. PECK, Jr., Department of Cell Biology, Baylor College of Medicine, Houston, Texas (335)

THORU PEDERSON, Worcester Foundation for Experimental Biology, Shrewsbury, Massachusetts (377)

A. M. C. REDEKER-KUIJPERS, Department of Genetics, Faculty of Sciences, Catholic University, Nijmegen, The Netherlands (81)

PAUL A. RICHMOND, Department of Anatomy, University of Florida, College of Medicine, Gainesville, Florida (75)

CARL L. SCHILDKRAUT, Department of Cell Biology, Albert Einstein College of Medicine, Bronx, New York (93)

[2] Present address: Department of Biology, Hobart and William Smith Colleges, Geneva, New York.

THOMAS C. SPELSBERG, Department of Molecular Medicine, Mayo Clinic, Rochester, Minnesota (303)

ERIC STAKE, Mayo Medical School, Mayo Clinic, Rochester, Minnesota (303)

GARY S. STEIN, Department of Biochemistry and Molecular Biology, University of Florida, Gainesville, Florida (293)

GRETCHEN H. STEIN, Department of Molecular, Cellular, and Developmental Biology, University of Colorado, Boulder, Colorado (271)

JANET L. STEIN, Department of Immunology and Medical Microbiology, University of Florida, Gainesville, Florida (293)

ELTON STUBBLEFIELD, Department of Biology, The University of Texas System Cancer Center, M. D. Anderson Hospital and Tumor Institute, Houston, Texas (101)

DAVID WITZKE, Mayo Medical School, Mayo Clinic, Rochester, Minnesota (303)

D. T. ZAVA, Department of Medicine, University of Texas Health Science Center, San Antonio, Texas (325)

ALFRED ZWEIDLER, The Institute for Cancer Research, The Fox Chase Cancer Center, Philadelphia, Pennsylvania (223)

PREFACE

During the past several years considerable attention has been focused on examining the regulation of gene expression in eukaryotic cells with emphasis on the involvement of chromatin and chromosomal proteins. The rapid progress that has been made in this area can be attributed largely to development and implementation of new, high-resolution techniques and technologies. Our increased ability to probe the eukaryotic genome has far-reaching implications, and it is reasonable to anticipate that future progress in this field will be even more dramatic.

We have attempted to present, in four volumes of *Methods in Cell Biology*, a collection of biochemical, biophysical, and histochemical procedures that constitute the principal tools for studying eukaryotic gene expression. Contained in Volume 16 are methods for isolation of nuclei, preparation and fractionation of chromatin, fractionation and characterization of histones and nonhistone chromosomal proteins, and approaches for examining the nuclear-cytoplasmic exchange of macromolecules. This volume (Volume 17) deals with further methods for fractionation and characterization of chromosomal proteins, including DNA affinity techniques. Also contained in this volume are methods for isolation and fractionation of chromatin, nucleoli, and chromosomes. Volume 18 focuses on approaches for chromatin fractionation, examination of physical properties of chromatin, and immunological as well as sequence analysis of chromosomal proteins. In the fourth volume (Volume 19) enzymic components of nuclear proteins, chromatin transcription, and chromatin reconstitution are described. Volume 19 also contains a section on methods for studying histone gene expression.

In compiling these four volumes we have attempted to be as inclusive as possible. However, the field is in a state of rapid growth, prohibiting us from being complete in our coverage.

The format generally followed includes a brief survey of the area, a presentation of specific techniques with emphasis on rationales for various steps, and a consideration of potential pitfalls. The articles also contain discussions of applications for the procedures. We hope that the collection of techniques presented in these volumes will be helpful to workers in the area of chromatin and chromosomal protein research, as well as to those who are just entering the field.

We want to express our sincere appreciation to the numerous investigators who have contributed to these volumes. Additionally, we are indebted to Bonnie Cooper, Linda Green, Leslie Banks-Ginn, and the staff at Academic Press for their editorial assistance.

GARY S. STEIN
JANET L. STEIN
LEWIS J. KLEINSMITH

Erratum

METHODS IN CELL BIOLOGY, Vol. 16

Page 450 lines 15–17 should read:
matin is found in the interphase which is not collected; however, if chromatin is prepared by other methods this could vary and should be carefully controlled. The amount of chromatin in the upper and lower phases is deter-

Part A.   Isolation of Nuclei and Preparation of Chromatin. 11

Chapter 1

Procedures for Minimizing Protease Activity during Isolation of Nuclei, Chromatin, and the Histones

PETER HOFFMANN* AND ROGER CHALKLEY

*Department of Biochemistry,
College of Medicine, University of Iowa,
Iowa City, Iowa*

I.   Introduction

Of the many proteins in the eukaryotic cell nucleus, the highly basic proteins associated with DNA, the histones, rank among the most intensively studied. Early work (*1–3*) indicated that there might be a large number of different histones, which led to the idea that they may act as gene repressors (*4*). However, it was later realized that there were only five major classes of histones (*5–9*) and that the erroneous earlier estimates, based on the number of bands observed on a polyacrylamide gel, were the consequence of proteolytic degradation.

Preparation of nucleoprotein has followed one of two principles: (1) purification from disrupted whole tissue or (2) preparation from purified nuclei. The method described by Bonner *et al.* (*10*), based on the procedure of Zubay and Doty (*11*), illustrates the first principle and is still widely used. Although Zubay and Doty specifically designed their system to minimize nucleolytic degradation by removal of metal ions with ethylenediaminetetraacetic acid (EDTA) and operation at pH 8 (*11*), these conditions, un-

*Present address: Department of Experimental Therapeutics, Roswell Park Memorial Institute, Buffalo, New York.

fortunately, afford proteolysis of the histones. The second principle is typified by a rigorous purification of nuclei for the removal of cytoplasmic contaminants, which have been considered as a probable source of proteolytic activity. The use of sucrose solutions for the isolation of intact nuclei (*12*) and the washing of the nuclei with Triton X-100 for removal of the outer aspect of the nuclear membrane and adhering cytoplasmic components (*13*) are the main features of the method. An equally important inclusion has been sodium bisulfite as an inhibitor of histone proteolysis. There is no obvious difference between the two methods when histones are prepared from calf thymus, but when applied to tissues having a greater ratio of cytoplasm to nucleus, such as liver, the method of Bonner *et al.* (*10*) can yield histones which can be significantly degraded. The concerns of cytoplasmic protease contamination (*14*) have been substantiated (*15,16*) and one study (*16*) indicated that a "neutral chromatin-associated protease" (*17*) is, in fact, of mitochondrial origin.

Since the nature of the problem of proteolysis of nuclear proteins is currently best understood with regard to the histones, we have purposely limited the scope of this article to those proteins. However, there is no reason to believe that the nonhistone proteins are any less susceptible to proteolytic degradation, and in one report this has been shown to be so (*18*). The methods and comments outlined below are derived largely from extensive experience using calf thymus and rat liver. These two tissues are used almost exclusively for studies on the structure and function of chromatin. However, we suggest that the following method for minimizing proteolysis can be employed with some confidence, at least as a starting point, for the isolation of nuclei (*19,20*) and nucleoprotein from any source, with the proviso that the integrity of the nucleoproteins is assayed at every step.

II. Methods

A. Assay for Proteolytic Degradation of Nuclear Proteins by Polyacrylamide Gel Electrophoresis

The histones may be conveniently assayed on acid-urea gels (*6,7*). The loss of any of the major fractions and appearances of degradation products are readily apparent when compared to a standard of acid-extracted whole histone from calf thymus. A sodium dodecyl sulfate (SDS)–polyacrylamide gel system has been used to assay for proteolysis of the nonhistone proteins (*18*).

B. Purification of Nuclei and the Isolation of Chromatin (8,9)

1. STOCK SOLUTIONS

The solutions required are: (a) "Grinding medium" containing 0.25 M sucrose–10 mM MgCl$_2$–10 mM Tris-HCl–50 mM NaHSO$_3$. The sucrose–MgCl$_2$–Tris can be prepared as a 5-fold concentrated stock at pH 8.0 and stored at 4°. The NaHSO$_3$ is added as the solid just prior to use because of its hydrolysis to H$_2$SO$_3$ in solution. After addition of solid NaHSO$_3$, the medium is used without readjustment of the pH. (b) "Washing medium" contains all the components of grinding medium and in addition 0.2–1.0% Triton X–100. The concentration of Triton X–100 used depends on the fragility of the nuclei and must be determined for a particular tissue. For calf thymus and rat liver 1.0% Triton X–100 is used. This solution can also be prepared as a 5-fold concentrated stock (pH 8.0). Solid NaHSO$_3$ is again added just prior to use, without readjustment of the pH. (c) "Tris–EDTA–bisulfite" contains 10 mM Tris–HCl–20 mM EDTA–50 mM NaHSO$_3$. The Tris and EDTA are dissolved, and the solution is adjusted to pH 8.0. Solid NaHSO$_3$ is then added, and the solution is used immediately without further adjustment of the pH.

2. DISRUPTION OF TISSUE

All operations are performed at 4°. Isolation and purification of nuclei should be monitored with a light microscope and staining with acetocarmine. Fresh tissue is cut into small pieces and frozen at −15°C (also for storage until required). Freezing is necessary to cause cell breakage. The frozen tissue is added to ice-cold grinding medium (approximately 20 ml/gm tissue) and blended at maximum speed in a Waring blender for 3 minutes. The suspension is filtered through four layers of cheese cloth and then two layers of Miracloth (Chicopee Manufacturing Company, New Jersey). The filtrate is centrifuged at 480 g for 10 minutes, and the supernatant is discarded. Disruption of tissue is carried out in the absence of Triton X-100 (which would otherwise cause frothing) when maximum blending speed is used. For the isolation of nuclei from cultured normal or tumor cells, frozen cells are simply suspended directly into washing medium using a Thomas tissue homogenizer or, for larger quantities, a VirTis homogenizer at low speed (10–15 V).

3. PURIFICATION OF NUCLEI

The pellet of nuclei is washed by suspension in washing medium (10 ml or greater per gram of tissue) using either a Waring blender at low speed (20 V) for 3 minutes or a Thomas tissue homogenizer. The suspension is centrifuged at 480 g for 10 minutes, and the process is repeated until the supernatant is clear (three or more washings).

It is sometimes necessary to purify nuclei further by centrifugation through a concentrated sucrose solution. The need for this step can be determined by comparing the proteins (on polyacrylamide gels) extracted from nuclei which either have or have not been subjected to this treatment. Calf thymus nuclei do not require this step, whereas rat liver nuclei do.

The nuclear pellet is homogenized (several strokes with a Thomas tissue homogenizer) in about 10 volumes of washing medium containing 2.2 M sucrose, and then it is layered onto washing medium containing 2.4 M sucrose. The suspension of nuclei in 2.2 M sucrose can occupy up to 50% of the volume of the centrifuge tube. The interface is disrupted by stirring, and the material is centrifuged at 60,000 g for $2\frac{1}{2}$ hours at 4° in a preparative ultracentrifuge, preferably using a swinging-bucket rotor. The supernatant is discarded.

4. Disruption of Nuclei

The purified nuclei are disrupted by thorough homogenization with a hand homogenizer or blender in 40–50 volumes of Tris–EDTA–bisulfite solution. The sticky stringy mixture is sedimented at 4300 g for 10 minutes, the supernatant is discarded, and the procedure is repeated.

5. Swelling of the Chromatin Gel

The pellet from above is swollen to a gel by homogenization in 40–50 volumes of deionized, glass-distilled water, and then it is centrifuged at 12,000 g for 10 minutes. This process is repeated until the material forms a gel and will no longer form a pellet upon centrifugation. For calf thymus only one water wash is required. Homogenization of the pellet in a second portion of water yields a gel which will not sediment at 12,000 g.

Thus an appropriate second volume of water is added to the swollen pellet (after the first wash) and homogenized to yield a chromatin gel of the desired DNA concentration.

C.   Extraction of Histones

1. Extraction with Acid

Chromatin is diluted to a DNA concentration of about 0.25 mg/ml ($A_{260\,nm} = 5$) and then sheared at maximum speed in a VirTis homogenizer for 5 minutes, to yield a clear nonviscous solution. The solution is centrifuged at 27,000 g for 10 minutes, and a small pellet of membranous debris is discarded. Histones are extracted by adding 4 N H_2SO_4 to the supernatant to a final concentration of 0.4 N H_2SO_4 and stirring in the cold for at least 1 hour. After centrifugation (12,000 g for 20 minutes) to remove the precipitated DNA, histone is precipitated from the supernatant by dialysis against at least 6 volumes of 95% ethanol at 4° overnight. The histone is collected by centrifugation at 12,000 g for 15 minutes, washed twice with cold 95% ethanol, and then dried under vacuum. The solid is stored at −15°.

2. EXTRACTION WITH NaCl

The histones can be differentially extracted with NaCl. Only histone H1 is extracted between 0.3 M NaCl and 0.6 M NaCl. H2A and H2B are removed by 1.2 M NaCl, and 2 M NaCl is necessary to extract H3 and H4 (*21*). Thus, chromatin which has been depleted of one or three histones can be readily prepared. The viscous chromatin is only slightly sheared (20–30 V, VirTis homogenizer) and then treated with solid NaCl to the desired concentration, in the presence of a buffer (50 mM sodium cacodylate) at pH 6.0. The material is centrifuged at 27,000 g for 5 minutes to remove the air and any membraneous debris. The supernatant is then layered onto 20% sucrose (containing the same salt and buffer concentrations), and the DNA is removed by centrifugation at 90,000 g for at least 7 hours or overnight at 4°. The sucrose solution occupies 50% of the centrifuge tube. After centrifugation the top half of the tube contents and $\frac{1}{8}$–$\frac{1}{4}$ of the sucrose is removed and contains the histones.

Whole histone also can be extracted from nuclei in a similar manner. A suspension of nuclei is disrupted by addition of 2 M NaCl and buffer at pH 6.0. The DNA is then slightly sheared and removed as described above. The solution of histones in 2 M NaCl is stored frozen.

3. DISPLACEMENT WITH PROTAMINE

The following outlines the procedure that has been developed in the authors' laboratory with the specific aim of eliminating the exposure of the histones to high ionic strength (Hoffmann *et al.*, unpublished). Chromatin, at a concentration of 5 mg/ml ($A_{260 nm} = 100$), is treated with an equal volume of solution containing 0.4 M NaCl–0.1 M cacodylate buffer (pH 6.0)–0.1–NaHSO$_3$–80 mg/ml protamine sulfate. The mixture is vigorously mixed on a Vortex mixer and then stored at 4° for at least 4 hours. Alternatively, histone H1 can be selectively displaced by first treating chromatin to a final concentration of 3.5 mg/ml protamine in the same ionic environment as above. The mixture is centrifuged at 1000 g for 5 minutes, and the four histones are then obtained by treating the pellet with a final concentration of 40 mg/ml protamine in the same final ionic environment as above. After removal of the DNA–protamine complex by centrifugation (12,000 g for 10 minutes) the supernatant is concentrated in a dialysis tube against solid sucrose at 4°. A large portion of the excess protamine is then removed as a sticky sediment by standing the solution in ice water overnight. The residual protamine can be removed by passage through Sephadex G-50, or the histones can be further fractionated by chromatography on Sephadex G-100 in 0.2 M NaCl-50 mM cacodylate buffer (pH 6.0) (22).

4. Extraction with Detergents

Extraction of histones with detergents has been used as an analytical tool rather than on a preparative scale.

Samples of chromatin are made at least 2% sodium dodecyl sulfate (SDS) and can then be analyzed on SDS-containing gels (23).

The cationic detergent cetyltrimethylammonium bromide (CTAB) can also be used to extract histones. A suspension of nuclei is treated first with 4 M urea and then with 2% CTAB (23a). Histones in the supernatant can then be analyzed on neutral gels in the presence of CTAB or on acid–urea gels after addition of 1 M acetic acid (23a).

III. Conditions and Patterns of Proteolytic Degradation of Histones

A. Introduction

Although proteolytic degradation of nucleoprotein has long been recognized (24,25), many of these observations were probably of contaminating enzymes coming from other than the nucleus (14–16). The activities which do seem to be associated with the nucleus (26–29) are, like many other proteolytic enzymes (30), remarkably resilient. They are able to survive and, in some cases, even function in harshly denaturing conditions such as 5 M urea (18) or 0.4N H_2SO_4. Thus, it should be remembered that the potential for proteolytic degradation of nucleoprotein preparations is always present, and it should be considered as being, at best, reversibly inhibited.

B. Patterns of Histone Degradation

Figure 1 shows the usual pattern of acid-extracted calf thymus histones on an acid–urea polyacrylamide gel. The positions of bands which correspond to degradation products of the histones are indicated, and some examples of degraded samples are shown.

The bands representing degradation products are highly reproducible.

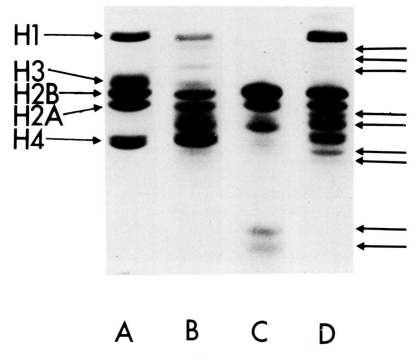

FIG. 1.   Electrophoresis of histones showing various degrees of proteolytic degradation. Electrophoresis was on acid–urea polyacrylamide gels, and migration was from top to bottom. The nine arrows on the right show the positions of fragments resulting from proteolytic degradation. (A) Acid-extracted undegraded whole histone standard. (B) Histones displaced by protamine from chromatin prepared by nuclease digestion. (C) Protamine-extracted H2A and H2B in 0.2 M NaCl–50 mM sodium cacodylate (pH 6.0) after storage at 4° for 1 month. (D) Histones displaced by protamine from sheared chromatin. For samples B and D, the protamine was removed by passage through Sephadex G-50 equilibrated in 0.15 M NaCl–50 mM NaHSO₃–50 mM sodium acetate (pH 5), and then the samples were stored at 4° for 2 weeks.

Three bands occur between H1 and H3 and are presumably derived from H1. Another band occurs just ahead of H2A, while an often seen degradation product runs just behind H4. The latter probably arises from H2A. On occasion, bands are sometimes observed migrating ahead of H4.

By referring to gels of acid-extracted histones from freshly prepared chromatin and the data of Fig. 1, proteolytic degradation can be readily detected. However, the loss of H3 should be interpreted with care and may not necessarily represent degradation. Histone H3, obtained from calf thymus, contains two cysteine residues and as such is susceptible to oxidation, either intramolecularly or intermolecularly. Intermolecular oxidation yields a dimer which migrates more slowly than H1 (6, 7), whereas intramolecular oxidation alters the shape of H3 such that it comigrates with H2B (8, 9). Accordingly, samples should be treated with 1 M 2-mercaptoethanol in 0.9 N acetic acid overnight before electrophoretic analysis.

C. Inhibition of Proteolysis and the Stability of Nucleoprotein Preparations

1. GENERAL

The general properties of the protease activity with respect to degradation of calf thymus nucleohistone have been documented (29). It is essentially inactive below pH 7.0, and the activity is decreased at lower ionic strengths. Thus the inhibiting effect of sodium bisulfite is probably partly due to the lowering of the pH. Addition of sodium bisulfite to grinding and washing medium lowers the pH to about 6.3. While the following outlines the stability of various preparations from calf thymus, it should only be used as a general guide for other tissues. Since different tissues are susceptible to proteolysis to varying degrees (29), the conditions for any particular tissue should be determined. In all cases it is wise to prepare material immediately prior to use.

2. NUCLEI

Nuclei should be used the same day on which they are prepared. Some proteolysis of histones can be detected after storage overnight.

The medium described by Hewish and Burgoyne (31) has also been used for the isolation of nuclei. Incubation, at 37°, of nuclei isolated in this way results in the rapid degradation of histones H3 and H2A.[1] However, if the mercaptoethanol in the medium is replaced with 20 mM sodium bisulfite (final pH 6.3), degradation is markedly reduced.[2] Also, if nuclei prepared in the presence of sodium bisulfite are given one wash with the medium in the absence of sodium bisulfite (and mercaptoethanol), the same specific pattern of DNA fragments is obtained upon digestion with micrococcal nuclease[3] as has been described previously (32).

3. CHROMATIN

One should endeavor to prepare the chromatin gel as rapidly as possible since the low ionic strength inhibits proteolysis. Chromatin at very low ionic strength ($I = 5 \times 10^{-4}$) can be stored at 2–4° for 5 days or longer before proteolysis is observed (29). While the histones remain associated with

[1] P. J. Hoffmann, unpublished observations.
[2] P. J. Hoffmann, unpublished observations.
[3] V. Jackson, personal communication.

the DNA, H1 and H3 are preferentially degraded, whereas the other three classes of histone are almost totally resistant to proteolytic attack (*29*).

4. ACID-EXTRACTED HISTONES

At least a portion of the proteolytic activity survives treatment with 0.4 N H_2SO_4, though it is virtually irreversibly inactivated by precipitation with ethanol and drying under vacuum. This if histones are extracted with 0.4 N H_2SO_4, dialyzed against water, lyophilyzed, and then dialyzed at neutral pH (or greater) at 4° for two or more days, substantial degradation is observed.[4,5]

5. NaCl-EXTRACTED HISTONES

Although high ionic strength increases the rate of proteolysis (*18,27,29*), there is little effect of ionic strength if the pH of the preparation is kept at pH 6.[6,7] Thus histones extracted with 2M NaCl [or other concentrations: 0.6M NaCl for extraction of H1; 1.2M NaCl for extraction of H1, H2A, and H2B (*21*)] at pH 6 are not degraded over a period of 7 days at 4°. Some slight degradation may be observed after this time. Histones in the presence of high concentrations of salt are best stored frozen for maximum stability. However, it should be remembered that the effect of freezing on the conformation of histones has not been determined.

If histones are extracted with salt above pH 7, there is significant degradation during the 7–16 hours of centrifugation necessary to remove the DNA.

Note that when the histones are no longer associated with DNA, the susceptibility of the various histone fractions to proteolysis is almost completely reversed when compared with intact nucleohistone; H1 is resistant to proteolysis while the other fractions are rapidly degraded (*29*).

6. PROTAMINE-DISPLACED HISTONES

We have fractionated histones extracted with protamine for the purpose of other studies but have noted their susceptibility to proteolysis. Like NaCl-extracted histones, fractionation and storage at pH 6 are the most important requirements for the minimization of proteolysis. After selective extraction of H1, the other four histones can be fractionated on Sephadex G-100 in 50 mM sodium cacodylate–50mM $NaHSO_3$–0.2M NaCl–0.02% sodium azide (pH 6). The fractionation is similar to that described previously (*22*). Two peaks are obtained; the first contains H3 and H4 and the

---

[4] V. Jackson, personal communication.
[5] J. K. Petell, personal communication.
[6] P. J. Hoffmann, unpublished observations.
[7] R. Chalkley, unpublished observations.

second H2A and H2B. Histones H3 and H4 and the previously obtained H1 seem completely stable and have been kept for up to 1 month at 4° without noticeable degradation. The H2A-H2B fraction is not so stable. After 3 weeks at 4° it is obvious that H2A is being proteolyzed, but it was difficult to determine whether H2B had been affected.

7. INHIBITORS OF HISTONE PROTEOLYSIS

Apart from the use of sodium bisulfite and of media at pH 6, phenyl-methylsulfonyl fluoride (PMSF) has been used most extensively as an inhibitor of proteolysis of nuclear proteins. Claims of its effectiveness have probably stemmed from the inhibition of contaminating proteases of cytoplasmic origin (16). In the authors' laboratory, PMSF has been used with only partial effectiveness. Incubation of purified nuclei at pH 7.5 with 10 mM PMSF afforded only partial protection against histone degradation and was no more effective than incubating the nuclei at pH 6.[8] (Nuclei represent a special case where significant degradation of histones is apparent even after only 24 hours of storage in washing medium, pH 6.3.) Nevertheless, there does appear to be a chromatin-associated protease which is activated by high ionic strength (at pH 7 or greater) and can be irreversibly inhibited by 1 mM PMSF and the alkylating reagent, carbobenzoxyphenylalanine chloromethylketone, in the presence of organic solvents (18). Application of these inhibitors may be especially useful in work with chromatin in solutions containing high concentrations of salt and/or urea.

Complete inhibition of proteolysis has been observed during experiments on the cross-linking of chromatin using the bifunctional reagent, dimethylsuberimidate.[9] This may indicate that the nuclear protease contains an active site lysine residue which therefore may also be inhibited by other reagents specific for the ε-amino group of lysine.

No proteolysis of histones has been observed in samples stored, prior to analysis, in the presence of denaturing detergents such as SDS.

IV.   Concluding Remarks

The use of purified nuclei, free of contaminating cytoplasmic protease(s) cannot be strongly enough recommended. Even with this precaution the potential for proteolytic degradation of nuclear proteins by nuclear pro-

---

[8] R. Hardison, personal communication.
[9] R. Chalkley, unpublished observations.

tease(s) still exists. Unfortunately, under conditions that are physiologic (pH 7.4, moderate ionic strength) and normally chosen for experimentation, nuclear proteolytic activity is apparent. It is essential, therefore, that this problem be recognized and that workers assay for the integrity of nuclei or nucleoprotein preparations at the conclusion of an experiment. Such an approach should help to eliminate misinterpretation of data from experiments where proteolytic activity may have been a significant contributing factor (33).

REFERENCES

1. Neelin, J. M., and Connel, G. E., *Biochim. Biophys. Acta* **31**, 539 (1959).
2. Shepherd, G. R., and Gurley, L. R., *Anal. Biochem.* **14**, 356 (1966).
3. Driedger, A., Johnson, L. D., and Marko, A. M., *Can. J. Biochem. Physiol.* **41**, 2507 (1963).
4. Huang, R. C. C., and Bonner, J., *Proc. Natl. Acad. Sci. U.S.A.* **48**, 1216 (1962).
5. Johns, E. W., *Biochem. J.* **92**, 55 (1964).
6. Panyim, S., and Chalkley, R., *Biochemistry* **8**, 3972 (1969).
7. Panyim, S., and Chalkley, R., *Arch. Biochem. Biophys.* **130**, 337 (1969).
8. Panyim, S., Bilek, D., and Chalkley, R., *J. Biol. Chem.* **246**, 4206 (1971).
9. Panyim, Sommer, K., and Chalkley, R., *Biochemistry* **10**, 3911 (1971).
10. Bonner, J., Chalkley, G. R., Dahmus, M., Fambrough, D., Fujimura, F., Huang, R. C. C., Huberman, J., Jenson, R., Marushige, K., Ohlenbusch, B., and Wisholm, J. in "Methods in Enzymology, "Vol. 12: Nucleic Acids (L. Grossman and K. Moldave, eds.), Part B, p. 3. Academic Press, New York.
11. Zubay, G., and Doty, P., *J. Mol. Biol.* **1**, 1 (1959).
12. Chaeveau, J., Moule, Y., and Rouiller, C., *Exp. Cell Res.* **11**, 317 (1956).
13. Ueda, K., Matsumura, T., Noboru, D., and Kawai, K., *Biochem. Biophys. Res. Commun.* **34**, 322 (1969).
14. Panyim, S., Jenson, R. H., and Chalkley, R., *Biochim. Biophys. Acta* **160**, 252 (1968).
15. Destree, O. H., D'Adelhart-Toorop, H. A., and Charles, R., *Biochim. Biophys. Acta* **378**, 450 (1975).
16. Heinrich, P. C., Raydt, G., Puschendorf, B., and Jusic, M., *Eur. J. Biochem.* **62**, 37 (1976).
17. Chong, M. T., Garrard, W. T., and Bonner, J., *Biochemistry* **13**, 5128 (1974).
18. Carter, D. B., and Chae, C-B., *Biochemistry* **15**, 180 (1976).
19. Taylor, C. N., Yeoman, L. C., and Busch, H., *Methods Cell Biol.* **9**, 349 (1975).
20. Muramatsu, M., *Methods Cell Physiol* **4**, 195 (1970).
21. Hnilica, L. S., "The Structure and Biological Function of Histones." CRC Press, Cleveland, Ohio, 1972.
22. Van Der Westhuyzen, D. R., and Von Holt, C., *FEBS Lett.* **14**, 333 (1971).
23. Hardison, R., Chalkley, R., This volume, p. 235.
23a. Hoffman, P. J., and Chalkley, R., Manuscript in preparation.
24. Dounce, A. L., and Umana, R., *Biochemistry* **1**, 811 (1962).
25. Stellwagen, R. H., Reid, B. R., and Cole, R. D., *Biochim. Biophys. Acta* **155**, 581 (1968).
26. Furlan, M., and Jericijo, M., *Biochim. Biophys. Acta* **147**, 135 (1967).
27. Furlan, M., and Jericijo, M., *Biochim. Biophys. Acta* **147**, 145 (1967).
28. Furlan, M., Jericijo, M., and Suhar, A., *Biochim. Biophys. Acta* **167**, 154 (1968).
29. Bartley, J., and Chalkley, R. *J. Biol. Chem.* **245**, 4286 (1970).
30. Perlmann, G. E., and Lorand, L., eds., "Methods in Enzymology,". Vol. 19: Proteolytic Enzymes. Academic Press, New York, 1970.

31. Hewish, D. R., and Burgoyne, J. A., *Biochem. Biophys. Res. Commun.* **52**, 504 (1973).

32. Noll, M., *Nature (London)* **251**, 249 (1974).

33. Chae, C-B., Gadski, R. A., Carter, D. B., and Efird, P. H., *Biochem. Biophys. Res. Commun.* **67**, 1459 (1975).

Chapter 2

Isolation of Nuclei from Animal Cells in Culture

G. D. BIRNIE

Wolfson Laboratory for Molecular Pathology,
The Beatson Institute for Cancer Research,
Bearsden, Glasgow, Scotland

I. Introduction

The objective of the ideal method for isolating nuclei is to obtain a preparation which consists solely of nuclear material and in which the nuclei are morphologically identical to those in undisrupted cells, with contents qualitatively and quantitatively identical to those of nuclei *in vivo*. The nucleus is a well-defined, easily recognized organelle, yet there have been published an enormous number of procedures designed to attain this objective. The reason for the plethora of methods is simply stated—the nucleus is a complex, rather fragile, and permeable organelle, and there is no method known at present which can be generally applied to all cell types in order to obtain preparations of uncontaminated, undamaged, and unchanged nuclei. Consequently, the first question that must be asked is—for what purpose are the isolated nuclei required? This question is not at all facetious. For example, for some purposes it is more important that the nuclei be undamaged than that they be completely free of cytoplasmic contamination; for others, quite severe morphological distortion is acceptable so long as the preparation is completely devoid of cytoplasmic material.

The state of the nuclei obtained by any procedure is dependent on many interacting, and all too often uncontrollable, factors including the type of cell or tissue from which the nuclei are isolated, the method used to lyse the cells and release the nuclei, the pH and composition of the isolation medium, and the technique used to separate the nuclei from the other cell constituents. The ways in which these factors affect nuclear isolation procedures which have been used with a variety of eukaryotic tissues and cells have been discussed in depth in recent reviews by Roodyn (*1*) and Smuckler *et al.* (*2*). A description of the effects of all of these variables, even so far as they pertain to the special case of cells cloned in tissue culture, is beyond the scope of this chapter. Instead, this article discusses briefly a few of the factors that are of more general importance in nuclear isolation procedures and the criteria that can be used to assess the integrity and purity of the isolated nuclei; it concentrates on describing two methods for isolating nuclei of which considerable experience has been gained in this laboratory as prelininary steps in the isolation of chromatin and chromatin constituents [see Rickwood and Birnie (3), MacGillivray (4); also see Chapter 2, Volume 18 and Chapter 28, Volume 19 of this series].

II. General Consideration of Nuclear Isolation Procedures

A. Problems Encountered

Nuclear isolation procedures essentially consist of two steps, namely, the lysis of the cells in a suitable medium and the separation of the nuclei from the unbroken cells and cell debris. The lysis of cells grown in tissue culture presents fewer problems than are encountered when dealing with whole tissues. One is not dealing with a mixture of cell types so there is no problem of differential susceptibility to lytic procedures, and there is no connective tissue or intracellular fibrous material which cause marked reductions in the efficiency of homogenizers. Consequently, there is no need to use harsh homogenization conditions like those obtained with, for example, whirling-blade homogenizers; instead, the more gentle and controllable Potter–Elvehjem and Dounce-type homogenizers function very well with cells grown in tissue culture. Even so, great care must be taken during the homogenization step. The clearance between the pestle and the wall of the homogenizer vessel must be chosen correctly, otherwise the proportion of cells lysed may be too low or, at the other extreme, a high proportion of the nuclei may be damaged. The choice of clearance depends both on the cell type and on the medium in which homogenization is done (see Section III). Also, it must be remembered that the rotation of the pestle within the homogenizer vessel produces considerable amounts of heat. Consequently, it is important not only to use precooled lysing media, but also to ensure that the homogenizer vessel is thoroughly chilled, not only before but also during the homogenizing process (in other words, the vessel should be kept in ice and not held in the hand). Finally, the homogenization process should be brief and should require only a few up-and-down strokes of the pestle. It is better to settle for less than total breakage of the cells than to risk damaging a high proportion of the nuclei.

The selection of a suitable homogenizing medium is, as indicated in Section I, a complex matter [see Smuckler et al. (2)], which will not be discussed here. However, the two methods described in Section III provide a sharp contrast of lysing media, ranging from one which uses a low pH medium with no osmotic stabilization to one in which the medium is osmotically stabilized at a slightly alkaline pH. As would be expected, the products obtained differ significantly (see Section III,D).

Separation of the nuclei from the other constituents of the cell is invariably done by centrifugation, advantage being taken of the facts that nuclei are larger than all other cell organelles and denser than most, and so will sediment faster and/or through denser media than nonnuclear material. There are a number of points in the centrifugation procedures used which require careful attention. If the separation is based on differences in sedimentation rate, the centrifugal force and the time for which it is applied must be carefully controlled—centrifugation for too long or at too high a speed may increase the yield of nuclei, but it increases contamination by nonnuclear material to a much greater extent. Also, it must be remembered that clumps of nonnuclear material may sediment as fast, or faster, than nuclei. Isolation of nuclei on the basis of density in practice means sedimentation through dense sucrose solutions (e.g., Chaveau et al. (5); Widnell and Tata (6); Blobel and Potter (7)], a simple procedure which, however, requires that considerable care be taken to control the concentration of the dense sucrose solution. Concentrated sucrose solutions are extremely viscous and, moreover, the viscosity of such solutions increases sharply even with a small increase in sucrose concentration [see Roodyn (1)]. Thus, small changes in the concentration of dense sucrose solutions have a profound effect on the rate at which particles sediment through them. Consequently, even a small increase in the concentration of the sucrose solution through which the nuclei are sedimented may cause a severe reduction in the yield of nuclei; on the other hand, use of a sucrose solution which is too dilute will result in contamination of the nuclei with cytoplasmic material. In this context, it should be noted that the viscosity of concentrated sucrose solutions increases dramatically with even small (1–2°C) decreases in temperature and, consequently, the temperature of the rotor must also be under very strict control if reproducible results are to be obtained. In addition to this, in methods in which the homogenate is layered on top of a pad of dense sucrose through which the nuclei are to be sedimented, the homogenate must be dilute (at least 10 ml or, better, 20 ml of medium per milliliter of cells), otherwise a large proportion of the nuclei are trapped in the plug of cell debris which collects on the top of the dense sucrose layer. Finally, it is worth noting that nuclei from some sources (e.g., regenerating liver) may not be dense enough to sediment through 2.2 M sucrose [see Smuckler et al. (2)].

After the nuclei have been isolated, it is usual to wash them to remove reagents which might interfere with the experiments for which the nuclei have been prepared—for example, to remove citric acid, or Triton X-100, or an excessive concentration of sucrose (see Section III). Obviously, the medium used to wash the nuclei must be chosen with care if damage to the nuclei or their contents is to be avoided (3).

B.   Assessment of Integrity and Purity of Isolated Nuclei

Examination of the isolated nuclei by high-resolution electron microscopy is far and away the best method of assessing the integrity of the nuclei and the purity of the preparation. Ideally, isolated nuclei should be seen to be identical in all respects to those in the whole cells, and there should be no trace of nonnuclear material in the preparation. Unfortunately, examination of nuclear preparations by electron microscopy is time-consuming; it is not a procedure that can readily be used routinely, though it is clearly the method of choice when a "new" isolation procedure is used, or when nuclei are being isolated from a cell line which has not been used previously by the experimenter. However, there is an excellent substitute for the electron microscope for routine monitoring of nuclear preparations—the phase-contrast microscope. It is unfortunate that biochemists are sometimes reluctant to make use of this instrument since it is extremely easy to use and gives an answer within minutes. Examination of a drop of either the homogenate or the nuclear suspension under the microscope immediately shows what proportion of the cells have been broken, the extent to which the nuclei are being lysed or distorted, and whether the nuclei have cytoplasmic tags or the suspension contains significant amounts of nonnuclear material. Consequently, the use of the phase-contrast microscope *at each stage* in a nuclear isolation procedure is to be recommended very strongly indeed since it provides an excellent guide to the progress of the procedure as well as to the acceptability of the final product. Moreover, microscopic examination of the initial homogenate is the simplest way of determining whether the clearance between the pestle and the wall of the homogenizer vessel lies within the range which gives good breakage of the cells without causing excessive lysis of the nuclei.

Chemical analyses can also be used to assess nuclear preparations though, for a variety of reasons, most of these analyses are of limited value only. Measurement of the relative proportions of DNA, RNA, and protein can be used to determine the reproducibility of a preparative procedure, but only strictly when the preparations are from the same cell line, grown under the same conditions. This approach is of doubtful value for comparing different isolation procedures; for example, if one procedure yields nuclei with a lower proportion of protein than another, it is difficult to determine whether the first method is more successful than the second because it removes cytoplasmic contamination more effectively, or less successful because it causes the loss of chromosomal proteins. More useful in this context is the determination of the proportion of histone to DNA in the nuclei, since this should be the same as in the whole cells. Similarly, the isolated nuclei should contain all the histones, in the same relative proportions as they exist in the whole cells. A crude, though useful, marker of cytoplasmic contamination is the ratio of RNA to DNA in the isolated nuclei—this should be much lower than in whole cells since 80 to 90% of the RNA (depending on cell type) in a cell is cytoplasmic in origin. Moreover, RNA extracted from nuclei should contain little or no 18 S ribosomal RNA (8), and the presence of significant amounts of RNA migrating at 18 S in polyacrylamide gels is indicative either of contamination with cytoplasm or of the presence of perinuclear ribosomes.

Enzymological analysis as a method of assessing nuclear preparations is less satisfactory. While the presence of significant amounts of succinic dehydrogenase and 5'-nucleotidase in a nuclear preparation will indicate contamination with mitochondria and plasma membranes, respectively, the absence of these enzyme activities could mean that the isolation procedure has inactivated them rather than removed the structures with which they are associated. Similar considerations apply to other marker enzymes. Assay of DNA polymerase and RNA polymerase activities as a means of determining the integrity of isolated nuclei is not particularly useful either. Not only are these enzymes among the most labile of nuclear proteins, but also their activity in isolated nuclei may be depressed simply because of a failure to use assay conditions which are optimal, or because the isolation procedure has removed a simple cofactor, such as a metal ion, from the nuclei.

III. Methods for Isolating Nuclei

A. Outlines of Two Diverse Methods

In this section, two variations of each of two methods are described in detail. The first of these methods is based on the pioneering work of Dounce and his co-workers [see Roodyn (1)], and in essence consists of lysing cells at low pH in a hypotonic solution of citric acid and isolating nuclei by differential sedimentation. The use of low-pH citric acid solutions is useful in that nucleases and proteases are inhibited, both by the low pH and by the action of the citric acid in chelating divalent cations. This method, and the sucrose–citric acid variation introduced by Busch and Smetana (9) is particularly useful for isolating nuclei which are to be used as a source of undegraded nucleoli, or RNA, or DNA. However, some modification is necessary when the nuclei are to be used for the preparation of chromatin or chromatin proteins since these are less tolerant of low pH. Consequently, the concentration of the citric acid is reduced approximately 10-fold in methods used to isolate nuclei for these purposes (10–12). The citric acid procedures described here have proved useful for isolating nuclei from a variety of animal cells grown in tissue culture (e.g., HeLa cells, and mouse LS, Friend, 3T3, and lymphoma L5178Y cells) as well as from whole mouse tissues (e.g., embryo, embryonic liver, adult liver, and brain).

The second of the methods is based on methods developed by Chaveau et al. (5), Widnell and Tata (6), and Blobel and Potter (7) using dense sucrose solutions at slightly alkaline pH values, combined with the use of detergent [e.g., Grunicke et al. (13)] to remove the outer nuclear membrane with its attached ribosomes and cytoplasmic tags. Methods like these can be considered to be more gentle than those using low pH in that they are more physiological (with respect to ionic composition of the isolation medium) and, consequently, they are less likely to cause translocation of nuclear components. They suffer from the disadvantage that the conditions under which the cells are lysed and the nuclei isolated are those in which degradative enzymes are active. Moreover, the conditions used in some (at least) of the methods cause rupture of the lysosomes. The only precaution taken to inhibit the activities of degradative enzymes is the use of low temperatures. The sucrose–detergent procedures described here have been used to isolate nuclei from the same cells and tissues as the citric acid methods [see Paul et al. (12), Rickwood and Birnie (3), and MacGillivray (4)].

B.   Method I: Isolation with Citric Acid

Two variations of the procedure using citric acid for isolating nuclei are summarized in Fig. 1. In both, the pellet of washed cells is suspended in 20 volumes of ice-cold 25 mM citric acid by vigorous shaking on a Vortex mixer, and the suspension is transferred to a glass Potter–Elvehjem homogenizer vessel which is buried almost to its rim in ice. The suspension is homogenized by 10 up-and-down strokes of a tight-fitting (0.1-mm clearance) cylindrical Teflon pestle, motor-driven at 2000 rpm. The homogenate is centrifuged at 200 g_{av} for 10 minutes at 4°C, and the supernatant fluid is carefully removed and discarded. Thereafter, in the first variation, the process of resuspension by Vortex mixing, homogenization, and centrifugation is repeated until the supernatant fluid is water-clear and the pellet, when

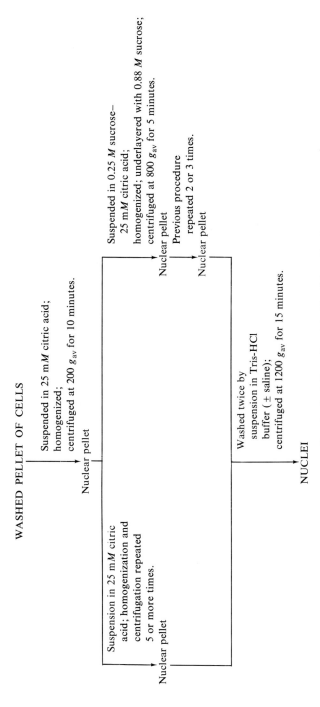

WASHED PELLET OF CELLS

Suspended in 25 mM citric acid;
homogenized;
centrifuged at 200 g_{av} for 10 minutes.

Nuclear pellet

Suspension in 25 mM citric
acid; homogenization and
centrifugation repeated
5 or more times.

Nuclear pellet

Suspended in 0.25 M sucrose–
25 mM citric acid;
homogenized; underlayered with 0.88 M sucrose;
centrifuged at 800 g_{av} for 5 minutes.

Nuclear pellet
Previous procedure
repeated 2 or 3 times.

Nuclear pellet

Washed twice by
suspension in Tris-HCl
buffer (± saline);
centrifuged at 1200 g_{av} for 15 minutes.

NUCLEI

FIG. 1. Isolation of nuclei with citric acid.

examined by phase-contrast microscopy, is seen to consist of clean nuclei without adherent tags of cytoplasm.

Although clean nuclei can be obtained in this way, it is frequently necessary to subject the nuclei to six or more cycles of homogenization and centrifugation to remove all the clumps of nonadherent cytoplasmic material which pellet with the nuclei. On occasion, some lysis of the nuclei occurs before cytoplasmic contamination has been reduced to acceptable levels. The removal of the clumps of cytoplasm is accelerated in the second variation of this method [after Busch and Smetana (9)]. In this, the pellet of nuclei, debris, and whole cells obtained by centrifugation after homogenization in 25 mM citric acid is resuspended by Vortex mixing and homogenized as before in ice-cold 0.25 M sucrose–25 mM citric acid. Centrifuge tubes are half-filled with the homogenate, which is then underlayered with an equal volume of 0.88 M sucrose–25 mM citric acid. The underlayering is conveniently done by injecting the heavy sucrose from a syringe through a horse-serum needle to the bottom of the centrifuge tube. The homogenate is then centrifuged at 800 g_{av} for 5 minutes at 4°C. The nuclei sediment through the denser sucrose, and debris collects at the interface between the 0.25 M and 0.88 M sucrose solutions. The supernatant fluid is removed by aspiration, care being taken to remove all the material at the interface before the 0.88 M sucrose solution is taken off, and the sides of the tubes are carefully wiped clean with sterile surgical gauze. The process of resuspension, homogenization, and centrifugation through 0.88 M sucrose is repeated until there is no debris visible in the supernatant fluid and the nuclei are seen in the phase-contrast microscope to be clean. Acceptable nuclei can be obtained in this way by three or, at most, four cycles of homogenization and centrifugation.

Finally, to remove the citric acid from nuclei prepared in either of these ways, the nuclei are washed twice, by suspension of the pellet by Vortex mixing in 10–20 volumes of ice-cold 0.2 M Tris-HCl (pH 7.5), or in 0.15 M NaCl–0.1 M Tris-HCl (pH 7.5). The suspension is allowed to stand in ice for 10–15 minutes, and nuclei are recovered by centrifugation at 1200 g_{av} for 15 minutes at 4°C. These nuclei can be used directly for the preparation of chromatin, or they can be stored frozen at −20°C in a tightly stoppered container.

Both of these procedures are simple and very tolerant of variations in conditions. However, it is essential that homogenizer pestles of the correct sizes are used, and that the progress of the preparations is monitored at each step with the phase-contrast microscope. In the sucrose–citric acid variation of the method, careful wiping of the sides of the centrifuge tube is particularly important.

C. Method II: Isolation with Sucrose and Detergent

The two procedures described in this section are again essentially variations of the one theme, and both start with a pellet of washed cells. In the first (summarized in Fig. 2), the cells are suspended by Vortex mixing in an ice-cold hypotonic buffer consisting of 10 mM NaCl–1.5 mM MgCl$_2$–10 mM Tris-HCl (pH 7.4) at a final concentration of about 4×10^7 cells/ml. After the suspension has stood in ice for 10–15 minutes to allow the cells to swell, it is homogenized in a glass Potter–Elvehjem homogenizer vessel by 8–10 up-and-down strokes of a cylindrical Teflon pestle, motor-driven at 2000 rpm.

WASHED PELLET OF CELLS

4×10^7 cells/ml swollen in
10mM NaCl–1.5mM MgCl$_2$–10mM Tris-HCl (pH 7.4);
homogenized;
sucrose added to 0.25 M;
centrifuged at 800 g_{av} for 10 minutes.

Nuclear pellet

Washed in 0.25 M sucrose
in buffer A (3 mM CaCl$_2$ –10 mM Tris-HCl (pH 7.4);
centrifuged at 800 g_{av} for 10 minutes.

Nuclear pellet

Homogenized in 2.2 M sucrose
in buffer A;
centrifuged at 40,000 g_{av} for 1 hour.

Nuclear pellet

Homogenized in 0.25 M
sucrose in buffer A;
centrifuged at 1200 g_{av} for 10 minutes.

Nuclear pellet

Homogenized in 0.25 M sucrose
in buffer A containing 1% Triton X-100;
centrifuged at 1200 g_{av} for 10 minutes.

Nuclear pellet

Washed twice in 0.25 M sucrose
in buffer A;
centrifuged at 1200 g_{av} for 10 minutes.

PURIFIED NUCLEI

FIG. 2. Isolation of nuclei with sucrose and detergent. From Rickwood and Birnie (3). Reproduced with permission.

The pestle should be looser than that used in Method I, a clearance of 0.13–0.15 mm being suitable for most cells. It is particularly important to examine the homogenate by phase-contrast microscopy at this stage since, if cell lysis is incomplete, homogenization can be continued. With most types of cell, however, it is inadvisable to use more than 20 strokes of the pestle, otherwise a significant proportion of the nuclei are ruptured. Immediately after homogenization is complete, sufficient 2 M sucrose is added to bring the concentration of sucrose to 0.25 M, and the mixture is centrifuged at 800 g_{av} for 10 minutes at 4°C. The pellet, which consists of nuclei, debris, and unbroken cells, is washed by resuspension and homogenization in 0.25M sucrose–3mM CaCl$_2$–10mM Tris-HCl(pH 7.4) followed by centrifugation at 800 g_{av} for 10 minutes. The crude nuclear pellet so obtained is then homogenized in 10 volumes of 2.2 M sucrose–3mM CaCl$_2$–10mM Tris-HCl (pH 7.4) using 10 up-and-down strokes of a tight-fitting Teflon ball homogenizer pestle (clearance 0.10 mm), motor-driven at 2000 rpm. The homogenate is centrifuged at 40,000 g_{av} for 1 hour at 5°C, preferably in a swing-out rotor though, for convenience, with large-scale preparations, a fixed-angle rotor can be used. In the latter, a considerable proportion of the nuclei pellet as a smear along the centrifugal wall of the centrifuge tube.

Nuclei prepared in this way have no gross cytoplasmic tags, and the preparation is free of contamination by cytoplasmic material. However, the outer nuclear membrane is still intact, and to remove this and the attached perinuclear ribonucleoprotein particles, treatment with detergent is required. This treatment also serves to remove the last traces of cytoplasm which may still be attached to some of the nuclei. The nuclei are first washed by being suspended in 10–20 volumes of 0.25 M sucrose–3mM CaCl$_2$–10 mM Tris-HCl(pH 7.4) using a few strokes of the Teflon ball pestle (just sufficient to ensure that all clumps are dispersed), and collected by centrifugation at 1200 g_{av} for 10 minutes at 4°C. The pellet is suspended in the same way in the same medium containing 1% (v/v) Triton X-100, and the nuclei are again collected by centrifugation at 1200 g_{av} for 10 minutes at 4°C. The nuclei are washed free of detergent by two cycles of suspension and centrifugation under the same conditions as before in 0.25 M sucrose–3 mM CaCl$_2$–10mM Tris-HCl(pH 7.4) or in whatever buffer is suitable for the use to which the nuclei are to be put. These nuclei can either be used directly for the preparation of chromatin, or they can be stored frozen at −20°C in a tightly stoppered vessel.

The second procedure using dense sucrose and detergent, summarized in Fig. 3, is a new one developed by Dr. R. S. Gilmour of the Beatson Institute and is yet to be characterized as fully as the first. However, chromatin satisfactory for transcriptional studies has been prepared from nuclei isolated in this way (see Chapter 28, Volume 19 of this series) and, moreover,

WASHED PELLET OF CELLS

> Homogenized in 5 mM MgCl$_2$–
> 0.5 mM dithiothreitol–10 mM Tris-HCl
> (pH 7.5)–1% Triton X-100;
> overlaid on 2.2 M sucrose–1 mM MgCl$_2$–
> 0.5 mM dithiothreitol–0.28 M NaCl–10 mM Tris-
> HCl (pH 7.5);
> centrifuged at 18,000 g_{av} for 1 hour.

Nuclear pellet

> Washed with 0.28 M NaCl;
> centrifuged at 2700 g_{av} for 10 minutes.

NUCLEI

FIG. 3. Isolation of nuclei with sucrose and detergent (short procedure).

since it is a rapid procedure, it may well prove to be eminently suitable in many circumstances. The cells are suspended in 20 volumes of ice-cold 5 mM MgCl$_2$–0.5 mM dithiothreitol–10 mM Tris-HCl (pH 7.5)–1% (v/v) Triton X-100, and the suspension is homogenized by five up-and-down strokes of a loose-fitting (0.13–0.15 mm clearance) cylindrical Teflon pestle in a glass Potter–Elvehjem homogenizer vessel. The pestle is slowly driven by hand only. The homogenate (10 ml) is then layered on top of 20 ml of 2.2 M sucrose–0.28 M NaCl–1 mM MgCl$_2$–0.5 mM dithiothreitol–10mM Tris-HCl (pH 7.5) and centrifuged in a swing-out rotor at 18,000g_{av} for 1 hour at 5°C. The top layer is removed by aspiration, care being taken to remove all of the cell debris which has collected at the interface between the buffer and the 2.2M sucrose, and the sucrose layer is decanted. The pellet, which is somewhat gelatinous in appearance, is then drained by up-ending the centrifuge tube on a piece of absorbent paper. The pellet is then washed by resuspending it in 10 volumes of 0.28 M NaCl, either by Vortex mixing or by a few strokes of a hand-driven homogenizer pestle, and the nuclei are recovered by centrifugation at 2700 g_{av} for 10 minutes. Nuclei isolated in this way are used directly for the preparation of chromatin.

Neither of these procedures is as tolerant of error as the citric acid procedures, and more care must be taken to monitor each step and to follow closely operational conditions found to give a satisfactory product. In neither method are steps taken to inhibit proteases and nucleases; consequently, it is particularly important to keep the material chilled thoroughly at all stages and to work as rapidly as possible. In the first of the two procedures, it is particularly important to control very carefully the concentration of the dense sucrose in which the nuclei are homogenized and the tempera-

ture at which the centrifugation is done (see Section II,A), and to add strictly
10 volumes of this solution to the nuclear pellet since the final concentration
of sucrose in the mixture should be no more than 2.0 M (or less than 1.8 M).
Similar considerations apply to the second sucrose–detergent method. In
addition, in this method it is important not to use too low a volume of buffer
for the initial homogenization, otherwise the yield of nuclei is reduced
(Section II,A).

D.   Comparison of the Products of Methods I and II

The morphology of nuclei prepared in citric acid differs from that of
nuclei in whole cells. The difference is quite marked when high concentrations (0.25 M) of citric acid are used, as described by Busch and Smetana (9),
in that the nucleoli appear to be enlarged, with hazier borders, and the
chromatin is clumped. Most of these changes can be attributed to extraction
of chromosomal proteins, in particular the more basic proteins, and to precipitation of some of the proteins by the low pH of the medium (9, 14). However, with the milder conditions (25 mM citric acid) described in Section
III,B, the morphological changes seen are much less marked, and the nuclei
do not appear to be severely distorted. Electron micrographs of nuclei prepared in this way show that the outer nuclear membrane is effectively removed by homogenization in citric acid, and that preparations of nuclei are
devoid of cytoplasmic contaminants, especially when the sucrose–citric acid
variation of this method is used. Nuclei prepared by the sucrose–detergent
methods are less distorted, and they retain morphological features more
closely resembling those seen in nuclei in the whole cells. However, nuclei
from some cells do show clumping of the chromatin even when they have
been isolated by these gentle procedures [see Smuckler et al. (2)]. Nuclei
prepared by similar methods omitting the detergent retain the outer nuclear
membrane, together with the perinuclear ribosomes, while treatment with
Triton X-100 removes these structures without causing the nuclei to lyse (2),
indicating that the inner nuclear envelope is quite resistant to mild treatment
with detergent. However, more drastic treatment, such as with Tween 40
plus sodium deoxycholate (8), does cause the nuclei to lyse.

MacGillivray et al. (15) have compared the chromatin proteins from
nuclei isolated in citric acid with those from nuclei prepared by sucrose–
detergent methods. The ratio of protein to DNA in "citric acid" nuclei was
very similar to that in "sucrose" nuclei which had been washed with Triton
X-100. In contrast, nuclei prepared by the double-detergent method of
Penman (8) yielded chromatin which was severely depleted in protein. The
"citric acid" nuclei appeared to be somewhat deficient in histone Hl while
Triton-washed "sucrose" nuclei did not. Polyacrylamide gel electrophoresis

in sodium dodecyl sulphate (SDS) showed that the chromatin nonhistone proteins from Triton-washed "sucrose" nuclei contained a relatively smaller proportion of high-molecular-weight species than those from "citric acid" nuclei. However, it must be noted that these analyses were done with proteins which had been separated from the DNA by hydroxylapatite column chromatography of chromatin dissociated in 2 M NaCl–5 M urea, and MacGillivray *et al.* (*15*) found that the recovery of protein from the column was much lower with material from "citric acid" nuclei. The low recovery of protein from nuclei isolated in citric acid appeared to be nonspecific and, thus, is not indicative of specific losses caused by the use of citric acid in the isolation procedure. However, this point should be borne in mind if nuclei are being prepared for the purpose of isolating nonhistone proteins (*4*).

Nuclei isolated by either Methods I or II contain only a small proportion of RNA [less than 10% of the total nucleic acids (*15*)], very little of which is 18 S RNA. This confirms the conclusion drawn from the microscopic examinations of these preparations, namely, nuclear preparations obtained by these procedures are virtually devoid of cytoplasmic contamination and perinuclear ribosomes. High-molecular-weight nuclear RNA is better preserved in "citric acid" nuclei (particularly those isolated in 0.25 M citric acid) than in those prepared by sucrose–detergent procedures (*9,14*), indicating that nucleases are considerably more active in nuclei isolated by the latter methods. Functionally, nuclei prepared in citric acid are inactive whereas those prepared by the first of the sucrose–detergent methods (Fig. 2) actively incorporate thymidine 5'-monophosphate (TMP) and uridine 5'-monophosphate (UMP) into DNA and RNA, respectively, when incubated *in vitro* under appropriate conditions. Nuclei isolated by the second sucrose–detergent method (Fig. 3) have not yet been tested for functional activity. However, chromatin prepared from "citric acid" nuclei and from Triton-washed "sucrose" nuclei is actively transcribed *in vitro* by *E. coli* RNA polymerase and, moreover, chromatin from both sources appears to retain much (if not all) of its tissue-specific template activity (see, for example refs 10–12, 16–18).

ACKNOWLEDGMENTS

The Beatson Institute is supported by grants from the Medical Research Council and the Cancer Research Campaign. I am indebted to Dr. R. S. Gilmour of the Beatson Institute for details of the method summarized in Fig. 3 and for permission to publish these details. I am grateful to Butterworths, London, for permission to republish Fig. 2.

REFERENCES

1. Roodyn, D. B., *in* "Subcellular Components: Preparation and Fractionation" (G. D. Birnie, ed.), 2nd ed. p. 15. Butterworths, London, 1972.

2. Smuckler, E. A., Koplitz, M., and Smuckler, D. E., *in* "Subnuclear Components: Preparation and Fractionation" (G. D. Birnie, ed.), p. 1. Butterworths, London, 1976.

3. Rickwood, D., and Birnie, G. D., *in* "Subnuclear Components: Preparation and Fractionation" (G. D. Birnie, ed.), p. 129. Butterworths, London, 1976.

4. MacGillivray, A. J., *in* "Subnuclear Components: Preparation and Fractionation" (G. D. Birnie, ed.), p. 209. Butterworths, London, 1976.

5. Chaveau, J., Moulé, Y., and Rouiller, C., *Exp. Cell Res.* **11**, 317 (1956).

6. Widnell, C. C., and Tata, J. R., *Biochem. J.* **92**, 313 (1964).

7. Blobel, G., and Potter, V. R., *Science* **154**, 1662 (1966).

8. Penman, S., *J. Mol. Biol.* **17**, 117 (1966).

9. Busch, H., and Smetana, K., "The Nucleolus." Academic Press, New York, 1970.

10. Paul, J., and Gilmour, R. S., *J. Mol. Biol.* **16**, 242 (1966).

11. Paul, J., and Gilmour, R. S., *J. Mol. Biol.* **34**, 305 (1968).

12. Paul, J., Carroll, D., Gilmour, R. S., More, I. A. R., Threlfall, G., Wilkie, M., and Wilson, S., *Karolinska Symp. Res. Methods Reprod. Endocrinol.* **5**, 277 (1972).

13. Grunicke, H., Potter, V. R., and Morris, M. P., *Cancer Res.* **30**, 776 (1970).

14. Busch, H., Choi, Y. C., Daskal, I., Inagaki, A., Olson, M. O. J., Reddy, R., Ro-Choi, T. S., Shibata, H., and Yeoman, L. C., *Karolinska Symp. Res. Methods Reprod. Endocrinol.* **5**, 35 (1972).

15. MacGillivray, A. J., Cameron, A., Krauze, J., Rickwood, D., and Paul, J., *Biochim. Biophys. Acta* **277**, 384 (1972).

16. Paul, J., Gilmour, R. S., Affara, N., Birnie, G. D., Harrison P. R., Hell, A., Humphries, S., Windass, J. D., and Young, B. D., *Cold Spring Harbor Symp. Quant. Biol.* **38**, 885 (1974).

17. Gilmour, R. S., Harrison, P. R., Windass, J. D., Affara, N. A., and Paul, J., *Cell Diff.* **3**, 9 (1974).

18. Gilmour, R. S., Windass, J. D., Affara, N., and Paul, J., *J. Cell Physiol.* **85**, 449 (1975).

Chapter 3

Preparation of Chromatin from Animal Tissues and Cultured Cells

WILLIAM T. GARRARD

Department of Biochemistry,
University of Texas Health Science Center,
Dallas, Texas

AND

RONALD HANCOCK

Swiss Institute for Experimental Cancer Research,
Lausanne, Switzerland

I. Introduction

Most laboratories engaged in studies of chromatin have developed their own preferred procedures for its isolation. Although these methods may appear numerous, they depend on a limited number of basic principles. Because recent advances in our understanding of the molecular structure of chromatin now allow a critical evaluation of many existing methods, we briefly review these concepts here.

More than 80% of the chromatin of higher cells consists of a flexible chain of subunits *(1–6)* [nucleosomes *(2)*], each consisting of about eight histone molecules [possibly two each of histones 2A, 2B, 3, and 4 *(3,4)*], and about 200 base pairs of DNA *(2,5)*. This DNA is believed to be wrapped around the outside of the histone cluster *(6,7)* with a packing ratio of about 6:1 *(2)* and a constraint equivalent to an average of one negative superhelical turn *(8)*. A subclass of nucleosomes may exist which contain, in addition, nonhistone proteins *(9,9a)*. Histone 1 appears to be located outside the nucleosome, and apparently it can form both intrachain *(2)* and interchain *(10)* cross-links under appropriate ionic strength conditions. DNA in the nucleus may be of chromosome length *(11)*, and regions remote from sites of replication or transcription are probably packed into specific

27

higher order structures (*12, 13*). Interactions between nucleosomes, and thus the conformation of and interactions between Chromatin molecules, are exquisitely sensitive to the ionic environment (*14–16*). It follows from these structural considerations that the ionic conditions employed during chromatin isolation are important in determining the properties of the product; and also that some degree of fragmentation of the giant chromatin molecules is necessary to obtain material suitable for experimental manipulations *in vitro*.

The nuclear envelope, which is made up of a pore–fibril network and of membranes (*17*), confines the chromatin of animal cells, except during mitosis. Thus, membrane contacts must be disrupted during the preparation of chromatin. We know of no convincing evidence that the lipids found in certain chromatin preparations (*18, 19*) are not derived from the nuclear envelope. Chromatin can be isolated as an independent, discrete structure after detachment of these surface layers (*20, 20a*). Transcription of RNA occurs at the periphery of condensed chromatin in the nuclear interior(*21*), and purification of chromatin requires elimination of the free ribonuclear-protein (RNP) products of transcription. The survival of the nucleolus as a structural entity depends on the isolation conditions; in view of the very active transcription of nucleolar genes the presence of nucleolar-derived regions in purified chromatin deserves careful evaluation.

We describe here a number of representative procedures with which we have direct experience, and which depend on somewhat different principles, for purification of chromatin from animal cells, starting from either whole cells (direct methods) or purified nuclei. The rationales, advantages and disadvantages, and appropriate uses of these procedures are discussed.

II.   Rationales of Methods

Procedures to isolate chromatin are based on its solubility properties, its size, its density, and/or its sensitivity to nucleases. Chromatin or isolated nucleosomes form precipitates in the presence of either moderate concentrations of divalent metals (1 to 10 mM Mg^{2+} or Ca^{2+})(*16,22*), or monovalent cations (40 to 250 mM Na$^+$) (*14–16,22*). Histone 1 appears to be responsible for the precipitation of chromatin caused by monovalent cations (*23*). Below these concentrations of cations, unfragmented chromatin exists as a gel. The size of chromatin DNA is of little importance when chromatin is prepared as a precipitate by differential centrifugation, but it is of paramount importance when chromatin is isolated as a gel. Depending on the preparation method, fragmented chromatin under low ionic strength conditions may not be a monodisperse solution of independent fibers, but rather a network of molecules with occasional fiber–fiber associations (24).

The buoyant density of chromatin is heterogeneous and intermediate between protein and DNA, approximately 1.4 to 1.5 gm/cm³ (9, 25–27). Nonionic media have been applied to fractionate chromatin by density, and they are discussed elsewhere (9). Ionic gradients are also useful in studying chromatin, but they require prior fixation of proteins to DNA, most commonly with formaldehyde (25–27).

A brief description of the rationale of using nucleases to prepare chromatin is presented later (Section VI,B).

III.   Technical Considerations

A.   Homogeneity of Starting Cell Populations

Chromatin is often prepared from a heterogeneous population of cells. Ideally, investigations of changes in chromatin during the growth cycle would best be performed using synchronized cells, studies on liver regeneration with purified parenchymal cells, and studies of globin synthesis with purified reticulocytes. Although they are beyond the scope of this article, we note that techniques exist to fractionate cell classes from tissues (28,29) and from mixed cell populations (30–32) and to synchronize cultured animal cells (33–35). Since inhibitors used to synchronize cells can cause unbalanced biological situations, appropriate controls should be performed using cells obtained by natural mitotic selection (33).

B.   Yield

As a general rule, the more steps involved in chromatin preparation, the lower the yield. Yields range from 30 to 90%, depending on methods; techniques of nuclei isolation rarely give better than 80% yields. An important unanswered question is whether selective losses of certain DNA sequences may occur; for example, methods based on chromatin precipitation require histones to be associated with DNA, and regions of free DNA (if they exist) could thus be lost using such techniques. Recent experiments based on hybridization, however, suggest that transcribed sequences are probably not lost during routine methods of chromatin isolation (36).

C. Proteolysis

Workers in the chromatin field are plagued with the problem of proteolytic degradation of proteins during experimental manipulations. This problem should receive paramount attention, but unfortunately has not, and the literature thus contains results that may be attributed to proteolysis artifacts [e.g., during incubation of chromatin for *in vitro* RNA synthesis, the template activity increases with time (37)]. The question of whether or not proteases reside within nuclei *in vivo* is not resolved; however, isolated nuclei (38) or chromatin (38,39) always contain some protease activity, regardless of the techniques employed, but this varies considerably depending on the cell source.

Three methods are used to overcome proteolysis problems: inhibitors, pH control, and ionic strength control. The best inhibitor is diisopropyl fluorophosphate (DFP); this compound is a volatile neurotoxin, and great care should be taken when it is used. Chromatin may be treated with DFP at a very early stage in purification to irreversibly inactivate endogenous proteases (40); since DFP at high concentrations lyses nuclei, only direct chromatin isolation methods are amenable to this procedure. Another inhibitor, phenyl methyl sulfonyl fluoride (PMSF), is less dangerous. Fresh stock solutions are made containing 0.1 M PMSF in isopropyl alcohol; this stock solution is diluted 1/100 in all buffer solutions used for nuclei and chromatin preparation (with rapid mixing) immediately before use. PMSF does not lyse nuclei; it does not completely inhibit proteases, but it does considerably reduce ($>50\%$) most problems (22).

Sodium bisulfite at 50 m M is a very effective protease inhibitor (39), but it can only be used in methods where chromatin exists as a precipitate since a high concentration is required to be completely effective and considerable alkali must be added to maintain a neutral pH. Bisulfite inhibition is not irreversible (22,39); it is chemically reactive and could possibly modify chromatin [even by catalysis leading to protein–DNA cross-links (41)], and it is also a disulfide reagent.

Endogenous proteases usually have pH optima around 8.0, and proteolysis can be greatly reduced by working at pH 6.5. Calf thymus nuclear preparations show complete inhibition of proteolysis at pH 6.5, while at pH 7.4 they contain fragmented histones (22). When extreme pHs are unadvisable, even pH 7.0 compared with 7.4 produces noticeable improvements (22). In general, very low protease activities are observed below 1 m M salt while extensive activity exists at 0.15 M salt (38,39). In practice, a combination of these factors is of great value in minimizing proteolysis; working rapidly and at 0°C are necessities, particularly when moderate ionic strength conditions are used.

SDS GEL   UREA GEL

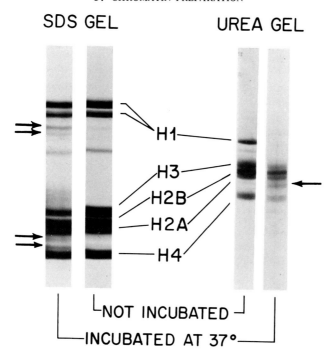

NOT INCUBATED

INCUBATED AT 37°

FIG. 1.   Assay of histone proteolysis. The left pair of SDS gels show typical early proteolysis products of calf thymus histones (arrows). The right pair of urea gels show histones from early regenerating rat liver before and after incubation of the chromatin for 4 hours at 37°C.

The simplest reliable assay for endogenous proteases in chromatin preparations is to perform gel electrophoresis of the histones prior to and after a 2- to 4-hour incubation at 37°C, using either urea (42) or sodium dodecyl sulfate (SDS) gel electrophoresis (4). Histone 1 is most susceptible, followed by histone 3; bands between histones 2A and 4 are indicative of proteolysis (Fig. 1) (38,39).

D.   DNA Size

Nucleases in chromatin preparations can often cause considerable fragmentation of DNA. In mouse and rat liver, which contain an endogenous Ca^{2+}, Mg^{2+}-activated endonuclease (43), preparation methods using Ca^{2+} or Mg^{2+} as stabilizing agents are not recommended, for internucleosomal DNA becomes fragmented during nuclear isolation (22) resulting in formation of nucleosome monomers. In an excellent technique for preparation of nuclei from tissues with high activities of endogenous nucleases (43), polyamines are used as stabilizing agents instead of divalent cations.

E. Miscellaneous

A number of other technical problems should be brought to the reader's attention. The question of contamination, either cytoplasmic or nucleoplasmic, is the most obvious and is discussed in Section XIII. Others include conformational changes in nucleosome structure (44–46), rearrangement of proteins [sliding and exchange between sites (20a,)], and protein loss during chromatin preparation. Dephosphorylation of phosphorylated histones can occur (47), oxidation-reduction or disulfide exchange reactions may lead to formation of fiber-fiber cross-links during aeration (48). The advantages of fresh versus frozen tissue should be considered, since freezing may cause structural changes (49). Since chromatin is one of the most complex structures in the cell, its study is subject to every problem and possibility of artifact encountered in other areas of biochemistry.

IV.   Direct Methods of Chromatin Isolation

We use the term "direct" to indicate that chromatin is isolated from a cell lysate rather than from purified nuclei; such procedures may also be used to isolate chromatin from mitotic cells (48,50). These methods usually employ isolation media at pH 8–8.5 where removal of nonchromatin proteins is improved (51), although proteolytic activity may not be minimal.

A.   Intermediate Ionic Strength Methods

Mechanical homogenization is usually necessary to rupture cells in tissues containing an intercellular matrix. Chromatin is also fragmented during this procedure (unless conditions are such that nuclei remain intact), and its subsequent recovery requires an ionic environment in which chromatin fragments interact to form a precipitate separable from the homogenate by low-speed sedimentation. The most extensively studied method of this type, that of Bonner et al. (52), evolved from earlier methods (14,51) and has been applied to a number of animal tissues. Slightly modified procedures are used by different investigators; one is given below for frozen rat liver.

Method 1.

Starting material: Tissue rapidly frozen, stored at −80°C, and broken into small fragments.

Solutions: A. Homogenization medium: 0.075 M NaCl–0.024 M Na$_2$EDTA (pH 8.0).
B. Washing and resuspension medium: 0.01 M Tris–0.2 mM EDTA–0.1% Triton X-100 (pH 8.0) at 0°C.
C. 0.01 M Tris–0.2 mM EDTA (pH 8.0) at 0°C.
D. 1.7 M sucrose–0.01 M Tris–0.2 mM EDTA (pH 8.0) at 0°C.

The procedure is shown in Scheme 1.

A prodecure developed from the same origins but for frozen cultured cells (50) is useful when cells must be stored or accumulated over a period of time before use. The sucrose centrifugation step (Scheme 1) may be added.

Comments: Proteases are fairly active under these conditions; DFP treatment at an early step therefore may be adopted as a routine procedure (40). Alternatively, it is recommended that 1 mM PMSF be added to all solutions (Section III,C).

Chromatin attaches readily to glass under these ionic conditions, and care is necessary to avoid losses with small samples. Intermolecular interactions occur at this ionic strength, and further fragmentation may be necessary to obtain manageable material for further studies. Intermolecular transfer of histone 1 and possible other histones (20a, 24) may occur in methods of this class; they therefore are not suitable for studies of chromatin assembly. Nonchromatin material (RNP complexes and membrane fragments) may be trapped between interacting chromatin molecules; these are generally [but not always (32)] removed by the sucrose gradient step. Fragments of membrane are seen by electron microscopy in chromatin prepared without sucrose centrifugation (49); inclusion of Triton X-100 reduces this contamination. Ethylenediaminetetraacetic acid (EDTA) reduces nuclease activity and chromatin aggregation. Vigorous homogenization can induce disulfide bond formation between H3 molecules (48).

B. Low Ionic Strength Methods

Chromatin may be isolated directly from cells of a number of cultured lines as "chromatin structures" (20), whose integrity may reflect the existence of giant topologically constrained DNA molecules (53). The preservation of these fragile structures requires gentle manipulations and absence of mechanical homogenization. Cells are lysed by a nonionic detergent in a medium of very low ionic strength containing EDTA to suppress intermolecular interactions. This procedure has recently been described in detail (20).

SCHEME 1

METHOD 1. DIRECT ISOLATION OF CHROMATIN FROM TISSUES AT INTERMEDIATE
IONIC STRENGTH[a]

15 g frozen $(-80°)$ rat liver in 200 ml solution A + 0.5% (v/v) 2-octanol in precooled Waring blender. Homogenize 1 minute, 85 V, then 4 minutes, 45 V.[b] Filter through two layers of miracloth.[c]

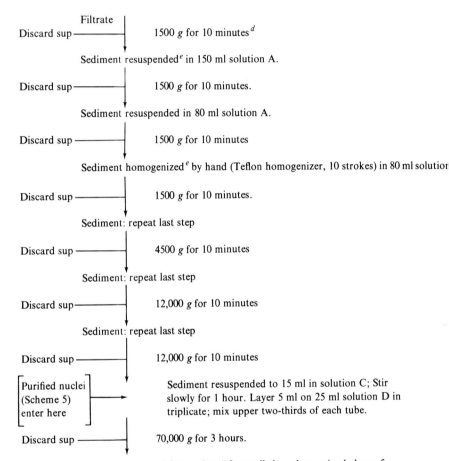

Filtrate

Discard sup ———— 1500 g for 10 minutes[d]

Sediment resuspended[e] in 150 ml solution A.

Discard sup ———— 1500 g for 10 minutes.

Sediment resuspended in 80 ml solution A.

Discard sup ———— 1500 g for 10 minutes

Sediment homogenized[e] by hand (Teflon homogenizer, 10 strokes) in 80 ml solution

Discard sup ———— 1500 g for 10 minutes.

Sediment: repeat last step

Discard sup ———— 4500 g for 10 minutes

Sediment: repeat last step

Discard sup ———— 12,000 g for 10 minutes

Sediment: repeat last step

Discard sup ———— 12,000 g for 10 minutes

⌈Purified nuclei⌉   Sediment resuspended to 15 ml in solution C; Stir
⎢(Scheme 5) ⎥──────► slowly for 1 hour. Layer 5 ml on 25 ml solution D in
⌊enter here ⌋   triplicate; mix upper two-thirds of each tube.

Discard sup ———— 70,000 g for 3 hours.

Sediment removed from tubes (after walls have been wiped clean of debris) and resuspended[e] in solution C (Teflon homogenizer) to desired concentration. Residual sucrose may be removed by two centrifugations (30,000 g for 20 minutes) or by dialysis.

Time required ≃ 7 hours   Yield (as DNA) ≃ 50%

---

[a] All operations are at 4°C.
[b] Voltages for 110 V instruments.
[c] Chicopee Co., Miltown, New Jersey.
[d] All g values in this article represent g_{max}.
[e] Use a spatula to first completely resuspend pellet in a small volume, then dilute.

Method 2.

Starting material: Cells in suspension culture, or detached from mono-layers by trypsin followed by blocking with trypsin inhibitor (*20*).

Lysing solutions: E. 0.1 *M* sucrose (Schwarz–Mann, RNase-free) in 0.2 m*M phosphate buffer (prepared by titrating NaH₂PO₄* with KOH to pH 7.5, to give a final phosphate con-centration of 0.2m*M*).

F. 0.5% (v/v) Nonidet P-40 (NP40: Shell Chemical Co.) in 0.2m*M* EDTA, pH 7.5. The NP-40 is added drop by drop, with stirring.

Washing solution: G. As solution E, but pH 8.5.

SCHEME 2

METHOD 2. ISOLATION OF CHROMATIN STRUCTURES FROM CULTURED CELLS[a]

Pellet of cells washed once in growth medium. Resuspended in solution E at ~5 × 10⁶ cells/ml (gentle vortex).

Discard sup ——————————| 1500 *g* for 7 minutes

Resuspend cells in solution E at 1–5 × 10⁷ cells/ml (gentle Vortex). Add drop by drop equal volume of solution F (gentle Vortex). Layer lysate onto 5 volumes G (wide-bore polystyrene pipette[b]).

Remove upper part of sup from surface by pipette; wipe inside tube and remove remainder of sup by pipette and discard

3000 *g* for 15 minutes (swing-out rotor)

Take up pellet intact in same volume of solution E used above; relayer on solution G as above in Dounce homogenizer[c] (pipette as above).

Remove sup as above ——————— 3,000 *g* for 15 minutes and discard

Resuspend chromatin pellet in 0.2 m*M* EDTA–0.2 m*M* dithiothreitol by 5–20 strokes of tight piston.

Time required ≃ 1.5 hours   Yield (as DNA) ≃ 80–90%

[a]All operations are at 4°C.
[b]Falcon Plastics, Oxnard, California.
[c]Kontes Glass Co., Vineland, New Jersey.

Resuspension medium: 0.2 mM EDTA–0.2 mM dithiothreitol (pH 7.2). It is preferable to sterilize these solutions by membrane filtration. The entire procedure is shown in Scheme 2.

Comments: Mild fragmentation of chromatin structures yields linear chromatin fragments which show virtually no intermolecular interaction by biochemical or electron microscopical criteria (20), and which are thus suitable for electron microscopical visualization of transcription and also for separation of regions of chromatin-bearing nascent RNA (54). Numerous samples of relatively small numbers of cells are easily processed, and this procedure is thus convenient for studies of biosynthesis and assembly of chromatin (55). This method has been applied to frozen cells (56), but some interchain transfer of histone 1 may be detectable after freezing and thawing (49). Higher order conformations are probably unfolded during this procedure, since only chromatin fibers of ~120 Å diameter are seen in sections of chromatin structures (20).

V.   Chromatin Isolation from Nuclei

A.   Isolation of Nuclei

Other contributors to these volumes have discussed nuclear isolation methods; however, we describe briefly here procedures forming an integral part of methods for chromatin isolation which have proved very reliable for calf thymus and for cultured cells, essentially eliminating proteolysis of histones (22).

1.   TISSUES (METHOD 3)

Starting material: Fresh calf thymus tissue, transported to the laboratory at 0° in 200 ml of solution H (see below). Remove membrane and connective tissue and mince 25 gm in 50 ml of solution H.

Solutions: H. Homogenization medium: 0.3 M sucrose– 0.05 M triethanolamine–0.025 M KCl–5 mM MgCl$_2$–1 mM PMSF (pH 6.5) at 4°C.

I. 2.15 M sucrose–0.05 M triethanolamine–0.025 M KCl–5 mM MgCl$_2$ –1 mM PMSF (pH 6.5) at 4°C.

The procedure is shown in Scheme 3.

Comments: The method is based on that of Blobel and Potter (57), but triethanolamine is employed instead of Tris to allow the use of cross-linking reagents which react with amino groups. For tissues which contain endogenous nucleases (Section III,D), other methods should be used for isola-

SCHEME 3

METHOD 3. ISOLATION OF CALF THYMUS NUCLEI FOR CHROMATIN PREPARATION[a]

---

25 gm fresh minced calf thymus in 200 ml of solution H in Waring blender (precooled). Homogenize at 40 V for 5 minutes.[b] Filter through two layers of cheesecloth.

↓

Filtrate

Discard sup ——————| 800 *g* for 10 minutes.

↓

Sediment resuspended[c] in 40 ml of solution H and filtered through cheese-cloth to remove any lysed nuclei.

To filtrate, add 2.5 volumes of solution I. Mix, distribute to centrifuge tubes leaving room for a 20% (by volume) underlay with solution I; add underlay.

Discard sup, ——————| 116,000 *g* for 90 minutes
wipe tube

↓

Purified nuclei

Time required ≃ 3.5 hours   Yield (as DNA) ≃ 80%

---

[a]All steps are performed at 4°C.
[b]Voltage for 110 V instruments. Foaming can be minimized by inserting a plastic container tightly into the blender vessel to just cover the solution–air interface.
[c]First use a spatula with a small volume of buffer; then add more buffer and use a Teflon piston that fits very loosely into plastic centrifuge tubes for gentle homogenization.

tion of nuclei (*43*). When using liver, cell lysis is best performed on finely minced tissue using a motor-driven Teflon homogenizer (*57*) rather than a Warning blender. Attempts to wash calf thymus nuclei with Triton X-100 (0.1%) have led to lysis (*22*).

2. CULTURED CELLS

Methods to isolate nuclei from cultured cells differ from those used for tissues. Blenders are not commonly used for cell lysis, a hypotonic swelling step is usually necessary, and on occasion problems of aggregation of cells or nuclei may occur.

Method 4a.

Starting material: Mouse P815 cells (*50*) suspended in solution J (below) at 2 to 7 × 10⁷ cells/ml.

SCHEME 4

METHODS 4a and 4b. ISOLATION OF NUCLEI FROM CULTURED CELLS FOR CHROMATIN PREPARATION[a]

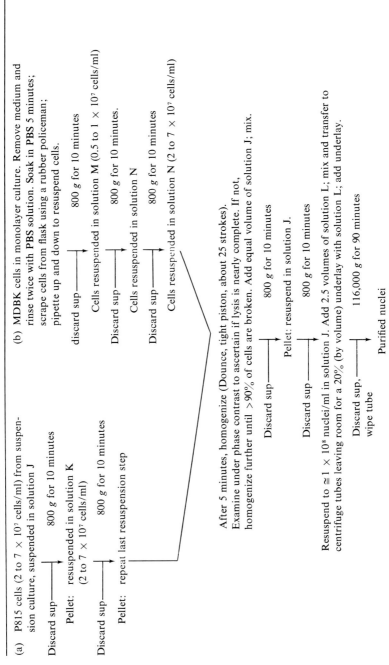

(a) P815 cells (2 to 7 × 10⁷ cells/ml) from suspension culture, suspended in solution J

Discard sup ⟶ 800 g for 10 minutes

Pellet: resuspended in solution K (2 to 7 × 10⁷ cells/ml)

Discard sup ⟶ 800 g for 10 minutes

Pellet: repeat last resuspension step

(b) MDBK cells in monolayer culture. Remove medium and rinse twice with PBS solution. Soak in PBS 5 minutes; scrape cells from flask using a rubber policeman; pipette up and down to resuspend cells.

discard sup ⟶ 800 g for 10 minutes

Cells resuspended in solution M (0.5 to 1 × 10⁷ cells/ml)

Discard sup ⟶ 800 g for 10 minutes.

Cells resuspended in solution N

Discard sup ⟶ 800 g for 10 minutes

Cells resuspended in solution N (2 to 7 × 10⁷ cells/ml)

After 5 minutes, homogenize (Dounce, tight piston, about 25 strokes). Examine under phase contrast to ascertain if lysis is nearly complete. If not, homogenize further until >90% of cells are broken. Add equal volume of solution J; mix.

Discard sup ⟶ 800 g for 10 minutes

Pellet: resuspend in solution J.

Discard sup ⟶ 800 g for 10 minutes

Resuspend to ≅1 × 10⁸ nuclei/ml in solution J. Add 2.5 volumes of solution L; mix and transfer to centrifuge tubes leaving room for a 20% (by volume) underlay with solution L; add underlay.

Discard sup, ⟶ 116,000 g for 90 minutes
wipe tube

Purified nuclei

Time required ≃ 5 hours   Yield (as DNA) ≃ 70%

Solutions: J. Solution H (Method 3), but pH 7.4 at 4°C.
K. 0.01 M triethanolamine–0.01 M KCl–1.5 mM MgCl$_2$ (pH 7.4) at 4°C.
L. Solution 1 (Method 3), but pH 7.4 at 4°C.

The procedure is shown in Scheme 4.

Method 4b.

Starting material: Monolayer cultures of bovine kidney cells (line MDBK).

Solutions: PBS (Dulbecco phosphate buffered saline minus Ca^{2+} and Mg^{2+}).
NaCl 8.0 gm/liter, KCl 0.2 gm/liter, Na$_2$HPO$_4$ 1.15 gm/liter, KH$_2$PO$_4$ 0.2 gm/liter, pH 7.4.
J, L (above).
M. 0.3 M sucrose–0.05 M triethanolamine–0.025 M KCl–1.5 mM CaCl$_2$–1 mM PMSF (pH 7.4) at 4°C.
N. 0.01 M triethanolamine–0.01 M KCl–1.5 mM CaCl$_2$ (pH 7.4) at 4°C.

The procedure is also shown in Scheme 4.

Comments: MDBK cells aggregate in solutions containing Mg^{2+} ions, but substitution by Ca^{2+} completely eliminates this problem. The buoyant density of nuclei varies in different cell lines, and the concentration of sucrose in solution L (Method 4a) may require modification.

B. Chromatin Isolation

For chromatin isolation from nuclei, an abbreviated version of the intermediate ionic strength method (Method 1) is used; fewer washes are necessary due to the homogeneity of the starting material.

Method 5.

Starting material: Purified nuclei, washed twice by gentle resuspension in solution J (Method 4a) and centrifugation (800 g for 10 minutes at 4°C).

Solutions: As in Method 1. The procedure is shown in Scheme 5.

Comments: The volumes of buffers A–C (method 1) are related to the amount of material; 25 ml/10 mg DNA is a reasonable ratio. Proteolysis problems are dealt with as described in Method 1.

VI. Fragmentation of Chromatin

Purified chromatin preparations contain fragments associated by inter-molecular interactions, or very long chromatin molecules; further fragmen-

SCHEME 5

METHOD 5.   ISOLATION OF CHROMATIN FROM PURIFIED NUCLEI[a]

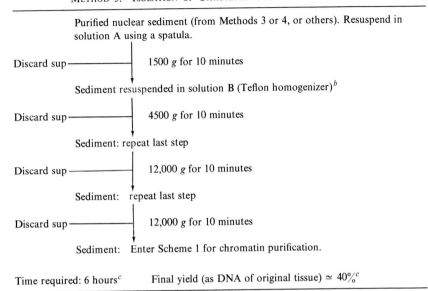

Purified nuclear sediment (from Methods 3 or 4, or others). Resuspend in solution A using a spatula.

Discard sup ——————— | 1500 g for 10 minutes

Sediment resuspended in solution B (Teflon homogenizer)[b]

Discard sup ——————— | 4500 g for 10 minutes

Sediment: repeat last step

Discard sup ——————— | 12,000 g for 10 minutes

Sediment: repeat last step

Discard sup ——————— | 12,000 g for 10 minutes

Sediment: Enter Scheme 1 for chromatin purification.

Time required: 6 hours[c]   Final yield (as DNA of original tissue) $\simeq 40\%$[c]

[a]All steps are at 4°C.
[b]See note c, Scheme 3.
[c]To chromatin.

tation (shearing) may be necessary to obtain material suitable for further manipulations. Many published studies have employed mechanical fragmentation, but more gentle fragmentation by nucleases will probably replace these methods in the future, especially for structural studies.

A. Mechanical Fragmentation

To suppress intermolecular ionic interactions, resuspension and mechanical fragmentation is carried out in a low ionic strength buffer (< 1 mM) containing a chelating agent (EDTA or EGTA, 0.2 mM) and an agent to reduce S—S bonds (β-mercaptoethanol or dithiothreitol, 0.2 mM); dialysis against the same solution removes ions carried over from previous steps. Mechanical fragmentation of chromatin has been studied in less detail than that of DNA, but the same procedures have been employed (Table I). The length and structures of the fragments generated are very dependent on the conditions used; it is recommended that the procedure employed be calibrated by examination of the product in the electron microscope [e.g., by the simple technique employed by Oudet et al. (2), by determination of the

TABLE I

LENGTH OF DNA IN CHROMATIN FRAGMENTED BY DIFFERENT METHODS

Method	Reference	DNA length[a] (double-stranded) as		
		Daltons	Base pairs	Nucleo-somes[b]
A. Mechanical				
VirTis homo-	(14)	8×10^6	12,000	60
genizer or				
Waring blender	(10)	12×10^6	18,000	90
	(15)	2.5×10^6	3,800	19
	(58)	3 to 5×10^6	6,000	30
	(59)	3×10^6	4,500	23
Sonication	(59)	0.7×10^6	1,000	5
	(60)		200	≥ 1
			(and greater)	
French Press	(61)	2×10^6	3,000	15
Hand homogenization:				
Teflon-glass	(3)	5 to 10×10^7	76,000 to	380 to
			150,000	750
Dounce	(20)	1.5 to 2×10^6	2,700	14
(tight piston)				
B. Nuclease				
DNase II	(15)	0.5×10^6	800	4
	(9a)		≥ 500	≥ 3
Micrococcal	(2, 16, 44–		200	1
nuclease	46, 62, 63)		(and	(and
			multiples)	multiples)
DNase I	(64)			
Restriction	(59)		Variable, depending on	
enzymes	(65)		enzyme and conditions.	

[a]Unless otherwise indicated, these values are weight average sizes.
[b]Approximate figure assuming 200 base pairs of DNA per nucleosome, and that all DNA is nucleosomal.

DNA fragment length by velocity sedimentation in alkaline sucrose gradients with appropriate markers (66), or by gel electrophoresis (67). The limiting size attainable by mechanical fragmentation represents [with one exception (60)] a length corresponding to three to five nucleosomes; single nucleosomes are usually prepared by enzymic cleavage (Table I). Depending on the preparative procedure, solutions of fragmented chromatin may still contain fragments of nucleoli, nuclear pore complex, and debris; these may be removed by centrifugation (at 15,000 g for 15 minutes).

Mechanical fragmentation results in detectable modification of certain properties of chromatin. A reduction (1.5°C) in the melting temperature

(T_m) of DNA in sonicated chromatin has been interpreted in terms of an increased number of end regions (59); the effects of sonication depend on the ionic conditions (68). Different degrees of fragmentation (VirTis homogenizer) yield products with small differences in circular dichroism (CD) spectra and in T_m of their DNA (69). Homogenization of chromatin may render the nucleosome-associated DNA more accessible to digestion by micrococcal nuclease (44–46), although such chromatin is transcribed by E. coli RNA polymerase with tissue-specific discrimination (58). The possible release of nonhistone proteins during fragmentation has been discussed (70). Intermolecular association is greater in fragmented chromatin than in unsheared chromatin, which suggests the generation of sites available for interactions (49). These subtle modifications will probably be interpretable with better understanding of the relevant molecular structures. Sonication conditions employed for chromatin can break ribosomal RNA molecules (71) and thus may fragment template-bound nascent RNAs.

B.   Enzymic Fragmentation

An area of intensive present research and great future potential for studies of chromatin rests on the use of nucleases, whose action have been used for separation of template active and inactive regions (9a, 15) and for release of single nucleosomes and multimers (5, 16, 44–46, 62, 72) (Table I). Micrococcal nuclease and DNase II have a strong preference to form double-strand breaks in internucleosomal DNA (22, 44). The former enzyme produces fewer cuts in nucleosomal DNA during brief exposure, but both nucleases also degrade nucleosomal DNA after prolonged treatment (5, 16, 22, 44–46). Micrococcal nuclease requires Ca^{2+} ions, while DNase II is inhibited by divalent metals. Thus, chromatin is insoluble during enzyme treatment with the former nuclease, but addition of EDTA to stop digestion solubilizes the fragmented chromatin. The contrary is true for DNase II. DNase I, which requires Mg^{2+} ions, shows little discrimination for internucleosomal DNA, and rapidly forms single-strand nicks within nucleosomal DNA (64).

Since restriction endonucleases show high sequence specificity and since histones appear to be associated randomly with specific DNA sequences (65), these enzymes look exceptionally promising to fragment chromatin selectively (67). The fragment sizes in limit digests of chromatin are larger than those from purified DNA, owing to the random occurrence of histone-covered restriction sites (65).

We have found the following method extremely reliable for preparation of polynucleosomes from nuclei of tissues and cultured cells (73).

SCHEME 6

METHOD 6. ISOLATION OF CHROMATIN AS NUCLEOSOMES AND OLIGOMERS[a]

Nuclei suspended in solution 0 at 5 mg DNA/ml are preincubated for 5 minutes at 37°C. One ninety-ninth volume of micrococcal nuclease solution is added with rapid mixing (200 units/ml final).

Incubate 20 sec to 15 min at 37°C, cool rapidly to 0°C

Discard sup; — 800 g for 10 minutes
wipe inside of
tube to remove
residual solution 0 Resuspend in solution P to original volume. Disperse pellet with the tip of a wide-bore Pasteur pipette without suction; then suck back and forth until completely solubilized.[b]

Discard — 10,000 g for 10 minutes
pellet

Supernatant

Time required ≃ 1 hour   Yield (as acid-precipitable DNA) = 88 to 96%

[a]All operations are at 4°C unless otherwise indicated.
[b]Care is necessary to avoid losses by attachment of chromatin to glass. Nuclei from some cell lines do not lyse immediately upon exposure to solution P. It may be necessary to dialyze overnight against solution P, or solution P plus 0.05% (v/v) NP-40, or to wash nuclei before digestion in solution 0 plus 1% Triton X-100.

Method 6.

Material: Isolated nuclei, washed twice by centrifugation (800 g for 10 minutes in solution 0 (below). Micrococcal nuclease (Worthington) dissolved at 20,000 units/ml in solution O (below) minus PMSF.

Solutions: O. 0.3 M sucrose–0.05 M triethanolamine–0.025 M KCl–4 mM MgCl$_2$–1 mM CaCl$_2$–1 mM PMSF (pH 7.0 at 37°C). P. 2 mM EDTA (pH 7.2 at 4°C).

The procedure is shown in Scheme 6.

Comments: The method yields chromatin at > 4 mg/ml (as DNA) in high yield, which can be fractionated further by gel electrophoresis (73), sucrose gradient centrifugation in normal (5) or zonal (62, 72) rotors, or gel filtration (16,62,63). Very brief digestion (20 seconds) yields giant polynucleosomes, while after 15 minutes about 60% of the product is monomer size (Fig. 2). The trace of DNA left in the pellet is predominantly of high molecular weight in brief digests; when digestion proceeds too long subnucleosomes precipitate due to charge neutralization, as nucleosomal DNA is degraded. It is advisable to cool to stop enzyme action (73), rather than

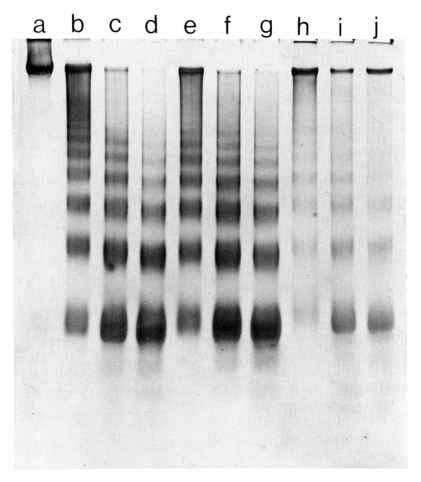

FIG. 2. DNA fragment patterns of enzymically prepared calf thymus chromatin, minutes of Wells (b), (c) and (d) are fragment patterns of total nuclear DNA after 2-, 8-, and 15-minutes of digestion, respectively (Method 6). Upon lysis of the nuclei, the DNA of the resulting solubilized chromatin gives the patterns shown in (e), (f), and (g), respectively, while the pellet DNAs (h), (i), and (j) consist predominantly of high-molecular-weight material. Well (a) is DNA isolated from nuclei after a 15-minute incubation in the absence of added enzyme. The amounts of acid-soluble DNA in (b), (c), and (d) are 3%, 10%, and 13% of the total, respectively.[73]

to add EDTA (44), since nuclei from some sources lyse upon EDTA addition; approximately 30 seconds of digestion occur during cooling. PMSF does not inhibit the nuclease at this high concentration of nuclei, but may do so at lower nuclear concentrations. The amount of acid-soluble DNA ranges from 1 to 14% over the time period used (73). The majority (~65%) of the nonhistone nuclear protein, as well as most of the nuclear RNA (~65%), remains in the pellet fraction (22).

VII. Storage

Conditions for storage of purified chromatin have been determined empirically only for certain specific purposes. Storage in 25–50% glycerol at −20°C maintains transcriptional activity (74) and conserves the principal features of nucleosomal structure as seen in the electron microscope (2). Storage at 0–4°C (without freezing) results in an increase in template activity for *E.coli* RNA polymerase (52) and in an apparent increase in the length of internucleosomal DNA (75). The general rule is not to store chromatin before the parameters under study have been established for fresh chromatin, to determine if they are affected by storage.

VIII. Criteria of Purity

There is at the present time no absolute criterion of purity for chromatin. Obvious possible impurities include cytoplasmic proteins and membrane material; these can be evaluated by assays of the marker enzymes glucose-6-phosphatase, 5'-nucleotidase, and cytochrome c oxidase (76), and by measurement of phospholipid content (18). The question becomes more complex (and perhaps semantic) when proteins are considered which may be in equilibrium *in vivo* between the nucleoplasm and chromatin; methods using exhaustive intermediate ionic strength washes may result in their loss from the chromatin (20). RNA-binding proteins, for example, may be bound to nascent RNA still attached to the chromatin template as well as to nucleoplasmic RNA. Low ionic strength methods may allow adsorption of nonchromatin nucleoplasmic proteins. Reconstruction experiments tend to indicate that the majority of the proteins of chromatin originate from the chromatin *in vivo* (50,74,76,77,79). However, for such reconstruction experiments to give rigorously valid conclusions, possible adsorption sites must not all be occupied, and/or adsorbed contaminants must be able to exchange with the exogenously added test components. The degree to which these criteria are met is not known.

TABLE II

COMPARISON OF SOME PROCEDURES FOR CHROMATIN ISOLATION

Procedure	Method number in this article	Starting material and quantity	Appropriate applications	Limitations[a]
I. Direct Methods A. Intermediate ionic strength	1	Tissue (10–20 gm) frozen, or frozen cultured cells	General studies of components (10,24,40,52); transcription in vitro (52); separation of template-active regions (9a,15)	Interchain interactions occur (H1 transfer possible) (20a,24, 49); possible effects of vigorous mechanical homogenization (44,48); extrachromatin protein and membrane contamination possible (32).
B. Low ionic strength	2	Cultured cells (5×10^6 to 5×10^8 fresh)	Separation of regions bearing nascent RNAs (54); chromatin assembly (55, 56); similar method (53) used to study DNA conformation and folding	Higher-order structures unfolded (20); extrachromatin nonhistone contaminants possible.
II. Methods from nuclei A. Intermediate ionic strength	5	25 gm fresh tissue or 2×10^8 to 3×10^9 fresh cultured cells	As for I,A, above, but decreased likelihood of extrachromatin protein contamination	As for I,A, above, but cytoplasmic protein contamination excluded; DFP treatment limited by nuclear lysis at high concentrations (22).
B. Enzymic (nucleases)	6	As in II,A above	Structure and composition studies of nucleosome monomers and multimers (2,9a,15,16, 44–46,62,65,72,73); separation of template-active regions (9a,15)	Residual nuclease may remain (22), so EDTA must be present throughout (immobilized DNAase will soon be available); nucleoplasmic contamination in crude digest.

IX.   Criteria of Structural Preservation

Absolute criteria cannot yet be proposed since we are just beginning to know the fine details of chromatin structure. The length of internucleosomal DNA may show an apparent increase during manipulation of chromatin (75,78), possibly due to partial detachment from the nucleosomal surface. Although the micrococcal nuclease digestion pattern may be modified in chromatin exposed to excessive mechanical fragmentation (44), this effect is not seen after milder fragmentation (80). The relative sensitivity of histones to tryptic digestion is also modified in extensively fragmented chromatin (44).

X.   Choice of Methods

The method of choice for chromatin preparation depends heavily on the problem to be investigated; it is probable that no single isolation method is appropriate for studies of all aspects of chromatin structure and function. To purify histones, the obvious major concerns are to eliminate nonhistone contamination and proteolysis rather than to maintain nucleosomal structure. To study the assembly of chromatin proteins onto newly made DNA, the major concern is to eliminate protein rearrangement. To isolate a specific protein or enzyme from the nonhistone chromatin protein fraction, this protein must be retained during the preparation procedure. Some appropriate applications and limitations of the procedures described here are summarized in Table II. Further procedures appropriate for specific problems are described elsewhere in these volumes.

NOTE ADDED IN PROOF

Chromatin has been isolated from rat ascites tumor cells and from rat liver using a modification of Method 2 (81,82). Such chromatin shows a less heterogeneous thermal melting profile than chromatin prepared at higher ionic strengths, and a markedly sharper micrococcal nuclease digest pattern.

ACKNOWLEDGMENTS

Supported by NIH, a Basil O'Connor Starter Research Grant from The National Foundation–March of Dimes, The Texas Affiliate of the American Heart Association (W.T.G.), and the Fonds National Suisse de la Recherche Scientifique (R.H.), and the Hoffmann-La Roche Foundation. The expert technical assistance of Mr. S. C. Albright is gratefully acknowledged.

References

1. Olins, A. L., and Olins, D. E., *Science* **183**, 330 (1974).
2. Oudet, P., Gross-Bellard, M., and Chambon, P., *Cell* **4**, 281 (1975).
3. Kornberg, R. D., and Thomas, J. O., *Science* **184**, 865 (1974).
4. Thomas, J. O., and Kornberg, R. D., *Proc. Natl. Acad. Sci. U.S.A.* **72**, 2626 (1975).
5. Noll, M., *Nature (London)* **251**, 249 (1974).
6. Baldwin, J. P., Boseley, P. G., Bradbury, E. M., and Ibel, K., *Nature (London)* **253**, 245 (1975).
7. Pardon, J. F., Worcester, D. L., Wooley, J. C., Tatchell, K., and Van Holde, K. E., *Nucleic Acids Res.* **2**, 2163 (1975).
8. Germond, J. E., Hirt, B., Oudet, P., Gross-Bellard, M., and Chambon, P., *Proc. Natl. Acad. Sci. U.S.A.* **72**, 1843 (1975).
9. Paul, J., and Malcolm, S. *Biochemistry* **15**, 3510 (1976).
9a. Gottesfeld, J. M., Garrard, W. T., Bagi, G., Wilson, R. F., and Bonner, J., *Proc. Natl Acad. Sci. U.S.A.* **71**, 2193 (1974).
10. Chalkley, R., and Jensen, R. H., *Biochemistry* **7**, 4380 (1968).
11. Kavenoff, R., Klotz, L. C., and Zimm, B. H., *Cold Spring Harbor Symp. Quant. Biol.* **38**, 1 (1973).
12. Carlson, R. D., and Olins, D. E., *Nucleic Acids Res.* **3**, 89 (1976).
13. Everid, A. C., Small, J. V., and Daines, H. G., *J. Cell Sci.* **7**, 35 (1970).
14. Zubay, G., and Doty, P., *J. Mol. Biol.* **1**, 1 (1959).
15. Marushige, K., and Bonner, J., *Proc. Natl. Acad. Sci. U.S.A.* **68**, 2941 (1971).
16. Honda, B. M., Baillie, D. L., and Candido, E. P. M., *J. Biol. Chem.* **250**, 4643 (1975).
17. Sheer, U., Kartenback, J., Trandelenburg, M. F., Streller, J., and Franke, W. W., *J. Cell. Biol.* **69**, 1 (1976).
18. Tata, J. R., Hamilton, M. J., and Cole, R. D., *J. Mol. Biol.* **67**, 231 (1972).
19. Jackson, V., Earnhardt, J., and Chalkley, R., *Biochem. Biophys. Res. Commun.* **33**, 253 (1968).
20. Hancock, R., Faber, A. J. and Fakan, S., *Methods Cell Biol.* **15**, 127 (1977).
20a. Hancock, R., *J. Mol. Biol.* **86**, 649 (1974).
21. Fakan, S., and Bernhard, W., *Exp. Cell Res.* **67**, 129 (1971).
22. Garrard, W. T., Todd, R. D., Boatwright, D. T., and Albright, S. C., Unpublished material.
23. Davies, K. E., and Walker, I. O., *Nucleic Acids Res.* **1**, 129 (1974).
24. Jensen, R. H., and Chalkley, R. J., *Biochemistry* **7**, 4388 (1968).
25. Hancock, R., *J. Mol. Biol.* **48**, 357 (1970).
26. Ilyin, Y. V., and Georgiev, G. P., *J. Mol. Biol.* **41**, 299 (1969).
27. Brutlag, D., Schlehuber, C., and Bonner, J., *Biochemistry* **8**, 3214 (1969).
28. Berry, M. N., and Friend, D. S., *J. Cell. Biol.* **43**, 506 (1969).
29. Pretlow, T. G., Glick, M. R., and Reddy, W. J., *Am. J. Pathol.* **67**, 259 (1972).
30. Lam, D. M. K., Furrer, R., and Buuce, W. R., *Proc. Natl. Acad. Sci. U.S.A.* **65**, 192 (1970).
31. Cross, M. E., *Biochem. J.* **128**, 1213 (1972).
32. Harlow, R., and Wells, J. R. E., *Biochemistry* **14**, 2665 (1975).
33. Tobey, R. A., Anderson, E. C., and Petersen, D. C., *J. Cell. Physiol.* **70**, 63 (1967).
34. Tobey, R. A., and Ley, K. D., *Cancer Res.* **31**, 46 (1971).
35. Tooze, J., "The Molecular Biology of Tumor Viruses," p. 75. Cold Spring Harbor Lab., Cold Spring Harbor, New York, 1973.
36. Lacy, D., and Axel, R., *Proc. Natl. Acad. Sci. U.S.A.* **72**, 3978 (1975).
37. Dupras, M., and Bonner, J., *Mol. Cell. Biochem.* **3**, 27 (1974).

38. Chong, M. T., Garrard, W. T., and Bonner, J., *Biochemistry* 13, 5128 (1974).
39. Bartley, J., and Chalkley, R., *J. Biol. Chem.* 245, 4286 (1970).
40. Douvas, A. S., Harrington, C. A., and Bonner, J., *Proc. Natl. Acad. Sci. U.S.A.* 72, 3902 (1975).
41. Shapiro, R., *in* "Aging, Carcinogenesis, and Radiation Biology" (K. D. Smith, ed.), p. 225. Plenum, New York, 1976.
42. Panyim, S., and Chalkley, R., *Arch. Biochem. Biophys.* 130, 337 (1969).
43. Hewish, D. R., and Burgoyne, L. A., *Biochem. Biophys. Res. Commun.* 52, 504 (1973).
44. Noll, M., Thomas, J. O., and Kornberg, R. D., *Science* 187, 1203 (1975).
45. Sollner-Webb, B., and Felsenfeld, G., *Biochemistry* 14, 2915 (1975).
46. Axel, R., *Biochemistry* 14, 2921 (1975).
47. Gurley, L. R., Walters, R. A., and Tobey, R. A., *J. Biol. Chem.* 250, 3936 (1975).
48. Garrard, W. T., Nobis, P., and Hancock, R., *J. Biol. Chem.* 252, 4962 (1977).
49. Hancock, R., unpublished material.
50. Hancock, R., *J. Mol. Biol.* 40, 457 (1969).
51. Dingman, D. W., and Sporn, M. B., *J. Biol. Chem.* 239, 3483 (1964).
52. Bonner, J., Chalkley, G. R., Dahmus, M. E., Fambrough, D., Fujimara, F., Huang, R. C., Huberman, J., Jensen, R., Marushige, K., Ohlenbusch, H., Olivera, B., and Wildholm, J., *in* "Methods in Enzymology," Vol. 12: Nucleic Acids (L. Grossman, and K. Moldave, eds.), Part B, p. 3. Academic Press, New York, 1968.
53. Cook, P. R., and Brazell, I. A., *J. Cell Sci.* 19, 261 (1975).
54. Faber, A. J., *Methods Cell Biol.* 16, 447 (1977).
55. Hancock, R., *Methods Cell Biol.* 16, 459 (1977).
56. Jackson, V., Granner, D. K., and Chalkley, R., *Proc. Natl. Acad. Sci. U.S.A.* 72, 4440 (1975).
57. Blobel, G., and Potter, V. R., *Science* 154, 1662 (1966).
58. Axel, R., Ceder, H., and Felsenfeld, G., *Proc. Natl. Acad. Sci. U.S.A.* 70, 2029 (1973).
59. Reeck, G. R., Simpson, R. T., and Sober, H. A., *Proc. Natl. Acad. Sci. U.S.A.* 69, 2317 (1972).
60. Senior, M. B., Olins, A. L., and Olins, D. E., *Science* 187, 173 (1975).
61. McConaughy, B. L., and McCarthy, B. J., *Biochemistry* 11, 998. (1972).
62. Bakayev, V. V., Melnickov, A. A., Osicka, V. D., and Varshavsky, A. J., *Nucleic Acids Res.* 2, 1401 (1975).
63. Shaw, B. R., Herman, T. M., Kovacic, R. T., Beaudreau, G. S., and Van Holde, K. E., *Proc. Natl. Acad. Sci. U.S.A.* 73, 505 (1976).
64. Noll, M., *Nucleic Acids Res.* 1, 1573 (1974).
65. Polisky, B., and McCarthy, B., *Proc. Natl. Acad. Sci. U.S.A.* 72, 2895 (1975).
66. Noll, H., *in* "Techniques in Protein Biosynthesis" (P. N. Cambell and J. R. Sargent, eds.), Vol. 2, p. 101. Academic Press, New York, 1967.
67. Pfeiffer, W. Horz, W., Igo-Kemenes, T., and Zachau, H. G., *Nature (London)*, 258, 450 (1975).
68. Lisharskaya, A. I., and Mosevitsky, M. I., *Biochem. Biophys. Res. Commun.* 62, 822 (1975).
69. Johnson, R. S., Chan, A., and Hanlon, S., *Biochemistry* 11, 4347 (1972).
70. Murphy, R. F., and Bonner, J., *Biochim. Biophys. Acta* 405, 62 (1975).
71. Tolstoshev, P., and Wells, J. R. E., *Biochemistry* 13, 103 (1974).
72. Hjelm, R. P., Kneale, G. G., Suau, P., Baldwin, J. P., Bradbury, E. M., and Ibel, K. *Cell* 10, 139 (1977).
73. Todd, R. D., and Garrard, W. T., *J. Biol. Chem.* 252, 4729 (1977).
74. Murphy, E. C., Hall, S. H., Shepherd, J. H., and Weiser, R. S., *Biochemistry* 12, 3543 (1973).
75. Oudet, P., and Chambon, P., personal communication.
76. Garrard, W. T., and Bonner, J., *J. Biol. Chem.* 249, 5570 (1974).
77. Bhorjee, J. S., and Pederson, T., *Proc. Natl. Acad. Sci. U.S.A.* 69, 3345 (1972).

78. Pignatti, P. F., Cremisi, C., Croissant, O., and Yaniv, M., *FEBS Lett.* **60**, 369 (1975).
79. Peterson, J. L., and McConkey, E. H., *J. Biol. Chem.* **251**, 548 (1976).
80. Woodhead, L., and Johns, E. W., *FEBS Lett.* **62**, 115 (1976).
81. Yaneva, M., and Dessev, G., *Eur. J. Biochem.* **66**, 535 (1976).
82. Tsanev, R. G., and Petrov, P., *J. Microsc. Biol. Cell* **27** (in press).

Chapter 4

Methods for Isolation of Nuclei from Cultured Mammalian Cells; Conditions for Preferential Retention of Selected Histones

MARGARIDA O. KRAUSE

Department of Biology,
University of New Brunswick, Fredericton,
New Brunswick, Canada

I. Introduction

One of the basic requirements for comparative studies on chromosomal proteins concerns the integrity and purity of the nuclear preparations obtained from a variety of tissues and cellular types. Since different tissues respond differently to a given method in terms of yield of uncontaminated whole nuclei, a variety of methods are currently being employed. Even within a single cell type, researchers tend to select one or another method on the basis of such criteria as yield, ease of preparation, purity, or a combination of the three. Although they seldom mention it, most are aware that once cells are lysed in aqueous media several components can leave the nucleus or, conversely, cytoplasmic components may adhere to or enter it. Such problems can greatly complicate interpretation of the results obtained in studies on the composition and metabolic activity of chromosomal proteins. Unfortunately there is, at present, a lack of appropriate criteria that can be used to test not just the purity but also the integrity of the nuclear preparation.

Histones are the only class of chromosomal proteins which have so far been well characterized chemically, are known to be present only in the nucleus, and can be easily identified and quantitated. They can therefore serve as useful standards to assay for leakage of nuclear components as a function of the composition of the media used for cell lysis and isolation of nuclei.

Studies in our laboratory have revealed that none of a variety of aqueous methods commonly used for cell lysis prior to isolation of nuclei and chromatin conserves all of the histones and that arginine-rich histones in particular are poorly conserved by methods which use chelating agents, such as ethylenediaminetetraacetic acid (EDTA) or citrate, in the absence of appropriate concentrations of divalent cations (1–3).

Furthermore, selective leakage was found to depend on cell cycling (1) and degree of acetylation of the arginine-rich fraction (2).

If, in fact, no single method of nuclear isolation can prevent leakage of one or another protein species, it is important that one select the method that appears most appropriate for the purpose of each study, that is, one resulting in optimal retention of one histone class at the partial sacrifice of another. In this case a comparison of more than one method becomes imperative whenever more than one class of proteins is under study.

Conditions for optimal retention of lysine-rich histones are generally well known. The lability of these fractions, particularly H1, to changes in pH and salt concentration is to be expected since their binding in chromatin depends largely on electrostatic interactions with DNA (4). Studies carried out in our laboratory showed that fine changes in pH can result in selective leakage of H1 depending on whether cells are progressing through the cell cycle or whether they are in a G_1-arrested state. Table I shows that when Mg^{2+} and Ca^{2+} are present in low concentrations in the lysing medium, a pH of 2.75

TABLE I

RELATIVE AMOUNTS (%) OR HISTONE FRACTIONS RETAINED IN NUCLEI
ISOLATED FROM EXPONENTIAL (E) AND STATIONARY (S) L-CELLS WITH
1 mM Mg^{2+} AND Ca^{2+a}

		3		5		10	
mM Citrate		3		5		10	
pH (calculated)		2.85		2.75		2.62	
Ionic strength ($\times 10^{-3}$)		6.9		7.3		8.6	
		E	S	E	S	E	S
Lysine-rich	H1	21	20	8	16	6	8
Slightly	H2A	15	16	17	16	19	17
lysine-rich	H2B	27	29	33	32	33	31
Arginine-rich	H4	18	19	20	22	22	21
	H3	19	16	22	17	20	23

^{a}All nuclei were isolated from monolayer cultures of mouse L-cells in medium containing 0.1% Triton X-100 and 0.1 M sucrose, in addition to $MgCl_2$, $CaCl_2$, and citrate in quantities indicated above. Histones were extracted from isolated nuclei with 0.25 N HCl at 4°C and fractionated in urea–acetic acid gels (5). Quantitation of histone fractions was carried out as described (1).

will allow for leakage of almost two-thirds of the H1 fraction from exponential phase cells but not from G_1-arrested (stationary phase) cells. A slightly lower pH (2.62) allows the same two-thirds leakage of H1 from both exponential and stationary phase cells with no apparent loss of the slightly lysine-rich fractions H2A and H2B.

At this level of divalent cation concentration, recovery of arginine-rich histones was found to be only partial, regardless of pH. Table II shows the relative amounts of histones recovered when the lysing medium contained 10 times the concentration of Mg^{2+} and Ca^{2+} and varying amounts of citric acid. Maximum recovery of arginine-rich fractions appears to occur at low pH with 26 mM citrate under conditions where all of the H1 molecules leak out of the nuclei. Again preferential recovery of H4 over H3 or vice versa depends on whether cells are stationary or in a logarithmic phase of growth.

While the low pH method results in an increased quantitative recovery of arginine-rich histones at the expense of the H1 fraction, we were concerned about the possible leakage of the more highly acetylated H3 and H4 molecules whose interaction with DNA is presumed to be weakened. We therefore labeled mouse L-cells with [³H]acetate and compared the recovery of acetylated histone fractions from nuclei isolated at different pH and divalent cation concentrations. The results are illustrated in Fig. 1. The concentra-

TABLE II

RELATIVE AMOUNTS (%)[a] OF HISTONE FRACTIONS RETAINED IN NUCLEI ISOLATED FROM EXPONENTIAL (E) AND STATIONARY (S) L-CELLS WITH 10 mM Mg^{2+} AND Ca^{2b}

mM Citrate		0		26		52		104	
pH (calculated)		7.00		2.37		2.22		2.09	
Ionic strength ($\times 10^{-3}$) –		60.0		64.7		66.5		68.6	
		E	S	E	S	E	S	E	S
Lysine-rich	H1	22 ± 5	22 ± 4	—	—	—	—	—	—
Slightly	H2A	17 ± 2	16 ± 9	17 ± 1	16 ± 1	19 ± 4	20 ± 2	17 ± 1	20 ± 1
lysine-rich	H2B	35 ± 2	32 ± 4	30 ± 1	30 ± 2	34 ± 1	31 ± 1	33 ± 1	33 ± 1
Arginine-rich	H4	17 ± 4	11 ± 2	35 ± 1	26 ± 2	38 ± 2	22 ± 3	30 ± 3	22 ± 2
	H3	10 ± 3	18 ± 4	19 ± 0	28 ± 3	20 ± 3	26 ± 4	20 ± 3	25 ± 3
Total	H3	27	29	54	54	48	48	50	48
arginine-rich	H4								

[a]Standard deviations indicated.

[b]All nuclei were isolated from monolayer cultures of mouse L-cells in medium containing 0.1% Triton X-100 and 0.1 M sucrose in addition to $MgCl_2$, $CaCl_2$, and citrate in the quantities of histone fractions were carried out as indicated for Table I.

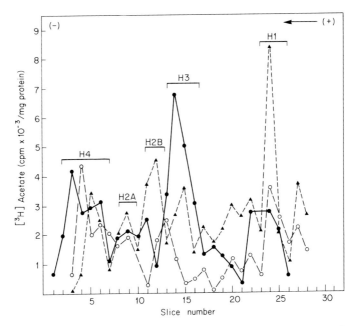

FIG. 1. Comparison of the amounts of [³H]acetate in histone fractions of log-phase L-cell nuclei isolated in media containing 0.1% Triton X-100, 0.1 M sucrose, and varying concentrations of citric acid, $MgCl_2$, and $CaCl_2$. Histones were fractionated in acetic acid–urea gels (5). Gels were scanned, sliced, and assayed for radioactivity as described (2). (●) from nuclei isolated in the presence of 10 mM Mg^{2+}–10 mM Ca^{2+}–26 mM citric acid (pH 2.4). (○) from nuclei isolated in the presence of 1 mM Mg^{2+}–1 mM Ca^{2+}–10 mM citric acid (pH 2.5). (■) from nuclei isolated in the presence of 10 mM Mg^{2+}–10 mM Ca^{2+} (pH 7.0) without citric acid.

tion of divalent cations appears to be the critical factor since more acetylated histones are recovered at the higher divalent cation concentration at both neutral and low pH. However, with the exception of acetylated H1, which is almost completely lost in the presence of 26 mM citrate (pH 2.4) and some loss of acetylated H2A and H2B, there is a definite increase in the recovery of acetylated H4 and a very substantial increase in the recovery of acetylated H3 at this low pH. It is possible that the function of citrate in this case is to act as a chelator of Mg^{2+} and Ca^{2+} so as to keep the divalent cation concentration at an optimum level. Presumably this ionic environment influences the conformation of the molecules so as to strengthen intermolecular interactions, thus preventing leakage (6). While the low pH method appears to be the best choice for studies on arginine-rich histones, in studies involving lysine-rich and slightly lysine-rich histone it is important to keep the pH near neutrality in order to ensure complete recovery of these

TABLE III

RECOVERY OF HISTONE FRACTIONS FROM NUCLEI ISOLATED
IN THE PRESENCE AND ABSENCE OF DIVALENT CATIONS
(10 mM Mg^{2+} AND 10 mM Ca^{2+})[a]

Number of nuclei extracted:	Control	Cations added	
	5 × 10^7	5 × 10^7	
Histone fraction	μg protein	μg protein	Increased recovery (×)
H1	80	105	1.3
H3	124	218	1.7
H2B	216	232	1.0
H2A	70	92	1.3
H4	50	153	3.0
Total histone recovered	540	800	1.5

[a]All nuclei were isolated from monolayer cultures of mouse L-cells in 1% Triton X-100–20 mM EDTA–80 mM NaCl–50 mM NaHSO$_3$ (pH 7.2). Histones were extracted overnight at 4°C in 0.25 N HCl and electrophoresed in acetic acid–urea gels (5). Quantitation of histone fractions was carried out as described (3) using calf thymus histone as a standard.

species. We therefore developed a neutral pH method which, at the same time, minimizes leakage of arginine-rich histones through the addition of 10 mM Mg^{2+} and Ca^{2+}. The effect of the divalent cation addition on the recovery of the various histone species at neutral pH is presented in Table III. Although Mg^{2+} has been claimed to favor proteolysis (7), there is no evidence of this in nuclei isolated under our conditions, whether at low or neutral pH.

On the basis of this background the following methods are recommended.

II.   Method A: For Studies on Lysine-Rich and Slightly Lysine-Rich Histones

Cultured cells must be washed three times in phosphate-buffered saline solution in order to remove any traces of serum. Aliquots ranging between 10^7 and 10^8 saline-washed cells are then resuspended at 4°C in 10 ml of medium containing 1% Triton X-100–20 mM ethylenediaminetetraacetic acid (EDTA)–80 mM NaCl–10 mM MgCl$_2$–10 mM CaCl$_2$–50 mM NaHSO$_3$ (pH 7.2).

Lysis is carried out by rapidly drawing the suspension in and out of a syringe equipped with a 22-gauge needle. This procedure normally requires two to three drawings through the syringe in order to yield 90–100% clean nuclei. However, some cell lines, in particular nontransformed diploid populations, appear more resistant to lysis than transformed cells; therefore the procedure should be monitored in a phase-contrast microscope and repeated as necessary as quickly as possible. Time is an important factor since, although the procedure is carried out in the cold and in the presence of a proteolytic inhibitor, decreased yields of histone are common whenever care is not taken to minimize preparation time. This is the reason for selected lysing with a syringe and needle rather than by repeated mixing on a Vortex and centrifugation.

The isolated nuclei are then pelleted by centrifugation at 1200 g for 5 minutes and washed twice in 10 ml of medium containing 0.15 M NaCl–0.01 M Tris (pH 8.0)–10 mM CaCl$_2$–10 mM MgCl$_2$–50 mM NaHSO$_3$ in order to wash out the detergent and remove nuclear sap proteins. The whole procedure including lysis and washing of the nuclei should take only 20–30 minutes.

It is not necessary to lyse the nuclei in order to ensure a thorough extraction of histones. However, it is important to resuspend the nuclear pellets thoroughly by mild Vortex mixing in the leftover drop of solution prior to adding new solution during the isolation and washing procedure, and especially before addition of acid whenever acid extraction of histones is to be carried out. If clumps are formed at this time, homogenization in a glass homogenizer with loose-fitting pestle is recommended in order to ensure complete histone recovery.

III.   Method B: For Studies on Arginine-Rich Histones

Aliquots ranging between 10^7 and 10^8 saline-washed cells are resuspended at 4°C in 10 ml of medium containing 0.1% Triton X-100–0.1 M sucrose–10 mM CaCl$_2$–10 mM MgCl$_2$–50 mM NaHSO$_3$–25 mM citric acid (pH 2.4). The same procedure as in Method A, involving two to three rapid drawings of the solution through a syringe and 22-gauge needle, is normally sufficient to ensure a 90–100% recovery of clean nuclei. Again the suspension should be monitored by phase-contrast microscopy since some cell populations appear more resistant to lysis by this procedure than others. The isolated nuclei are pelleted by centrifugation at 1200 g for 5 minutes and washed twice in medium containing 0.2 M sucrose–10 mM MgCl$_2$–10 mM CaCl$_2$–50 mM NaHSO$_3$.

Despite the loss of H1, nuclei prepared in this way show not only an increased content of H3 and H4 histones, as compared with those obtained at neutral pH Fig. 2, but they are also enriched in acetylated species (Fig. 1).

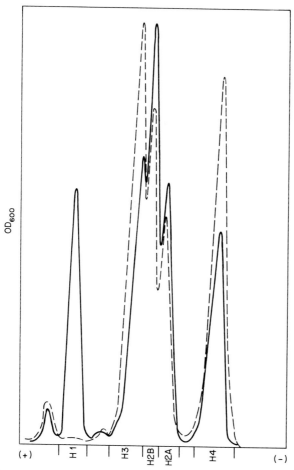

FIG. 2. Comparison of absorbance scans of histone gels extracted from nuclei of log-phase WI-38 human diploid fibroblasts isolated with 80 mM NaCl–20 mM EDTA (pH 7.2) (solid line) and from nuclei isolated in the presence of 10 mM Mg^{2+}–10 mM Ca^{2+}–26 mM citric acid (pH 2.4). The same number of nuclei were extracted in each case (8).

References

1. Krause, M. O., Yoo, B. Y., and MacBeath, L., *Arch. Biochem. Biophys.* **164**, 172 (1974).
2. Krause, M. O., and Inasi, B., *Arch. Biochem. Biophys.* **164**, 179 (1974).
3. Krause, M. O., and Stein, G. S., *Exp. Cell Res.* **92**, 175 (1975).
4. Johns, E. W., *in* "Histones and Nucleohistones" (D. M. P. Phillips, ed.). Plenum, New York, 1971.
5. Panyim, S., and Chalkley, R., *Biochemistry* **8**, 3972 (1969).
6. Bradbury, M., Carpenter, B. G., and Rattle, H. W. E., *Nature (London)* **241**, 123 (1973).
7. Douvas, A. S., Harrington, C. A., and Bonner, J., *Proc. Natl. Acad. Sci. U.S.A.* **72**, 3902 (1975).
8. Krause, M. O., *Exp. Cell Res.* **100**, 63 (1976).

Chapter 5

Methods for Isolation of Nuclei from Spermatozoa

YASUKO MARUSHIGE

AND

KEIJI MARUSHIGE

Laboratories for Reproductive Biology and Department of Biochemistry,
Division of Health Affairs,
University of North Carolina,
Chapel Hill, North Carolina

I. Introduction

Comparative studies of spermatozoa have shown an enormous diversity in their structural features and in the properties of their chromatin (*1–6*). Some of these differences may be attributed to the varying requirements of fertilization such as in reaching the egg, in penetrating structures surrounding the egg, and in fusing with the egg cell membrane. Needs for safe and effective transport of genes in the male gamete and for appropriate changes following syngamy must have made effective use of natural selection following countless mutations, since only one successful spermatozoon shares in creating the offspring after severe competition with other spermatozoa.

In general, the sperm nucleus is a compact chromatin mass enclosed by a firmly attached nuclear envelope. In contrast to the remarkable similarity of somatic histones of various organisms (*7*), proteins associated with sperm DNA vary widely in different species. Bloch (*1*), in his "A Catalog of Sperm Histones," classified these proteins into several groups. Since then more detailed biochemical data characterizing sperm histones of many species have become available.

Sperm histones of eutherian mammals are rich in arginine and stabilized by disulfide bonds formed from their cysteine residues. Protamines contain arginine residues in amounts up to two-thirds of the total amino-acid content and have been found in some metatherian mammals, birds, reptiles, (amphibians), fish, and molluscs. Sperm histones, which are very similar to somatic histones, have been found in some amphibians, fish, echinoderms, and molluscs. A variety of basic proteins with features intermediate between protamines and somatic histones have been found in some amphibians and molluscs. Basic proteins of some insects, such as crickets and grasshoppers, appear to be rich in arginine. The nuclei of these species are very compact. In contrast to these basic proteins, the proteins associated with the DNA of some crabs and crayfish are acidic. Their nuclei are less condensed than somatic nuclei, and their spermatozoa lack tails.

Since many properties of sperm chromatin depend on their histones, methods for isolation of sperm nuclei differ from case to case. Structural and chemical differences between spermatozoa of different species and at different stages make necessary many modifications in methods for isolation of nuclei. In principle, spermatozoa are isolated from seminal ejaculate or reproductive tissue; sperm heads are separated from sperm tails, after which nonnuclear materials of the heads are removed. Usually the sperm nucleus is surrounded by a very thin perinuclear layer of cytoplasm, and the acrosome covers the anterior part of the sperm head. However, some components surrounding sperm nuclei are firmly attached, so that isolated sperm nuclei often contain remnants of nonnuclear materials. Processes for removing nonnuclear material may extract some of the nuclear proteins.

In this article we describe some examples of methods for isolating sperm nuclei appropriate for organisms with various types of sperm histones. We then mention some modifications suitable for other species, based on our own experience and on methods reported by many other workers. All operations are carried out at 0–4 °C unless otherwise specified. It is very helpful to monitor all steps in the preparation with a light microscope. There are some species in which the isolation of sperm chromatin has been achieved but attempts to isolate sperm nuclei have failed. There are many species for which methods for isolation of sperm nuclei have not been established. We shall mention some available information which may help lead to suitable methods. Nuclei of species which share similarities in properties of sperm chromatin with those for which satisfactory methods are available might also yield if subjected to similar procedures.

II. Spermatozoa of Eutherian Mammals

Sperm histones of many, perhaps all, eutherian mammals contain about 50–60 mol % of arginine and 8–16 mol % of cysteine residues (5,8–12). The chromatin of mature bull spermatozoa consists principally of DNA and sperm histones. Essentially all the cysteine residues are present in disulfide form (13). The replacement of somatic histones by sperm histone is achieved in the testis during late stages of spermatids (*14, 15*). The sperm chromatin is further stabilized by cross-linking of histone molecules through disulfide bond formation during passage through the epididymis (*14, 16*). The tail of mammalian spermatozoa possesses a row of fibrous sheath, in addition to the $9 + 9 + 2$ pattern of microtubules, in contrast to the simple axonemal $9 + 2$ patterns of primitive spermatozoa. Some components of sperm tails such as outer fibers and fibrous sheath are also stabilized by disulfides in the epididymis (*10*).

A. Spermatozoa in Semen

Ejaculated semen of large domestic animals and of the human comprise a useful source of spermatozoa. Semen contains spermatozoa and seminal plasma, which is the acellular portion of the ejaculate. Seminal plasma consists of secretions produced by accessory genital glands. It includes many organic substances and enzymes (*17*), particles of various sizes, and coagulum. Remnant cytoplasmic droplets are also found in semen (*18*). A method for isolation of sperm nuclei from bull semen has been developed on the basis of the method of Coelingh *et al.* (*19*). This method is described first. We then mention some modifications suitable for isolation of sperm nuclei from other mammals.

Ten grams of frozen bull semen is thawed and diluted to 35 ml with 10 mM Tris-HCl (pH 8). The suspension is centrifuged at 1500 g for 10 minutes. The sediment containing spermatozoa is washed twice with the same buffer under the same centrifugal conditions. The washed sediment is next resuspended in 15 ml of the Tris buffer with the aid of a hand-operated glass–Teflon homogenizer. It is then sheared with a VirTis-45 homogenizer at 85 V for 15 minutes. The resulting suspension is layered on an equal volume of 0.6 M sucrose–10 mM Tris-HCl (pH 8) and centrifuged at 1500 g for 20 minutes. The supernatant containing small pieces of broken tails is discarded. Although the sperm heads and the broken tails do not sediment as two clearly separated layers at this step, some of the tail pieces can be removed from the upper edge of the sediment. The sediment is resuspended thoroughly in the Tris buffer and centrifuged again through 0.6 M sucrose. The resulting sediment at the bottom is composed of the tightly packed sperm head fraction. The loosely packed upper fraction contains broken tails. The upper fraction is washed away by gentle swirling with the buffer. The sperm heads thus obtained are suspended in 30 ml of 10 mM Tris-HCl (pH 8)–1% Triton X-100, kept in an ice bath for 30 minutes, and then centrifuged at 1500 g for 10 minutes. The sediment is next resuspended in 30 ml of 10 mM sodium deoxycholate, incubated at 37°C for 1 hour, and centrifuged at 1500 g for 10 minutes. The sediment is finally washed four times with the Tris buffer by repeated resuspension and centrifugation. Ten grams of bull semen yields approximately 25 mg of DNA. The weight ratio of protein to DNA has been found to be 2.7 in the whole spermatozoa, 1.1 in the sperm head, and 0.8 after washing of the sperm head with detergents. The ratio of sperm histone to DNA is 0.7 (13). Little or no RNA is found in this preparation.

The concentrations of spermatozoa, the sizes and abundance of acellular particles, and the states of coagulation in semen vary from species to species and, to a less extent, from individual to individual. A proper dilution and sedimentation of semen at initial stages of preparation are therefore essential for obtaining sperm nuclei in satisfactory purity. Acellular materials larger than spermatozoa can be removed by centrifugation at a very low speed, leaving the majority of spermatoza in the supernatant. For human spermatozoa (20), ejaculate with a minimum of debris is selected, diluted with 2 volumes of a washing solution (120 mg $CaCl_2$, 300 mg KCl, 150 mg $NaHCO_3$, 3000 mg NaCl, 500 mg glucose, and 1 gm Dextran, Sigma type 200C, M 204.000, per 500 ml, pH 8.0), and centrifuged at 225 g for 5 minutes in order to sediment acellular materials of large size. Spermatozoa are then sedimented from the supernatant by centrifugation at 500 g for 15 minutes. Although Pedersen (20) has diluted semen with only twice the volume of washing medium, probably to prevent loss of acrosomal materials, we have found that, for preparation of sperm nuclei, acellular materials of large size in human semen can be removed effectively by a greater dilution (10 volumes) with 0.25 M sucrose–10 mM Tris-HCl (pH 8) and a mild homogenization with a Dounce homogenizer followed by centrifugation at 100 g for 5 minutes. The spermatozoa are then sedimented at 650 g (10 minutes), resuspended in 0.25 M sucrose, and further purified by centrifugation at 100 g and 650 g.

Rabbit semen forms a thick, coagulated lump. After freezing and thawing, a liquid portion of the semen containing spermatozoa separates from the coagulum and can be pipetted out simply. Ram semen has been diluted with 4 volumes of sucrose–citrate (0.275 M sucrose–16 mM citric acid, adjusted to pH 6.5 with NaOH) and centrifuged in a swinging-bucket rotor at 600 g for 15 minutes (21). Spermatozoa of the stallion have been sedimented from semen at 10,000 g for 10 minutes without dilution and washed several times with 0.9% NaCl (22).

Because of the chimical stability of mammalian sperm tails, mechanical avulsion of the tail is usually essential for isolation of mammalian sperm nuclei. Instead of a VirTis-45 homogenizer described above, a Bühler homogenizer (15 minutes at the maximum speed) has also been used for bull, ram, boar, and stallion spermatozoa (19, 23). Sonication appears to be more useful than shearing for spermatozoa with the short tails (human) or spermatozoa with extremely strong tails (murine rodents). In subsequent separation of the sperm heads from the tails, sedimentation through 0.6 M sucrose, as described above, is a simple operation and is capable of handling large amounts of material. In case a satisfactory separation of the sperm heads is not obtained by this centrifugation, sedimentation through dense sucrose solution can be employed. A sample in 10 ml of 1.6 M sucrose is layered on a discontinuous sucrose gradient consisting of 10 ml each of 1.8 M and 2.0 M sucrose and 5 ml of 2.2 M sucrose and spun in a Spinco SW-27 rotor at 15,000 rpm (30,000 g) for 45 minutes (24). The sperm heads are obtained in the pellet. In this centrifugation, an overloading of the sample results in a trapping of the sperm heads by broken tails and acellular contaminants at the interphases.

B. Spermatozoa in Cauda Epididymis

In small mammals, the cauda epididymis is a useful source of mature spermatozoa. The long epididymal duct is enclosed in a capsule of connective tissue. The cauda in its capsule is sliced with a razor blade or minced with scissors and homogenized in saline–ethylenediaminetetracetic acid (EDTA) [75 mM NaCl–24 mM EDTA (pH 8)] with a motor-driven glass–Teflon homogenizer (TRI-R Model S63C, setting 5). The connective tissue should be occasionally wiped from the pestle during this homogenization. Use of a VirTis homogenizer at this step has resulted in considerable contamination of the sperm head fraction by pieces of connective tissue. Since spermatozoa in cauda epididymis adhere to each other (4), special caution must be taken to minimize the intensity and duration of homogenization when preparing spermatozoa with long, thick tails, such as those of the rat. Excessive homogenization tends to twist these tails together so they become extremely resistant to mechanical disruption at a later step. The homogenate is appropriately diluted and strained through two layers of cheesecloth. The use of cloth with a finer mesh, such as Miracloth, is not recommended since spermatozoa, especially those with long tails, stick in the fine meshes and are lost. The strained homogenate is then centrifuged at 1500 g for 10 minutes. The sediment is successively washed twice with saline–EDTA containing 0.5% Triton X-100 (pH 8) and thrice with 10 mM Tris-HCl (pH 8) by resuspension and centrifugation (1500 g, 10 minutes). The washed sediment is resuspended in the Tris buffer and subjected to high-speed shearing or sonication followed by sedimentation through sucrose and deoxycholate treatment as described for spermatozoa in semen. Rabbit sperm nuclei can be obtained satisfactorily by shearing with a VirTis-45 homogenizer followed by the sedimentation through 0.6 M sucrose. Sonication and sedimentation through dense sucrose solution have proved to be suitable for preparation of rat sperm nuclei.

C. Spermatozoa in Caput Epididymis

The procedure for isolation of sperm nuclei from caput is the same as that described for spermatozoa in cauda epididymis. Spermatozoa in caput epididymis are still in process of stabilization. The sperm heads are resistant to strong mechanical shearing, but the sperm tails are easily broken into small pieces, even in the case of rat spermatozoa. Separation of sperm nuclei by centrifugation through 0.6 M sucrose is usually satisfactory. Some of the cysteine residues of sperm histone in this portion of epididymis are present as free sulfhydryls (14, 16). Therefore, it is possible for these sulfhydryls to become oxidized during preparation of sperm nuclei unless an appropriate blocking reagent (e.g., iodoacetamide) is used throughout the procedure.

D.   Testicular Spermatozoa and Spermatids at Late Stages

The spermatozoa detached from the Sertoli cells move rapidly into the epididymis. The population of spermatozoa in the testis is therefore extremely small. Late spermatid nuclei with shapes similar to those of mature sperm nuclei can be isolated from the testes of the basis of their resistance to mechanical shearing (14). Frozen testes are thawed and the tunicae albugineae removed. The testes are then homogenized in saline–EDTA in a Waring blender (85 V for 1 minute and then at 45 V for 3 minutes). They are then centrifuged at 1500 g for 10 minutes. The sediment is next washed twice with saline–EDTA containing 0.5% Triton X-100, and once with 10 mM Tris-HCl (pH 8). The sediment is homogenized in the Tris buffer with a hand-operated Teflon homogenizer and then spun at 10,000 g for 10 minutes. This step is repeated once. The resulting gelatinous sediment is resuspended in the Tris buffer, sheared in a Waring blender at 100V for 3 minutes, and centrifuged at 17,000 g for 20 minutes. The sediment is resuspended in the buffer and sheared again in a VirTis-45 homogenizer at 85 V for 15 minutes. The suspension is then layered on an equal volume of 0.6 M sucrose containing 10 mM Tris-HCl (pH 8), and centrifuged for 20 minutes at either 1500 g for the oval-shape nuclei (rabbit) or 10,000 g for the thin hooked-shape nuclei (rat). This centrifugation results in two layers of sediments. A loosely packed upper layer consists of partially condensed spermatid chromatin, broken tails, and some cell debris. A tightly packed bottom layer contains late spermatid nuclei. The upper layer is removed by gentle swirling with the Tris buffer. The bottom layer is homogenized thoroughly and sedimented again through 0.6 M sucrose in the same manner as before. Late spermatid nuclei thus obtained may be purified further by sedimentation through a dense sucrose solution similar to that described for isolation of sperm nuclei from semen. Essentially all the cysteine residues of sperm histone found in the testis are in the form of free sulfhydryls (14). Artificial oxidation of these cysteine residues can be prevented by the presence of 5 mM iodoacetamide at all steps of the process.

III.   Spermatozoa with Protamine

Since the first isolation of protamine from spermatozoa of Atlantic salmon by Friederich Miescher in 1874, sperm nuclei of various species of fish (25–30), of rooster (31–35), and of squid (36) have been reported to contain proteins similar to salmon protamine. Cytochemical evidence suggests that some metatherian mammals and reptiles are also in this category (1).

A.   Fish

The teleost fish belongs to a rare group of animals whose spermatozoa lack an acrosome. The spermatozoon reaches the egg through a micropile prior to fertilization (6, 37). Each teleost sperm tail contains an axonemal complex of the 9 + 2 pattern and is not stabilized by protein disulfides (38). The sperm nuclei of salmonid and related fish remain unaltered in water while the cytoplasm and the tail are disrupted under these conditions (26). Thus isolation of teleost sperm nuclei with protamine principally involves repeated washings of the spermatozoa with a hypotonic solution.

The spermatozoa can be obtained from milt or from testes of fish ready to spawn. Milt can be diluted with saline–EDTA and the spermatozoa then sedimented by centrifugation at 1500 g for 10 minutes. To obtain a cell fraction from testes highly enriched with spermatozoa, the whole testis or one coarsely minced with scissors can be squeezed in a sack composed of eight layers of cheese cloth. A milky suspension of spermatozoa is expressed and can be allowed to flow into saline–EDTA. If such a squeezing is not effective, minced testes can be homogenized briefly in saline–EDTA with a Teflon homogenizer and filtered through four layers of cheese cloth. The suspension is then centrifuged. The sediment of spermatozoa thus obtained is washed twice with saline–EDTA containing 0.5% Triton X-100 and thrice with 10 mM Tris-HCl (pH 8) by resuspension and centrifugation at 1500 g for 10 minutes.

Gelatinous sediments indicate that the original sperm suspension has been considerably contaminated with younger germinal cells. Centrifugation at 17,000 g for 10–20 minutes may be necessary. In such cases, the final preparation of sperm nuclei will be contaminated with condensed chromatin of younger spermatids.

The washed sediment is resuspended in the Tris buffer and homogenized in a VirTis-45 homogenizer at 45 V (at 100 V in a Waring blender) for 3 minutes. The suspension is next centrifuged at 1500 g for 10 minutes, and the resulting sediment is finally washed three to four times with the Tris buffer. The nuclei of younger germinal cells or nongerminal cells are disrupted during the washings in the Tris buffer, giving rise to gelatinous aggregates of chromatin. These are removed by homogenization and subsequent centrifugation at 1500 g. Trout sperm nuclei obtained by a method essentially identical to that described above possess a ratio of protamine to DNA by weight of 0.54 and contain little RNA and acid-insoluble protein (39).

Since the testis maturation of fish occurs seasonally (e.g., October to February in rainbow trout), testes in which germinal cells at a certain stage of development dominate the population can be obtained. Using immature trout, synchronized spermatogenesis can be induced by injection of salmon pituitary gonadotrophins (39,40). Trout testis cells at different stages of spermatogenesis have been separated by velocity sedimentation (41).

B.   Rooster

The nuclei of rooster spermatozoa are rod-shaped. The acrosome of rooster spermatozoa is said to be structurally rather similar to that of mammals (42). The sperm tail has a simple microtubule arrangement of the 9 + 2 pattern and is not stabilized by disulfides (38). Semen is the usual source of rooster spermatozoa. The nuclei may be isolated from rooster spermatozoa in a manner similar to that described for fish. Daly et al. (31) and Fischer and Kreuzer (32) have isolated well-formed nuclei from rooster semen by repeated washings of the spermatozoa with 0.14 M NaCl and then with 0.2% citric acid, which removes the acrosomal cap, the middle piece, and the tail. It has been noted that in this procedure shreds of cytoplasm adhere to some of the nuclei; this is probably due to the use of citric acid (26).

IV.   Spermatozoa with Somatic Histone-Type Protein

Spermatozoa of the leopard frog (43,44), the carp (45), various species of the sea urchin, Arbacia punctulata (46,47), A. lixula (48–51), Strongylocentrotus purpuratus (52–55), Lytechinos variegatos (unpublished data of this laboratory), and the sea cucumber (51, 56) contain chromosomal basic proteins similar to somatic histones. In general, in these species the major difference between somatic histones and the histones of spermatozoa resides in the fraction corresponding to lysine-rich histone H1. Zirkin (57) has shown that the structure of sperm chromatin of the leopard frog, the sea urchin, and the gold fish is basically similar to that of somatic chromatins. Spadafora and Geraci (58) have further shown that sperm chromatin of the sea urchin *Paracentrotus lividus* exhibits a nuclease digestion profile similar to that observed in somatic chromatins.

A. Sea Urchin

Sea urchin spermatozoa can be collected by injection of 0.5 M KCl to induce spawning or by shaking the ripe gonads in filtered sea water or artificial sea water. Palau *et al.* (48) have isolated sperm heads in the following manner: A sperm suspension in sea water is centrifuged at 30 g for 10 minutes to remove coarse materials. The resulting supernatant is centrifuged at 2500 g for 10 minutes. The sediment is washed three times with sea water by resuspension and centrifugation. The spermatozoa thus obtained are briefly sheared (Type Berrens, Barcelona, Spain; top speed for 5 seconds and then at a half speed for 5 seconds) in sea water. The sperm heads are sedimented by centrifugation at 500 g for 10 minutes and washed twice with sea water. Saline–EDTA (46,53) or standard saline–citrate (52,55) has been used for washing of spermatozoa. Gibbons and Fronk (59) reported that Triton X-100 solubilizes sea urchin sperm membranes, matrix proteins, and mitochondria.

In our experience with *Lytechinos variegatos*, washing sea urchin spermatozoa with 20% glycerol–10 mM EDTA (H. Ozaki, personal communication) is more efficient than washing with sea water. Freezing and thawing seem to detach the middle pieces and the tails from the sperm heads in most spermatozoa. The frozen ejaculate is thawed, diluted with 10 volumes of 20% glycerol containing 10 mM EDTA (pH 8), and centrifuged at a slow speed (e.g., International clinical centrifuge Model CL, setting 1) for 5 minutes at room temperature in order to remove coarse materials. The supernatant is centrifuged again in the same manner. The spermatozoa are then sedimented from the supernatant at 1500 g (10 minutes) at room temperature. The sediment is washed once with the glycerol–EDTA under the same centrifugal conditions. The sediment is next resuspended in saline–EDTA [75 mM NaCl–24 mM EDTA (pH 8)] containing 0.5% Triton X-100 and centrifuged at 1500 g for 10 minutes. This step is repeated once. The sediment is finally washed thrice with saline–EDTA by resuspension and centrifugation (1500 g, 10 minutes). A light microscopic examination of sea urchin sperm nuclei thus obtained and subsequently stained with Lee's solution [methylene blue-basic fuchsin; (60)] showed that the nuclei are virtually free from contamination by mid pieces, tails, or acellular materials and that the surfaces of the heads are smoother and stained more lightly than those of the original sperm heads. Moreover, materials staining pink with Lee's stain at the tip of the sperm head (possibly acrosomal material) were removed during preparation.

B. Leopard Frog

Alder and Gorovsky (44) have isolated the nuclei from spermatozoa and late spermatids of the leopard frog (*Rana pipiens*). Although improvements are needed to reduce contamination by nuclei of younger spermatogenic cells, the application of a method essentially identical to that employed for isolation of somatic nuclei appears to be quite adequate.

The testes of a sexually mature leopard frog are macerated into 10% Holt-freter's solution (3.5 gm NaCl, 0.05 gm KCl, 0.10 gm $CaCl_2 \cdot 2H_2O$ and 0.20 gm $NaHCO_3$/liter) and allowed to sit for 5–10 minutes in order to release spermatozoa. The suspension of spermatozoa is next separated by decantation from the testicular materials settled at the bottom. The spermatozoa are then sedimented by low-speed centrifugation and quickly frozen. The frozen spermatozoa are homogenized (Waring blender) in 0.25 M sucrose containing 10 mM $MgCl_2$–10 mM Tris-HCl (pH 8)–50 mM $NaHSO_3$ at 30 V for 3 minutes and then at 60 V for 1 minute according to Panyim et al. (61). The homogenate is filtered through four layers of cheesecloth, then through two layers of Miracloth, and then centrifuged at 480 g for 10 minutes. The sediment is resuspended in the grinding medium containing 0.2–0.5% Triton X-100, and centrifuged at 480 g for 10 minutes. This step is repeated at least once. The sediment is next resuspended in approximately 10 volumes of 2.2 M sucrose containing the same ionic components as the grinding medium and layered on 2.4 M sucrose (approximately three-sevenths volume of the nuclear suspension) containing the same ionic components as the grinding medium. The nuclei are then sedimented by centrifugation at 60,000 g for 2.5 hours. Alternatively, the nuclei can be purified by resuspension of the washed sediment in 0.5 M sucrose containing the same ionic components as the grinding medium, followed by centrifugation at 8000 rpm for 10 minutes.

Ultrastructural and cytochemical studies have shown that sperm nuclei of other amphibians [*Bufo, Hyla, Xenopus*: (62); *Pleurodeles*: (63,64)] contain chromosomal basic proteins that are different from those of *Rana*.

V. Spermatozoa of Molluscs

The phylum *Mollusca* belongs to a different evolutionary branch than that of the vertebrates and embraces a broad variety of animals. Sperm histones of many species of molluscs have been studied by Subirana *et al.* (*36*). In *Gibbula, Haliotis, Loligo*, and *Octopus*, the main components are rather similar to salmonid protamines. A complex mixture of basic proteins, including ones similar to somatic histones and proteins intermediate in nature between protamines and somatic histones, are found in *Mytilus, Chilton, Ostrea, Spisula* and *Patella*. Spermatozoa of *Eledone* possess a mixture of basic proteins containing cysteine residues.

Although procedures for isolation of sperm nuclei from these organisms have not been well established, the methods described by Subirana *et al.* (*36*) for separation of spermatozoa from various species of molluscs might provide a basis for development of isolation procedures of their sperm nuclei and sperm chromatin. Spermatozoa of *Cryptochiton stellerii, Chiton olivaceus, Spisula solidissima*, and *Gibbula divaricata* can be released by shaking their ripe male gonads in sea water. The release of spermatozoa of *Mytilus edulis* requires an alternative treatment of single living individuals with cold (4°C) and warm (25°C) sea water at 15-minute intervals. Suspensions of spermatozoa of *Loligo pealeii, Octopus vulgaris*, and *Eledone cirrhosa* can be obtained by disruption of the spermatophores (the sacs containing spermatozoa) in sea water with the aid of a Dounce homogenizer followed by filtration through several layers of cheesecloth. The spermatozoa thus obtained can be sedimented at 3000 g (10 minutes). In *M. edulis, Ostrea edulis, G. divaricata, Haliotis tuberculata*, and *P. valgata*, ripe male gonads have been subjected to two cycles of homogenization in 0.25 M sucrose–3 mM CaCl$_2$ with a Waring blender and centrifugation at 3000 g for 10 minutes. The sediment of nuclei has then been washed successively with 0.1 M Tris-HCl (pH 8) and 0.15 M NaCl. The sperm nuclei prepared in this manner have been contaminated by variable amounts of nuclei from unripe gametes and other cell types present in the gonad. Nuclear samples containing more than 90% of sperm nuclei have been used for the protein extraction.

VI. Spermatozoa of Cricket

It has long been known from cytochemical studies that sperm nuclei of some insects such as grasshopper, cricket, coccid, katydid, and fruit fly contain tightly bound arginine-rich proteins (*1*). Tessier and Pallotta (*65*) have isolated the nuclei of spermatozoa (and late spermatids) of the house cricket (*Acheta domestica*) and analyzed their chromosomal basic proteins.

The frozen adult testes are homogenized in a grinding medium [0.25 M sucrose–2.5 mM CaCl$_2$–0.1% Triton X-100–50 mM Tris-HCl (pH 6.0)] with 10–15 strokes of a Dounce homogenizer and strained through four layers of cheesecloth. The homogenate is centrifuged at 1000 g for 10 minutes, and the sediment is washed once with the grinding medium under the same centrifugal conditions. The nuclear sediment thus obtained is either homogenized in 2.2 M sucrose–1.5 mM CaCl$_2$–50 mM Tris-HCl (pH 6.0) and centrifuged at 60,000 g for 1 hour or resuspended in the grinding medium, layered on 1 M sucrose–1.5 mM CaCl$_2$–50 mM Tris-HCl (pH 6.0), and centrifuged at 2000 g for 20 minutes after gentle stirring of the upper two-thirds. The purified nuclei are then washed once with 0.09 M NaCl–50 mM Tris-HCl (pH 6.0) by centrifugation (1000 g, 10 minutes), resuspended in distilled water with the aid of the Dounce homogenizer, and sheared in the VirTis homogenizer at maximum speed for 5 to 10 minutes. This homogenization leaves the nuclei of cricket spermatozoa (and late spermatids) intact while the nuclei of the spermatogonia, spermatocytes, and early spermatids are broken and the chromatin derived from these nuclei becomes solubilized. Alternatively, Kaye and McMaster-Kaye (66) have separated the nuclei of the spermatozoa and the late spermatids of the house cricket by velocity sedimentation at unit gravity as described by Meistrich and Eng (67). The nuclear fraction is layered on top of 5–12% sucrose gradient containing 0.9% Dextran and allowed to settle for 6 hours. The fraction which sediments most slowly contains only long, thin nuclei. Electrophoretic analyses of acid-soluble proteins by both Tessier and Pallotta (65) and Kaye and McMaster-Kaye (66) have shown that the nuclei of cricket spermatozoa and those of late spermatids contain two new histone fractions, one migrating between histone H1 and H3 and the other at a position similar to that of histone H3. In the nuclear preparation of Kaye and McMaster-Kaye (66), the new histones comprise approximately 80% of the total histone fraction, while, possibly due to contamination of the chromatins derived from younger spermatids, considerably larger amounts of somatic-type histones have been observed in the preparations of Tessier and Pallotta (65).

VII.   Spermatozoa of Crab

Cytochemical observations have shown that spermatozoa of many species of crabs and crayfish have nuclei in which chromatin is less condensed than in the nuclei of somatic cells. Moreover, these nuclei lack chromosomal basic proteins (1, 68–70). Since these nuclei are very fragile, and since these spermatozoa possess a well-developed acrosomal complex, it has not been possible to isolate the sperm nuclei from these organisms. Vaughn and Hinsch (*71*) have, however, succeeded in isolation of chromatin fraction of spermatozoa of the spider crab, *Libinia emarginata.*

The complete vas deferens of the crab is first minced and stirred in saline–EDTA $[50\,mM\,NaCl–100\,mM\,EDTA\,(pH\,8)]$ for 15 minutes to free spermatophores and spermatozoa. The suspension is filtered through four layers of cheesecloth, and the filtrate is then centrifuged at 90 *g* (800 rpm in the 870 rotor of the International B-20 centrifuge) for 10 minutes. The pellets are washed twice. The sediment containing spermatophores and spermatozoa is resuspended in 50 ml of saline–EDTA containing 1/200 volume of 2-octanol, homogenized in a VirTis-45 homogenizer at 85 V for 45 seconds, and centrifuged at 2500 *g* (6000 rpm in the same rotor) for 15 minutes. The sediment containing acrosomes and broken spermatophore walls is washed twice with saline–EDTA by resuspension and centrifugation. The combined supernatants are clarified by centrifugation at 45,000 rpm (133,000 *g*) in the Spinco 50 Ti rotor for 1 hour. The chromatin is then sedimented by centrifugation at the same speed for 14 hours. The gelatinous sediment is resuspended in saline–EDTA and sedimented again under the same centrifugal conditions. The washed sediment is twice resuspended in 50 m*M* Tris-HCl (pH 8), stirred magnetically for 1 hour, and pelleted. The chromatin sediment is next resuspended in 10 m*M* Tris-HCl (pH 8), stirred for 4 hours, and dialyzed against the same buffer overnight. The suspension is sheared in the VirTis homogenizer at 25 V for 90 seconds, followed by centrifugation at 10,000 *g* for 30 minutes. The supernatant contains sheared sperm chromatin. The chromatin thus obtained possesses a weight ratio of protein to DNA of 2.3–3.2 and exhibits a thermal denaturation profile identical to that of purified sperm DNA. The chromosomal protein is dissociable by salt and gives a ratio of basic to acidic amino acid residues of 0.43–0.48. An electron microscopic examination has revealed that the sperm chromatin has aggregated fibers with a diameter in the range of 30–95 nm or more. Pronase treatment of the thick fibers leads to formation of very fine fibers.

Acknowledgments

We are grateful to Dr. H. Stanley Bennett for his valuable discussion and to Dr. Hironobu Ozaki for his helpful information about spermatozoa of sea urchin. We thank Dr. Darrel W. Stafford for his generous gift of sea urchin spermatozoa. This work was supported by a grant from the Rockefeller Foundation to the Laboratories for Reproductive Biology, University of North Carolina and by United States Public Health Service Grant HD-08510 from the National Institutes of Health.

References

1. Bloch, D. P., *Genetics* **61**, Suppl. 1, 93 (1969).
2. Baccetti, B., "Comparative Spermatology." Academic Press, New York, 1970.

3. Fawcett, D. W., *Biol. Reprod.* **2**, Suppl. 2, 90 (1970).
4. Fawcett, D. W., *Devel. Biol.* **44**, 394 (1975).
5. Coelingh, J. P., and Rozijn, T. H., *in* "The Biology of the Male Gamete" (J. G. Duckett and P. A. Racey, eds.), p. 245. Academic Press, New York, 1975.
6. Baccetti, B., and Afzelius, B. A., "The Biology of the Sperm Cell." Karger, Basel, 1976.
7. DeLange, R. J., and Smith, E. L., *Annu. Rev. Biochem.* **40**, 279 (1971).
8. Coelingh, J. P., Monfoort, C. H., Rozijn, T. H., Leuven, J. A. G., Schiphof, R., Steyn-Parvé, E. P., Braunitzer, G., Schrank, B., and Ruhfus, A., *Biochim. Biophys. Acta* **285**, 1 (1972).
9. Kistler, W. S., Geroch, M. E., and Williams-Ashman, H. G., *J. Biol. Chem.* **248**, 4532 (1973).
10. Calvin, H. I., *in* "The Biology of the Male Gamete" (J. G. Duckett and P. A. Racey, eds.), p. 257. Academic Press, New York, 1975.
11. Calvin, H. I., *Biochim. Biophys. Acta* **434**, 377 (1976).
12. Bellvé, A. R., Anderson, E., and Hanley-Bowdoin, L., *Dev. Biol.* **47**, 349 (1975).
13. Marushige, Y., and Marushige, K., *Biochim. Biophys. Acta* **340**, 498 (1974).
14. Marushige, Y., and Marushige, K., *J. Biol. Chem.* **250**, 39 (1975).
15. Platz, R. D., Grimes, S. R., Meistrich, M. L., and Hnilica, L. S., *J. Biol. Chem.* **250**, 5791 (1975).
16. Calvin, H. I., and Bedford, J. M., *J. Reprod. Fertil. Suppl.* **13**, 65 (1971).
17. Mann, T., *in* "Fertilization" (C. B. Metz and A. Monroy, eds.) Vol. I, p. 99. Academic Press, New York, 1967.
18. Dott, H. M., and Dingle, J. T., *Exp. Cell Res.* **52**, 523 (1968).
19. Coelingh, J. P., Rozijn, T. H., and Monfoort, C. H., *Biochim. Biophys. Acta* **188**, 353 (1969).
20. Pedersen, H., *J. Reprod. Fertil.* **31**, 99 (1972).
21. Brown, C. R., Andani, Z., and Hartree, E. F., *Biochem. J.* **149**, 133 (1975).
22. Wagner, T. E., Mann, D. R., and Vincent, R. C. *J. Exp. Zool.* **189**, 387 (1974).
23. Monfoort, C. H., Schiphof, R., Rozijn, T. H., and Steyn-Parvé, E. P., *Biochim. Biophys. Acta* **322**, 173 (1973).
24. Calvin, H. I., Yu, C. C., and Bedford, J. M., *Exp. Cell Res.* **81**, 333 (1973).
25. Kossel, A., "The Protamines and Histones." Longmans, Green, New York, 1928.
26. Felix, K., *Adv. Protein Chem.* **15**, 1 (1960).
27. Ando, T., Iwai, K., Ishii, S., Azegami, A., and Nakahari, C., *Biochim. Biophys. Acta* **56**, 628 (1962).
28. Ando, T., and Suzuki, K., *Biochim. Biophys. Acta* **140**, 375 (1967).
29. Bretzel, G., *Z. Physiol. Chem.* **353**, 933 (1972).
30. Bretzel, G., *Z. Physiol. Chem.* **354**, 543 (1973).
31. Daly, M. M. Mirsky, A. E., and Ris, H., *J. Gen. Physiol.* **34**, 439 (1951).
32. Fischer, H., and Kreuzer, L., *Z. Physiol. Chem.* **293**, 176 (1953).
33. Nakano, M., Tobita, T., and Ando, T., *Biochim. Biophys. Acta* **207**, 553 (1970).
34. Nakano, M., Tobita, T., and Ando, T., *Int. J. Pept. Protein Res.* **8**, 565 (1976).
35. Nakano, M., Tobita, T., and Ando, T., *Int. J. Pept. Protein Res.* **8**, 579 (1976).
36. Subirana, J. A., Cozcolluela, C., Palau, J., and Unzeta, M., *Biochim. Biophys. Acta* **317**, 364 (1973).
37. Nakano, E., *in* "Fertilization" (C. B. Metz and A. Monroy, eds.), Vol. II, p. 295. Academic Press, New York, 1969.
38. Bedford, J. M., and Calvin, H. I., *J. Exp. Zool.* **187**, 181 (1974).
39. Marushige, K., and Dixon, G. H., *Dev. Biol.* **19**, 397 (1969).

40. Ingles, C. J., Trevithick, J. R., Smith, M., and Dixon, G. H., *Biochem. Biophys. Res. Commun.* **22**, 627 (1966).
41. Louie, A., and Dixon, G. H., *J. Biol. Chem.* **247**, 5490 (1972).
42. Tingari, M. D., *J. Reprod. Fertil.* **34**, 255 (1973).
43. Vendrely, R., *Arch. Julius Klaus- Stift. Vererbungsforsch. Sozialanthropol. Rassenhyg.* **32**, 538 (1957).
44. Alder, D., and Gorovsky, M. A., *J. Cell Biol.* **64**, 389 (1975).
45. Vendrely, R., Knoblock-Mazen, A., and Vendrely, C., *Biochem. Pharm.* **4**, 19 (1960).
46. Paoletti, R. A., and Huang, R. C. C., *Biochemistry* **8**, 1615 (1969).
47. Easton, D., and Chalkley, R., *Exp. Cell Res.* **72**, 502 (1972).
48. Palau, J., Ruiz-Carrillo, A., and Subirana, J. A., *Eur. J. Biochem.* **7**, 209 (1969).
49. Wangh, L., Ruiz-Carrillo, A., and Allfrey, V. G., *Arch. Biochem. Biophys.* **150**, 44 (1972).
50. Ruiz-Carrillo, A., and Palau, J., *Dev. Biol.* **35**, 115 (1973).
51. Subirana, J. A., *J. Mol. Biol.* **74**, 363 (1973).
52. Johnson, A. W., and Hnilica, L. S., *Biochim. Biophys. Acta* **224**, 518 (1970).
53. Ozaki, H., *Dev. Biol.* **26**, 209 (1971).
54. Senshu, T., *Biochim. Biophys. Acta* **243**, 323 (1971).
55. Johnson, A. W., Wilhelm, J. A., Ward, D. N., and Hnilica, L. S., *Biochim. Biophys. Acta* **295**, 140 (1973).
56. Phelan, J. J., Subirana, J. A., and Cole, R. D., *Eur. J. Biochem.* **31**, 63 (1972).
57. Zirkin, B. R., *J. Ultrastruct. Res.* **36**, 237 (1971).
58. Spadafora, C., and Geraci, G., *FEBS Lett.* **57**, 79 (1975).
59. Gibbons, I. R., and Fronk, E., *J. Cell Biol.* **54**, 365 (1972).
60. Bennett, H. S., Wyrick, A. D., Lee, S. W., and McNeil, J. H., *Stain Technol.* **51**, 71 (1976).
61. Panyim, S., Bilek, D., and Chalkley, R., *J Biol. Chem.* **246**, 4206 (1971).
62. Bols, N. C., and Kasinsky, H. E., *Can. J. Zool.* **50**, 171 (1971).
63. Picheral, B., and Bassez, T., *J. Microsc. (Paris)* **12**, 107 (1971).
64. Picheral, B., and Bassez, T., *J. Microsc. (Paris)* **12**, 441 (1971).
65. Tessier, A., and Pallotta, D., *Exp. Cell Res.* **82**, 103 (1973).
66. Kaye, J. S., and McMaster-Kaye, R., *in* "The Biology of the Male Gamete" (J. G. Duckett and P. A. Racey, eds.), p. 227. Academic Press, New York, 1975.
67. Meistrich, M. L., and Eng, V. W. S., *Exp. Cell Res.* **70**, 237 (1972).
68. Yasuzumi, G., *J. Biophys. Biochem. Cytol.* **7**, 73 (1960).
69. Chevaillier, P., *J. Cell Biol.* **32**, 547 (1967).
70. Vaughn, J. C., *J. Histochem. Cytochem.* **16**, 473 (1968).
71. Vaughn, J. C., and Hinsch, G. W., *J. Cell Sci.* **11**, 131 (1972).

Chapter 6

Manual Enucleation of Xenopus Oocytes

CARL M. FELDHERR AND PAUL A. RICHMOND

Department of Anatomy,
University of Florida,
College of Medicine,
Gainesville, Florida

I. Introduction

Manual isolation is, at present, the only means by which amphibian oocyte nuclei can be obtained. However, the technique is not only useful for experiments relating to this specific cell type but is appropriate for more general studies as well. The procedure can be mastered in a relatively short time and has several advantages over the more commonly used mass isolation methods. For example, nuclei can be removed in less than 30 seconds, without the use of homogenization and centrifugation, thus avoiding prolonged exposure of the organelles to chemical and physical conditions not normally encountered *in vivo*. The major disadvantage, of course, is that only a limited amount of material can be obtained.

II. Methods

A. Nuclear Isolation

Nuclear isolation techniques will be described for mature *Xenopus laevis* oocytes, but, with minor modifications, they should be applicable to other mature amphibian oocytes. Nuclei also can be obtained, albeit with more difficulty, from immature oocytes. The procedures are similar to those described previously by Gall (*1*).

Pieces of ovary are surgically removed from anesthetized females and rinsed well in Ringer's solution. Although it is not essential, there are advantages to defolliculating the cells prior to enucleation, especially if large numbers of nuclei must be obtained within a short time interval. When using defolliculated oocytes it is not uncommon for over 90% of the nuclei to remain intact during isolation. The yields for cells containing follicle layers are usually much lower, probably due to increased tension on the nuclei as they are being extruded. Defolliculation can be accomplished either manually, using watchmaker's forceps, or enzymatically, using Pronase (2) or collagenase (3).

Enucleation is performed in intracellular medium containing 102 mM KCl, 11.1 mM NaCl, 7.2 mM K_2HPO_4, and 4.8 mM KH_2PO_4 (pH 7.0 $\pm$ 0.1). A simpler solution described by Gall (1) can also be used; it consists of 5 parts 0.1 M KCl and 1 part 0.1 M NaCl ("5:1 medium"). Since these solutions are not physiological for whole oocytes, cells (or pieces of ovary) are stored in Ringer's and transferred to intracellular medium immediately before use. The oocytes are stored and enucleated in 60 × 15 mm Falcon culture dishes. All manipulations are carried out under a dissecting microscope at a magnification of about ×35.

The major steps in the isolation procedure are illustrated in Fig. 1A–C. Initially, a watchmaker's forceps (Dumont No. 5) having a rounded, blunt tip is positioned on the animal pole at a point approximately half-way between the equator and axis. The forceps is inserted into the oocyte to a depth sufficient to produce a tear, approximately 400 μm wide, in the cell surface (Fig. 1A). The same forceps is then used to hold the oocyte in place while a second pair of forceps is positioned at the opposite side and used to compress the cell as shown in Fig. 1B. Cytoplasm will begin to flow from the tear, and the nucleus will appear as a clear spherical structure, 300–400 μm in diameter. It is essential that pressure be applied very gently as the nucleus is being extruded. At this stage, nuclei can be transferred to an experimental solution or fixative using a fire-polished Pasteur pipette with a tip diameter of 0.8–1 mm (Fig. 1C). The nuclei tend to stick to clean surfaces and should not be allowed to settle in the dish or pipette. If necessary, adhering cytoplasm can be removed by drawing the nuclei in and out of the pipette several times. Care must be taken to avoid contact with the water–air interface, as this will rupture the nuclei.

The volume of intact oocyte nuclei is a function of the colloid osmotic pressure of the surrounding medium; thus, when nuclei are isolated in salt solution they will swell (regardless of the ionic strength) provided that the nuclear envelope is intact. This, in fact, is a useful way to check the integrity of the envelope following isolation. Swelling can be prevented by adding 2% bovine serum albumin (or an equal molar concentration of another non-toxic macromolecule) to the intracellular medium.

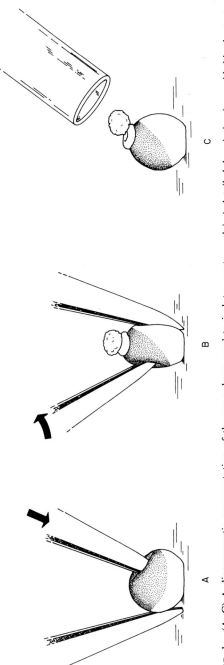

FIG. 1. (A–C) A diagrammatic representation of the procedures used to isolate oocyte nuclei. A detailed description is provided in the text.

The tendency for nuclei to adhere to plastic or glass surfaces can present a serious problem especially if the experimental procedure requires storing or incubating nuclei for any length of time. To avoid this difficulty one can cover the bottom of the dish with a thin layer of Sephadex G-15 beads which have been previously equilibrated with the appropriate solution. This does make it somewhat more difficult to visualize the nuclei under the dissecting microscope, but usually this is not a serious consideration.

B. Nuclear Fractionation

Isolated nuclei can be dissected into envelope and nucleoplasmic fractions by simply tearing the nucleus with a sharp instrument. In fresh preparations the nucleoplasm is a transparent gel and is easily separated from the envelope. This procedure can be performed on a layer of Sephadex G-15 beads using fine glass needles. Scheer (4) employed a different technique. After isolation, he transferred the nuclei to "5:1 medium" containing 10 mM $MgCl_2$ which coagulates the nucleoplasm and causes it to pull away from the envelope. When a pipette with a tip diameter slightly smaller than that of the nucleus is used, the envelope is ruptured and the nucleoplasm released. Still another method, which is especially useful when studying diffusable macromolecules, is to fix the nuclei in absolute ethanol immediately after isolation. This also causes the nucleoplasm to contract, and the envelope can be dissected away with watchmaker's forceps.

In immature amphibian oocytes the nucleoli are attached to, and can be isolated along with, the nuclear envelope (5). Aside from providing nucleolar-enriched fractions, this is a convenient system in which to separate nucleoli from other components of the nucleoplasm. Immature oocytes are also an excellent source of lampbrush chromosomes. The techniques for isolating and studying these structures have been reported in detail by Gall (1).

III.   Conclusions

Methods have been described for manually isolating and dissecting amphibian oocyte nuclei. These procedures are applicable to problems relating specifically to oocytes and are also useful for studying more general aspects of nuclear functions, especially when the experiments require minimum exposure to nonphysiological conditions. Despite the fact that only small amounts of material can be obtained, the methods described above have been used successfully to investigate the structure and function of lampbrush chromosomes (6), nucleocytoplasmic exchange (7,8), the structure of the nuclear envelope (9,10), the protein composition of various nuclear components (5), and aspects of RNA synthesis (11).

ACKNOWLEDGMENTS

The original investigations of the authors were supported, in part, by grant GM 21531 from the NIH.

REFERENCES

1. Gall, J. G., *Methods Cell Physiol.* **11**, 37 (1966).
2. Colman, A., *J. Embryol. Exp. Morphol.* **32**, 515 (1974).
3. Eppig, J. J., and Dumont, J. N., *In Vitro* **12**, 418 (1976).
4. Scheer, U., *Z. Zellforsch.* **127**, 127 (1972).
5. Maundrell, K., *J. Cell Sci.* **17**, 579 (1975).
6. Callan, H. G., *in* "Handbook of Molecular Cytology" (A. Lima-de-Faria, ed.), p. 540. North-Holland Publ. Co., Amsterdam, 1969.
7. Bonner, W. M., *J. Cell Biol.* **64**, 421 (1975).
8. Feldherr, C. M., *Exp. Cell Res.* **93**, 411 (1975).
9. Franke, W. W., and Scheer, U., *J. Ultrastruct. Res.* **30**, 288 (1970).
10. Franke, W. W., and Scheer, U., *J. Ultrastruct. Res.* **30**, 317 (1970).
11. Franke, W. W., and Scheer, U., *Symp. Soc. Exp. Biol.* **28**, 249 (1974).

Chapter 7

Macro and Micro Methods for Isolation and Injection of Cell Components from Polytene Tissues

N. H. LUBSEN AND A. M. C. REDEKER-KUIJPERS

Department of Genetics,
Faculty of Sciences,
Catholic University,
Nijmegen, The Netherlands

I. Mass Isolation of Salivary Glands from *Drosophila hydei*

The isolation of large quantities of cellular material is required for a biochemical analysis of the factors involved in gene expression. Such an analysis could extend the large number of cytochemical studies of gene induction in salivary glands of Drosophilae. For this purpose a method has been developed for the mass isolation of salivary glands and their subcellular components from *D. hydei* (*1*). These methods are also applicable to *D. melanogaster.*

A. Purification of Larvae

Synchronous 7-days-old larvae [cultured essentially according to Mitchell and Mitchell (*2*)] are washed free of the food under a stream of tepid tap water and collected in a plastic 1.1-mm mesh vegetable strainer. Residual food particles are removed by floating the larvae repeatedly on saline (80 gm NaCl/liter) in a 2-liter beaker. The larvae are then allowed to settle repeatedly through tap water to remove pupae and some further debris. Finally the larvae are collected in the strainer, rinsed with tap water, and blotted with filter paper. About 80 gm of larvae can be purified in one batch.

B. Salivary Gland Isolation

About 16 to 20 gm of larvae are placed on a clean glass plate measuring 40 $\times$ 100 $\times$ 0.3 cm. The larvae are wetted with a small amount of ice-cold Ringer's solution [6.5 gm NaCl, 0.14 gm KCl, 0.2 gm NaHCO$_3$, 0.204 gm MgCl$_2 \cdot$ 6 H$_2$O, 0.01 gm NaH$_2$PO$_4 \cdot$ H$_2$O per liter (pH 7.2)] and spread out over the plate with a paintbrush. A strong light at one end of the plate is used to orient the larvae with their heads away from the light.

A 33-cm long steel rod with a raised lip at each end such that the distance from most of the rod to the plate is 0.12 mm (for *D. melanogaster*, this distance should be shorter) is placed at the end of the plate closest to the light, which is now turned off to prevent heating the plate. The larvae are then squashed from behind by rolling the rod slowly down the plate. If the larvae stick to the rod while rolling, the rod may be wiped with a paintbrush during the rolling. After squashing, the larvae are washed from the plate with a stream of ice-cold Ringer's solution (from a wash bottle) into a 2-liter beaker and placed in an ice bath. The glass plate and the bar are wiped off with a paper towel. Up to four batches of squashed larvae can be collected before proceeding with the purification. All subsequent operations are performed at 0–4°C. The larvae are allowed to settle through the Ringer's, most of the supernate is decanted, and the sediment is strained and washed with a strong stream of Ringer's through a 1.1-mm mesh screen into a 2-liter beaker. This step removes the carcasses and large debris. The material is allowed to settle for a few minutes, and this supernate is decanted. The sediment is strained and washed through 0.85-mm mesh strainer into a 1-liter beaker, again with a strong stream of Ringer. After settling, the supernate is decanted and the sediment is resuspended in Ringer's and allowed to settle; the supernate is then decanted once more. At this point the supernatant should be virtually free of material; if not, the cycle of decanting and resuspension should be repeated.

The preparation now contains mainly salivary glands and intestines. The intestines are removed as follows: the preparation is poured into a plastic petri dish and swirled so that much material settles in the middle of the dish. Since the intestines are lighter than the salivary glands, they will settle on top and can be removed by gentle suction with a Pasteur pipette (attached to a water aspirator). During this process the preparation is examined continuously under a dissection microscope. After the top layer of intestines is removed, the preparation is resuspended in Ringer's and again allowed to settle while the plate is swirled. Another layer of intestines can then be removed. About 10 to 20 cycles of resuspension are required to remove most of the intestinal material. It is usually difficult to remove all without undue loss of salivary gland material, and the last pieces can best be removed individually with a pair of forceps. The final preparation still contains white malphighian tubes. According to Boyd and Presley (3), they may be removed by layering the gland preparation over 30% Ficoll. A gentle centrifugation pellets the malphighian tubes, while the glands float on the Ficoll solution.

C. Isolation of Subcellular Components

The isolation of nuclei and nucleoli from salivary glands from *D. melanogaster* has been recently reviewed by Boyd (4). Since the same procedures are used to isolate these components from *D. hydei*, they will not be described here. Chromatin has been prepared from salivary glands nuclei by Helmsing and van Eupen (5) and Elgin and Boyd (6).

II. Micromanipulation of Salivary Glands

The large size of the polytene cells of the salivary glands of Diptera allows the direct isolation of cellular components. Edström and his co-workers (7) were the first to take advantage of this possibility in their studies of RNA synthesis in Balbiani rings in *Chironomus tentans* (8). We have adapted their method, which has been reviewed (7,9), to study transcription in puff 2–48 BC in *Drosophila hydei* salivary glands (10).

A. Preparation and Fixation of Glands

Salivary glands are isolated by hand from late third instar larvae. The age of the larvae should be carefully selected so that the glands are transparent or almost so, and so that most of the mucopolysaccharides have already been secreted into the lumen and can be removed after fixation of the glands. The contents of younger glands become too sticky after fixation to allow micromanipulation.

The isolated glands are incubated in Poels medium (11) containing radioactive RNA procursors under appropriate puff-inducing conditions: e.g., 60 minutes at 37° or 4 hours in the presence of $5 \times 10^{-2} M$ vitamin B_6 for puff 2–48 BC. Glands are fixed for 15 to 20 minutes in freshly prepared absolute alcohol–acetic acid (3:1 v/v) on ice, rinsed twice for 10 minutes in ice-cold 70% alcohol, and stored for at least 90 minutes at $-20\,°C$ in absolute alcohol–glycerol (1:1 v/v). One or two glands are placed on a 32×5 mm piece of glass (cut from a cover slip, degreased with chloroform) and opened lengthwise with two needles to remove the secretion product, which should be visible as an opaque rod, from the lumen. The strip of glass is then placed in the oil chamber for micromanipulation.

B.   Preparation of Oil Chamber

The oil chamber is made from a thick piece of glass (the size of a microscope slide) with a transverse groove. A 32 × 5 mm piece of glass, cut from a coverslip (degreased with chloroform), is placed over the groove. On this coverslip is attached (with a drop of glycerol) a 1 mm² piece of coverslip (also degreased with chloroform), on which the isolated cellular components are collected (see Fig. 1). The piece of coverslip containing the salivary glands is placed next to this coverslip (Fig. 2). The area under these two glass strips is now filled with paraffin oil.

C.   Preparation of Needles for Micromanipulation

A glass rod with a diameter of 0.5–0.8 mm is pulled from a glass rod with a diameter of 3–5 mm after the latter is heated over a flame. The thin rod is heated over a small flame and then pulled out again. The rod is then cut to make a 10-cm-long needle with a 1.5-cm-long thin tip. This needle is placed in a microforge to form the required tips:

1. For the fixed needle (used to hold pieces of tissue): a tip is pulled perpendicular to the needle, and a second tip is pulled on this tip perpendicular to the plane of the needle and the first tip.

2. For the movable needle (this needle is used to cut out the desired cellular component): a short but sharp tip is pulled on the needle.

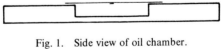

Fig. 1.   Side view of oil chamber.

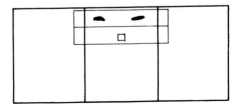

Fig. 2.   Top view of oil chamber.

D. Micromanipulation

The oil chamber is placed on the stage of a Wild microscope with phase-contrast optics. The needles are placed in a Fonbrune micromanipulator, and the desired pieces of tissue are isolated. For example, to isolate a part of the chromosome, the nuclei are first removed from the cell, then opened, and the chromosomes are spread out. The desired part is cut off and transferred to the 1-mm² piece of glass. Most cellular components will stick to glass after fixation.

The ease of micromanipulation is primarily dependent upon the state of the tissue (we have found that only late third instar glands are satisfactory in this respect) and on the cellular component desired. It is quite difficult to recognize some chromosome regions under these conditions. Furthermore, for unknown reasons, only in some glands is enough loose nucleoplasm present to make its isolation possible. Under optimal conditions, 100 puffs at 2–48 BC can be isolated in a day. Fifty puffs usually contain enough labeled RNA to allow an analytical gel electrophoresis.

E. Analysis of RNA from Isolated Cellular Components

The small piece of glass with the isolated components is removed from the oil chamber, transferred to a small petri dish, and rinsed several times with chloroform to remove the paraffin oil. It is then transferred directly to a small tube containing 25 μl of extraction medium [0.02 M Tris-HCl (pH 7.4), 2.5% sodium dodecyl sulfate (SDS), 0.1 mg/ml polyvinylsulfate, and 1 mg/ml $E. coli$ RNA], heated for 2 minutes at 90°, and directly analyzed by gel electrophoresis either on polyacrylamide or agarose gels.

III. Microinjection of Salivary Gland Cells

Microinjection of salivary gland cells has been used to study the gene-activating capacity (i.e., puff-inducing ability) of substances to which the cell is nonpermeable (12) or to study the localized effect of permeable substances [e.g., the effect of injecting ecdyson directly in the nucleus; Brady et al. (13)].

A. Isolation of Salivary Glands

Salivary glands are hand isolated from late third instar larvae of *Drosophila hydei*. All contaminating tissue, including the layer of fat cells, should be removed. Usually one of the pair of glands is injected with the material to be tested, while its sister gland is used as injection control.

B. Preparation of Needles

Needles are pulled from Pyrex capillary tubes with inner diameters of 1 mm with a Narishige microelectrode puller. The proper settings will have to be determined experimentally since the size and length of the needle pulled is not constant. The tip of the needle should have an outside diameter of about 2–3 μm (measured with an ocular micrometer) after breaking off the tip (see below), and the tip should be fairly long (about 10 to 15 nm). The needle is connected with a short piece of tubing (about 4 cm long) to a Fonbrune micromanipulator and with further tubing of a stationary 2-ml glass syringe. The syringe is mounted in a simple apparatus that allows the plunger action to be controlled by turning a screw nut.

The whole system is filled with paraffin oil. The needle is now placed under the microscope (Wild, with phase optics and a fixed stage), and the needle is opened by breaking off the tip with a pair of forceps. The needle can now be filled partially with oil. Avoid filling the tip since this will block the needle. A drop of injection fluid is placed on a siliconized slide on the microscope stage, and the tip of the needle is filled by placing the needle in this drop and slowly pulling back the plunger of the syringe. The paraffin oil and the injection fluid should be separated by air.

C. Injection of Salivary Glands

Salivary glands are fixed to a siliconized microscope slide as follows. A small square piece of Plasticine is placed on the slide and covered with a drop of Ringer solution [6.5 gm NaCl, 0.14 gm KCl, 0.2 gm NaHCO$_3$, 0.204 gm MgCl$_1$·6 H$_2$O, 0.01 gm NaH$_2$PO$_4$·H$_2$O/liter (pH 7.2)] or Poels (11) medium. One prong of a pair of forceps is pressed down on the Plasticine to make a V-shaped opening in one side of the square. The gland is placed in this opening, leaving the apical cells free. To hold the gland down firmly, the sides of the opening are pressed towards the gland and a strip of plasticine is placed on top. The slide is placed on the microscope stage, and the apical cells can now be injected. The volume of injected material can be determined by measuring the distance that the meniscus formed by the injection fluid has moved through the needle (with an ocular micrometer) and from the diameter of the needle. Usually about 1×10^{-7} μl is injected.

Injected glands can be incubated further in a suitable medium, such as Poels medium (11), and analyzed cytologically by standard methods. The injected nuclei can be marked by pulling out part of a chromosome by moving the needle up and down a few times. To analyze cells whose cytoplasm has been injected the noninjected part of the gland is cut off after fixation.

REFERENCES

1. Boyd, J. B., Berendes, H. D., and Boyd, H., *J. Cell Biol.* **38**, 369 (1968).
2. Mitchell, H. K., and Mitchell, A., *Drosophila Inf. Serv.* **39**, 125 (1964).
3. Boyd, J. B., and Presley, J. M., *Biochem. Genet.* **9**, 309 (1973).
4. Boyd, J. B., *Methods Cell Biol.* **10**, 135 (1975).
5. Helmsing, P. J., and van Eupen, O., *Biochim. Biophys. Acta* **308**, 154 (1973).
6. Elgin, S. C. R., and Boyd, J. B., *Chromosoma* **51**, 135 (1975).
7. Edström, J.-E., *Methods Cell Biol.* **1**, 417 (1964).
8. Daneholt, B., *Cell* **4**, (1975).
9. Lambert, B., and Daneholt, B., *Methods Cell Biol.* **10**, 17 (1975).
10. Bisseling, T., Berendes, H. D., and Lubsen, N. H., *Cell* **8**, 299 (1976).
11. Poels, C. L. M., *Cell Differ.* **1**, 63 (1972).
12. Sin, Y. T., *Nature (London)* **258**, 159 (1975).
13. Brady, T., Berendes, H. D., and Kuijpers, A. M. C., *Mol. Cell Endocrinol.* **1**, 249 (1974).

Chapter 8

Isolation of Nuclei and Chromatin from Phycomyces blakesleeanus

ROBERT J. COHEN

Department of Biochemistry and Molecular Biology,
University of Florida,
Gainesville, Florida

I. Introduction

Phycomyces blakesleeanus has long been studied as a model sensory system (*1,2*). Only recently has its potential as a model for the control of development and gene expression been suggested. *Phycomyces* belongs to one of the most primitive mold families, the Mucoraceae. The haploid nucleus has seven chromosomes with a total genome size of about 1.9×10^{10} daltons, approximately 6.7 times that of *E. coli* (*3*). Organogenesis is, however, already rather complex. The various stages of development are well defined (*1*). At several points in its development, external signals such as chemicals, heat, or light (*4–6*) determine the choice of morphology and differentiation. Mutants with lesions at most of the points of divergence are available.

The giant unicellular sporangiophore is the most remarkable organelle. This 3–5 cm long multinucleated fruiting body possesses a very sophisticated growth response system sensitive to light and nearby objects. It exhibits considerable regenerative powers (*7–11*) and even a few microliters of cytoplasm squeezed out onto a glass slide will regenerate mycelia and sporangiophores (*12,13*).

Control of the development of the sporangiophore at the level of gene expression may be relatively simple. Yet an important question is immediately posed. Nuclear replication and the regulation of gene expression and development are here not related to cell division. Yet they are exquisitely controlled by extrinsic factors. How, then, are the genome components of this multinucleated organelle regulated?

A preliminary investigation of gene components has been made (*14*). In it, chromatin isolated from various stages of synchronized sporangiophores was shown to contain nonhistone chromosomal proteins which exhibit considerable differences, both quantitative and qualitative, during sporangiophore development. Extracted chromatin contains a relatively large quantity of histones including H1, H2B, H2A, and H4; H3 was not detected. Hamkalo reports that the chromatin does contain nu bodies (private communication). Of interest would be to attempt to reconcile this information with the tetrameric histone complex proposed by Kornberg and Thomas (*15*).

II. Methods

The following methodology is used in our laboratory to grow mycelia and sporangiophores and to isolate intact nuclei from the latter. Other techniques are outlined in the papers indicated in the appropriate sections; particularly useful is Bergman *et al.* (*1*).

A. Growth of *Phycomyces blakesleeanus*

1. Mycelial growth is maintained in liquid minimal medium, GA, which consists of 30 gm glucose, 2 gm L-asparagine $\cdot$ H_2O, 0.5 gm $MgSO_4 \cdot 7 H_2O$, 1.5 gm KH_2PO_4, trace metals,[1] and 0.25 ppm thiamine/liter of H_2O, final pH 4.2. Yields are increased with the addition of 1–5 gm/liter yeast extract or casamino acids (Difco). Spores of *Phycomyces blakesleeanus* ($-$) (Burgeff) (NRRL1555) are heat shocked in 5 ml of sterile H_2O at 48° for 5 minutes to induce germination. In liquid, densities may range from 10^5–10^7 spores liter. Optimal growth occurs at 18–23°. Profuse growth occurs with oscillation and vigorous aeration. Yields are 15–30 gm/liter mycelia after about 4 days. Mycelia are easily harvested by filtration through a cold Büchner funnel. The mat may be washed with water and frozen or immediately processed. Cutting into 2–3 mm pieces and homogenizing is sufficient to release nearly all cytoplasmic material. Cessation of oscillation and aeration during incubation results in the initiation of sporangiophores at the air–liquid inter-

---

[1] R. P. Sutter (personal communication) has suggested that GA medium be modified to contain 10 μg Ca^{2+}/ml, 2 μg/ml citric acid $\cdot$ H_2O, 0.2 μg Fe^{3+}/ml, 0.2 μg Zn^{2+}/ml, 0.1 μg Mn^{2+}/ml, 0.02 μg Mo^{6+}/ml, and 0.01 μg Cu^{2+}/ml. Our laboratory routinely adds trace metals. Calcium is stored as 14% (w/w) $CaCl_2$ solution. Trace metal stock is prepared by dissolving *in order* in 100 ml of deionized water: (1) citric acid–H_2O, 2.0 g; (2) $Fe(NO_3)_3 \cdot 9 H_2O$, 1.5 g; (3) $ZnSO_4 \cdot 7 H_2O$, 1.0 g; (4) $MnSO_4 \cdot H_2O$, 0.3 g; (5) $CuSO_4 \cdot 5 H_2O$, 0.05 g; and (6) $NaMoO_4 \cdot 2 H_2O$, 0.05 g. The stock solution is stored at 4°.

face, especially during continuous illumination. This is a good technique to study initiation and to obtain large quantities of the complete organism.

2. Sporangiophores are best raised on either potato dextrose agar (Difco) supplemented with 0.2% yeast extract (PDA + Y), pH 5.6, or on commercial instant potato supplemented with 0.2% yeast extract. Four percent agar may also be added to the minimal media to provide solid support. For large quantities we usually use 50 cm × 30 cm stainless-steel trays with tempered glass covers. Agar-containing medium and trays are autoclaved separately; the medium is allowed to thicken slightly before pouring. The commercial potato medium is mixed to a smooth consistency. Trays containing the mix are autoclaved 40 minutes. The medium should then be somewhat wet and should flow slightly when tipped. In either case, after heat shocking, 5–10 spores/cm^2 are distributed. For the potato medium distribution is facilitated by suspending the spores in about 50 ml of H_2O per tray.

Trays are grown under fluorescent light for 48–60 hours until the first appearance of immature stage I sporangiophores. They are then transferred to a cold room overnight (4°) in the dark. The trays are removed and the glass covers opened to expose the fungus to room light and temperature. This procedure tends to synchronize development. Stage I sporangiophores are harvested 2–4 hours after their initial appearance, stages II and III, 8–9 hours, and stage IVb about 18 hours. The last should be the same height, from 3–5 cm. The degree of synchrony is from 85–100%; however, asynchronous sporangiophores can easily be plucked and discarded. Stages II and III cannot be visually distinguished easily and are usually collected together.

B. Isolation of Nuclei from Sporangiophores

Sporangiophores are returned to the cold room (4°) 30 minutes before harvesting. They are plucked, weighed, cut into 3-mm pieces, and gently squeezed with an 8-cm diameter roller into a flat stainless-steel tray containing 3–4 volumes of homogenizing buffer. Plucked sporangiophores always detach at their base. The roller may be a cylindrical glass bottle. Homogenizing buffer consists of 0.75 M sucrose–20 mM Tris-HCl–0.5 mM Na$_2$ethylenediaminetetraacetic acid (EDTA)–15 mM MgCl$_2$–40 mM KCl–25 mM ascorbate (pH 7.5). The ascorbate serves as an antioxidant. The solution should not be made up more than 24 hours beforehand since ascorbate will oxidize in solution.

The suspension is filtered through three layers of cheesecloth, and the filtrate is spun at 8000 rpm (7700 g) in a Sorvall SS-34 rotor for 15 minutes at 4°. The pellet is resuspended by mixing on a Vortex in "resuspension buffer" (the previous buffer, but lacking the ascorbate). Spores are removed from stage IVb extract by centrifuging 30 minutes at 100 g in a swinging-bucket rotor. The nuclear pellet is resuspended in 4 ml of resuspension buffer, and 1-ml aliquots are carefully layered over 4.5 ml of 2.2 M sucrose– 20 mM Tris-HCl–0.5 mM Na$_2$ EDTA–15 mM MgCl$_2$–40 mM KCl (pH 7.4). The sucrose suspension is centrifuged in a Beckman SW-50.2 swinging-bucket rotor at 35,000 rpm for 60 minutes. Cytoplasmic contamination remains at the interface. However, at high concentration of material, nuclei are also lost at the interface. Observation of the top, mid-portion, and bottom of the nuclear pellet by phase-contrast microscopy reveals that each region of the pellet is more than 95% free of cytoplasmic contamination. Treatment with detergent appears unnecessary. When spores are initially present, one sees only one spore per 2–5 × 10^3 nuclei. Nuclei are small and spherical, about 2 μm in diameter; spores are ellipsoidal, 5–10 μm in the long dimension.

C. Isolation of Chromatin

Nuclei are lysed by suspension in triple glass-distilled water with a few strokes of a Dounce homogenizer with a wide clearance. Chromatin is swelled for 3 hours at 4°, pelleted at 20,000 g for 15 minutes, resuspended in distilled water, and again pelleted at 20,000 g. In our hands, the ratios of histone:nonhistone chromosomal protein from *Phycomyces* chromatin were 0.80 for stage IVb, 0.84 for stages II–III, and 1.10 for stage I (*14*).

REFERENCES

1. Bergman, K., Burke, P. V., Cerdá-Olmedo, E., David, C. N., Delbrück, M., Foster, K. W., Goodell, E. W., Heisenberg, M., Meissner, G., Zalokar, M. Dennison, D. S., and Shropshire, W., *Bacteriol. Rev.* **33**, 99 (1969).

2. Cohen, R. J., Jan, Y. N., Matricon, J., and Delbrück, M., *J. Gen. Physiol.* **66**, 67 (1975).

3. Dusenbery, R. L., *Biochim. Biophys. Acta* **378**, 363 (1975).

4. Bergman, K., *Planta* **107**, 53 (1972).

5. Thornton, R. M., *Plant Physiol.* **51**, 570 (1973).

6. Thornton, R. M. *Am. J. Bot.* **62**, 370 (1975).

7. Burgeff, H., *Flora* **107**, 259 (1914).

8. Ootaki, T., Lightly, A. C., Delbrück, M., and Hsu, W.-J., *Molec. Gen. Genet.* **121**, 57 (1973).

9. Gruen, H. E., and Ootaki, T., *Can. J. Bot.* **48**, 55 (1970).

10. Ootaki, T., and Gruen, H. E., *Can. J. Bot.* **48**, 95 (1970).

11. Gruen, H. E., and Ootaki, T., *Can. J. Bot.* **50**, 139 (1972).

12. Weide, A., *Arch. Exp. Zellforsch.* **23**, 299 (1939).

13. Zaichkin, E. I., Orlova, S. A., and Fikhte, B. A., *Dokl. Akad. Nauk. SSSR* **225**, 1187 (1975).

14. Cohen, R. J., and Stein, G. S., *Exp. Cell Res.* **96**, 247 (1975).

15. Kornberg, R. D., and Thomas, J. O., *Science* **184**, 865 (1974).

Part B.   Chromosomes

Chapter 9

Mammalian Metaphase Chromosomes[1]

JOSEPH J. MAIO AND CARL L. SCHILDKRAUT

Department of Cell Biology,
Albert Einstein College of Medicine,
Bronx, New York

I.   Introduction

Recent reports indicate that isolated mammalian metaphase chromosomes may serve as carriers for the transfer of genetic information between mammalian cells. The genes transferred included those for functional cellular enzymes (*1–5*) and tumor viruses (*6–8*). Isolated metaphase chromosomes seem suited for this purpose since they may preserve natural linkage groups for other gene markers. Also, the DNA may be protected from extensive nuclease action during the transfer, and integration of the donor chromosomes with the recipient genome may be facilitated. The potential usefulness of isolated metaphase chromosomes for studying problems in the organization of the mammalian genome has not yet been fully realized. Such studies encompass the mechanisms of chromatin condensation and decondensation during the cell cycle, the subunit structure of nucleosomes in altered states of chromatin, and the association of specific types of DNA sequences and nuclear proteins with specific chromosomes. Fractionation of isolated metaphase chromosomes by velocity sedimentation (*9–13*) or by electronic particle-sorting techniques (*14*) offers an additional experimental approach to some of these problems and may prove useful in refining the gene transfer experiments.

Of the many methods that have been described for the isolation of mammalian metaphase chromosomes, all require the stabilization of chromosome structure by suitable buffer systems. The metaphase cells are then

[1] Supported by NIH grants NCI-CA 16790 and GM 19100 (J.J.M.), and by NSF grant GB 33691 and NIH grant GM 22332 (C.L.S.).

homogenized in the appropriate buffer to release the chromosomes, and the subsequent purification is through the removal of cytoplasmic and membranous debris and interphase nuclei by a variety of centrifugation techniques. Metaphase chromosome structure is stabilized at low pH (from pH 2.8 to about 5.6) (15–19), at neutral pH in the presence of divalent metal cations or organic solvents (20–23), and at pH 10.5 in the presence of Ca^{2+} and hexylene glycol (24). It is not our purpose to evaluate the advantages and disadvantages of each of these methods. Such an evaluation requires detailed knowledge of (1) the extent of DNA depurination, denaturation, or shearing; (2) the removal, displacement, exchange, or denaturation of intrinsic chromosomal proteins; (3) the binding of protein, DNA, or RNA as contamination artifacts; (4) the suitability or efficacy of the final product in the experimental design (e.g., gene transfer experiments); and (5) the preservation of chromosome morphology, especially if individual chromosome types are to be identified in the population of purified chromosomes. We present here the method with which we are most familiar (22). This method avoids the use of extremes of pH or organic solvents, it has been used in chromosome fractionation (11) and gene transfer experiments (1,4,5), and negligible chromosome protein exchange occurs during the purification (S. Kobayashi and C. Schildkraut, unpublished observations).

II.   General Procedures

A.   Cell Cultures

The method described here has been used successfully in the isolation of metaphase chromosomes from HeLa cells, L-cells, Syrian and Chinese hamster cells, mouse and human lymphoblasts, and African green monkey BSC-1 cells. All of these cell lines can be grown in suspension and are heteroploid, pseudodiploid, or adapted for long-term growth in culture. Chromosome isolation from diploid cells in monolayer culture has generally given disappointing results. This is because of the low percentage of metaphase cells obtained from such cultures and the consequent contamination of the final preparations by interphase nuclei. It is possible that suitable methods of synchrony (e.g., release from contact inhibition) could increase the percentage of metaphase cells in diploid cultures, but this approach has not been adequately explored.

For a typical preparation, 1 to 2 $\times$ 10^9 cells were harvested, of which at least half were in metaphase arrest. The final yield of purified chromosomes was usually 15 to 30% of the chromosomes in the starting homogenate as determined in a Petroff–Hausser bacterial counting chamber under phase-contrast optics. Despite the relatively low yields, such preparations from HeLa cells usually provided 60 to 75 mg dry weight of purified metaphase chromosomes.

B. Metaphase Arrest

Because different cell lines vary in their responses to metaphase arrest agents, preliminary studies are performed to determine the concentration of colchicine or vinblastine sulfate necessary to produce maximum metaphase arrest. For example, the S3 HeLa cell line undergoes metaphase arrest in the presence of 0.01 μg/ml of colchicine or vinblastine sulfate, whereas the Chinese hamster and L-cell lines require 5 to 10 μg/ml of these agents to produce a similar response. Colchicine or vinblastine sulfate are added under conditions such that the cell culture can complete at least one more division in the exponential growth phase.

In addition to the concentration, it is also advisable to determine the optimum length of the incubation period with the metaphase arrest agent for each cell line. This is especially important when working with cell lines that have a tendency to form micronuclei during prolonged incubation with vinblastine sulfate or colchicine (e.g., Chinese hamster cells). Micronuclei are very difficult to separate from the chromosomes by differential centrifugation and represent a loss in chromosome yield. Also, partial synchronization of the culture has been utilized for cell lines in which drugs do not produce a sufficiently enriched population of cells arrested in metaphase. For example, synchronization of Chinese hamster cells by a double thymidine block can be followed by treatment with colchicine. This increases the percent metaphase from less than 50% to about 80% (S. Kobayashi, unpublished results).

The percentage of cells in metaphase arrest may be determined by examining a drop of the cell suspension under a coverslip with phase-contrast optics. For this purpose, 1 to 2 ml of the culture of arrested cells is centrifuged at 500 g for 5 minutes in a small centrifuge tube with a conical bottom. The medium is poured off, and the cells are resuspended in 0.5 ml of TM buffer (described below). The cells expand in the hypotonic buffer over a period of about 10 minutes, but most of them remain opaque in phase-contrast. Ten microliters of a 5% solution of saponin (see below) are then added: the cells quickly become transparent, and metaphase chromosomes and interphase nuclei are easily scored.

III.   Isolation of Chromosomes

The cells are sedimented by centrifugation at 500 g for 15 minutes in the horizontal head of a PR-2 centrifuge (International Equipment Co., Needham Heights, Mass.) and washed twice in Earle's balanced salts solution. Throughout all subsequent operations the cells and extracts are maintained near 0°C.

The pellet of washed cells is resuspended in 13 volumes of TM (CaCl$_2$, MgCl$_2$, and ZnCl$_2$, each 0.001 M in 0.02 M Tris-HCl, pH 7.0). Hypotonic swelling and intracellular dispersion of metaphase figures occur over a period of about 10 minutes. Saponin (Fisher Scientific Co.; 5% solution previously filtered through two layers of Whatman No. 1 paper) is then added to a final concentration of 0.1%. After hypotonic swelling in TM, some cell lines such as Chinese hamster V79 cells are particularly difficult to break by Dounce homogenization. Cell breakage is more easily achieved when saponin is replaced by Triton X-100 (Rohm and Haas, Philadelphia, Pennsylvania) to 1% final concentration shortly before breaking the cells. However, we note that the use of Triton X-100 sometimes produces somewhat greater contraction of the purified metaphase chromosomes than does saponin. It is important that the saponin or Triton X-100 be added after the cells have had time to swell adequately in the hypotonic buffer. This usually requires 5 to 10 minutes at 0°C. After 5 minutes, aliquots of the cell suspension are transferred to a 40-ml capacity Dounce homogenizer equipped with a small-clearance pestle. The cells are broken and the chromosomes liberated by 20 to 60 strokes with the homogenizer. The extent of cell breakage and the effectiveness of the subsequent steps in the chromosome purification are followed by phase-contrast microscopy.

The homogenate is diluted in 2 volumes of TMS (TM containing 0.1% saponin) and 15-ml aliquots are distributed into 40-ml capacity, heavy-duty Pyrex centrifuge tubes (Corning No. 8340) with conical bottoms. The tubes can be siliconized before use. The distance from the surface of the liquid to the bottom of the tubes should not exceed 3 cm. Centrifugation for 5 minutes at 120 g in the horizontal head of a PR-2 centrifuge sediments nuclei and unbroken cells, leaving the chromosomes and fine cellular debris in suspension. The supernatant fraction, containing the chromosomes, is poured off and saved. Care is taken to stop the pouring-off as soon as it is apparent that pelleted material is about to contaminate the supernatant fraction. Chromosomes entrapped among the nuclei are extracted by resuspending the pellet in each tube to the original volume with TMS and repeating the centrifugation at 120 g. The extraction is repeated if significant quantities of chromosomes are still present in the pellets. Any residual nuclei in the combined supernatant fractions are removed by repeating the low-speed centrifugation as described above. The chromosomes in the pooled supernatants are sedimented by centrifugation in an angle-head centrifuge for 10 minutes at 2500 g, and the pellets are washed once by resuspension in 0.02 M Tris (pH 7.0) containing 0.1% saponin and recentrifugation at 2500 g for 10 minutes. Divalent metals are omitted during this step. The washed pellets are thoroughly resuspended in 54 ml of 2.2 M sucrose (sp gr ca. 1.28)–0.02 M Tris (pH 7.0)–0.1% saponin. Then 18-ml aliquots of this suspension are layered over 10 ml of the 2.2 M sucrose solution in three 1 in. × 3 in. tubes suitable for the SW 25.1 rotor of the Model L ultracentrifuge. Slight amounts of liquid entrapped in the pellet decrease the density of the suspension somewhat so that layering is easily accomplished. The interfaces are blended by stirring with a glass rod, but care should be taken to leave about 1 cm of sucrose solution undisturbed at the bottom of the tubes. The chromosomes are sedimented by centrifugation at 50,000 g for 1 hour. Cytoplasmic debris remains in suspension or rises to the surface of the sucrose to form a pellicle. A few milliliters of TM buffer are added to the chromosome pellet and the tube is left to stand in an ice bucket for 2 or 3 minutes before the chromosomes are resuspended. Sucrose is removed by washing the chromosomes in TM. The chromosomes can be stored in TM for at least 2 months at 0°C in ice buckets without alteration in morphology or staining properties. No significant amounts of radioactive material were released during this time from chromosomes whose nucleic acids and proteins were labeled with [^{32}P] and [^{3}H]leucine, respectively. For long-term storage at −20°C, it is necessary to add glycerol to a final concentration of 20% in TM in order to prevent chromosome breakage on freezing and thawing.

IV. Chromosome Aggregation

The purified metaphase chromosomes aggregate tenaciously in most stabilizing buffers. Such aggregates interfere with fractionation according to size classes and, presumably, with the uptake of individual chromosomes by recipient cells in gene transfer experiments. The presence of $ZnCl_2$ and $MgCl_2$ in the stabilizing buffer greatly increases the tendency of the isolated chromosomes to aggregate. $CaCl_2$ does not have this effect, although it maintains the compactly coiled state of the chromosomes when present at sufficiently high concentration (3 to 5 mM) (11). For some purposes, such as fractionation, $ZnCl_2$ and $MgCl_2$ are removed from the chromosomes after isolation by repeated washings in Tris–Ca buffer as described (11). However, for maximum stability during storage they should be transferred back to TM buffer.

Excessive centrifugal force also promotes aggregation. The lowest possible centrifugal force necessary to sediment all of the chromosomes should be used at all stages of the isolation, washing, or fixation procedures. In the absence of sucrose, this requires about 2500 g for 10 minutes.

Because some cells are difficult to break in TM buffer, and because the chromosomes aggregate excessively, we have also isolated metaphase chromosomes employing a modification of the TM procedure. Although this procedure gives good yields, the chromosomes are often considerably more contracted than in the standard method employing TM buffer and identification of individual chromosomes is difficult. The cells are washed in Earle's balanced salts solution as described, but hypotonic swelling occurs in 6 mM $CaCl_2$ –Tris-HCl (pH 7.0) for 10 minutes. Many cell lines can then be broken easily in this buffer without the addition of either saponin or Triton X-100, and the chromosomes show much less tendency to aggregate. Nuclei and cell debris are removed by low-speed centrifugation as described, employing Tris–$CaCl_2$ buffer at all stages, but centrifugation in 2.2 M sucrose is in the presence of 1.5 mM $CaCl_2$–Tris-HCl (pH 7.0). Ca^{2+} must be present at all stages in the isolation procedures and in subsequent manipulations or else the chromosomes will uncoil. Other divalent metals such as Mg^{2+} have also been employed to stabilize the chromosomes, but Ca^{2+} seems to give more satisfactory results.

REFERENCES

1. McBride, O. W., and Ozer, H., *Proc. Natl. Acad. Sci. U.S.A.* **70**, 1258 (1973).
2. Burch, J., and McBride, O. W., *Proc. Natl. Acad. Sci. U.S.A.* **72**, 1797 (1975).
3. Willecke, K., and Ruddle, F., *Proc. Natl. Acad. Sci. U.S.A.* **72**, 1792 (1975).
4. Wullems, G., van der Horst, J., and Bootsma, D., *Somatic Cell Genet.* **1**, 137 (1975).
5. Wullems, G., van der Horst, J., and Bootsma, D., *Somatic Cell Genet.* **2**, 359 (1976).
6. Shani, M., Huberman, E., Aloni, Y., and Sachs, L., *Virology* **61**, 303 (1974).
7. Shani, M., Aloni, Y., Huberman, E., and Sachs, L., *Virology* **70**, 201 (1976).
8. Ebina, T., Miao, R., and Watanabe, Y., *Exp. Cell Res.* **88**, 203 (1974).
9. Huberman, J. A., and Attardi, G., *J. Mol. Biol.* **29**, 487 (1967).
10. Mendelsohn, J., Moore, D. E., and Salzman, N. P., *J. Mol. Biol.* **32**, 101 (1968).
11. Maio, J., and Schildkraut, C., *J. Mol. Biol.* **40**, 203 (1969).
12. Schneider, E., and Salzman, N. P., *Science* **167**, 1141 (1970).
13. Skinner, L. G., and Ockey, C. H., *Chromosoma* **35**, 125 (1971).
14. Gray, J., Carrano, A., Steinmetz, L., Van Dilla, M., Moore, D., Mayall, B., and Mendelsohn, M., *Proc. Natl. Acad. Sci. U.S.A.* **72**, 1231 (1975).
15. Chorazy, M., Bendich, A., Borenfreund, E., and Hutchison, D. J., *J. Cell Biol.* **19**, 59 (1963).
16. Franceschini, P., and Giacomoni, D., *Atti. Assoc. Genet. Ital.* **12**, 248 (1967).
17. Cantor, K., and Hearst, J., *Proc. Natl. Acad. Sci. U.S.A.* **55**, 642 (1966).

18. Salzman, N., Moore, D. E., and Mendelsohn, J., *Proc. Natl. Acad. Sci. U.S.A.* **56**, 1449 (1966).
19. Huberman, J., and Attardi, G., *J. Cell Biol.* **31**, 95 (1966).
20. Somers, C. E., Cole, A., and Hsu, T. C., *Exp. Cell Res. Suppl.* **9**, 220 (1963).
21. Lin, H. J., and Chargaff, E., *Biochim. Biophys. Acta* **91**, 691 (1964).
22. Maio, J., and Schildkraut, C., *J. Mol. Biol.* **24**, 29 (1967).
23. Wray, W., and Stubblefield, E., *Exp. Cell Res.* **59**, 469 (1970).
24. Wray, W., Stubblefield, E., and Humphrey, R., *Nature (London) New Biol.* **238**, 237 (1972).

Chapter 10

Analytical Techniques for Isolated Metaphase Chromosome Fractions

ELTON STUBBLEFIELD AND SHIRLEY LINDE

Departments of Biology and Biomathematics,
The University of Texas System Cancer Center,
M. D. Anderson Hospital and Tumor Institute,
Houston, Texas

FRANCES K. FRANOLICH AND L. Y. LEE

Department of Biophysics,
The University of Houston,
Houston, Texas

I. Introduction

Metaphase chromosomes represent reproducible precise subfractions of the total genome of an organism. For those organisms where mitotic cells can be accumulated in sufficient quantities, isolated metaphase chromosomes provide material for the analysis of specific portions of the genetic apparatus, if pure preparations of a specific chromosome or group of chromosomes can be obtained. Chromosome separation techniques have so far failed to yield a pure preparation of a single chromosome, but this will undoubtedly be accomplished sooner or later. Stubblefield and Wray (*1*) reported preliminary evidence that the nonhistone proteins were different in the various chromosome fractions of the Chinese hamster. As this work has progressed, it has become apparent that reliable techniques are needed to evaluate the composition and purity of specific chromosome fractions.

The technology described in this report was developed to evaluate isolated chicken chromosomes in experiments designed to locate specific viral DNA sequences integrated into the chicken genome. In principle, however, the same approach can be taken to identify proteins peculiar to specific chromosomes, and we are also engaged in just such a study. We will limit this discussion to the analysis of chromosome fractions; the results of the study of DNA sequences in chicken chromosomes will appear elsewhere (2).

II. Chromosome Isolation

A variety of techniques for the isolation of metaphase chromosomes have been published (3–9). We have routinely used one of three simple procedures, depending on the goal of the experiment. For studies of chromosomal proteins, the chromosomes should be isolated at a pH near 7.0 to preclude extraction of either acid-soluble or alkali-soluble chromosomal proteins. For such studies we use the method of Wray and Stubblefield (8). If, on the other hand, one wants isolated chromosomes with high-molecular-weight DNA, then the hypotonic treatment must be done at pH 10.5 to inactivate the nucleases of the cell which would otherwise degrade the DNA. This procedure (9) causes some alteration of the high-molecular-weight proteins (1). However, both of these procedures require a rather dilute suspension of cells (about 2% by volume) so that large volumes are needed to handle large samples. In contrast, chromosomes can be isolated in 50% acetic acid in very thick suspensions, so for studies using nucleic acid hybridization to DNA isolated from the chromosome fractions the convenience of this approach is noteworthy. In this procedure (10) metaphase cells are accumulated in a growing population by treatment for 4–6 hours with Colcemid. The mitotic cells can be differentially removed from monolayer cultures, but in suspension cultures this is not so easily accomplished. Interphase cells produce isolated nuclei which contaminate the isolated chromosomes in all isolation procedures. However, the large difference in size and mass between a chromosome and a nucleus allows one to readily remove the nuclei by differential sedimentation or filtration.

All chromosome isolation procedures require a hypotonic treatment step to disrupt any spindle remnants and to make the cell susceptible to shear disruption. In the acetic acid procedure we treat the cells with either 0.075 M KCl or medium diluted 1:5 with water. Hypotonic treatment at room temperature for 10 minutes is usually adequate. The cells are then sedimented at room temperature in a clinical centrifuge. Refrigeration at this point should be avoided since it results in chromosome aggregates instead of free chromosomes when the cells are broken open later. The cell pellet is suspended in 5 ml of 50% acetic acid, and the cells are sheared open by rapid passage through a 22-gauge needle. As much as 1 ml of packed cells per 5 ml of acetic acid has been successfully prepared this way, and the treatment nicely parallels the technique used for chromosome squash preparations for karyotype analysis. Although the molecular DNA is broken or nicked and some proteins are probably lost, the chromosomes are structurally quite stable and the total DNA content is undiminished.

We have used the above procedure for the isolation of chicken chromosomes from a lymphoma cell line MSB-1 taken from a chicken with Marek's disease (11). The cells grow in stationary cultures but do not attach to the flask, so the isolation of mitotic cells from the culture is not readily possible. However, the cells grow rapidly, and after a 5–6 hour Colcemid treatment (0.06 μg/ml) about half of the cells are blocked in mitosis.

III. Fractionation of Isolated Chromosomes

Isolated chromosomes offer few advantages over isolated whole nuclei, unless some type of sorting of the individual kinds of chromosomes can be achieved. The most widely used method has been low-speed centrifugation (1, 7, 12). Centrifugation in a zonal rotor is preferable to centrifugation in tubes, since greater volumes can be handled and interaction of the chromosomes with the tube walls is avoided. We routinely sort chromosomes on a 1-liter 20% to 40% linear sucrose gradient in a Sorvall SZ-14 reorienting gradient zonal rotor. The linear gradient is generated with an LKB Ultragrad-11300 and loaded with the rotor at rest. The sucrose solutions for making the gradient were made up in the chromosome isolation buffer of Wray and Stubblefield (8) containing 1.0 M hexylene gylcol. The loaded rotor is slowly brought up to speed (3000 rpm) over a 10-minute interval by a special rate controller attached to the centrifuge. The isolated chromosomes (and contaminating nuclei) in 50% acetic acid are diluted with an equal volume of 20% sucrose, and the sample is loaded onto the gradient in the spinning rotor. Nuclei sediment rapidly to the rotor wall, but the largest chromosomes take longer, about 50 minutes for chicken and human chromosomes, but only about 25 minutes for Chinese hamster chromosomes. At the conclusion of the run the rotor is brought back to rest over a 10-minute interval using the rate controller to prevent too rapid a deceleration which might cause mixing of the gradient. The gradient is then pumped out dense-end first into 50-ml plastic centrifuge tubes. About 28 fractions of 40 ml each are collected, with larger chromosomes in the early fractions and smaller chromosomes in the later fractions. The contaminating nuclei stay in the rotor pelleted on the rotor wall. For this reason it is important not to use a dense cushion of 60% sucrose, as this causes the nuclei to be less tightly adherent, and they tend to come loose into the various fractions as the rotor is emptied.

IV.   Fraction Analysis

A.   Visual Inspection

Visual inspection of the various fractions with a phase-contrast microscope confirms the general impression that the chromosomes are distributed in the gradient according to size. However, for the gradient separation to be useful, we need to know with some precision the proportion of specific chromosomes in each fraction. In our earliest studies we made this analysis by visual microscopy, but the procedure is so time-consuming that we were reluctant to run such experiments. Better, faster assays are now available, but for some studies it is still important to examine the chromosome visually. We have used two approaches, sedimentation and filtration, with equal success.

In the sedimentation procedure a small circular coverslip is placed in the bottom of a centrifuge tube and covered with 2 ml of 20% sucrose. An aliquot of the sample is diluted to less than 20% sucrose and layered over the 20% sucrose in the tube. Centrifugation in a swinging-bucket rotor at 3000 g for 15 minutes is adequate to sediment all of the chromosomes onto the coverslip, which is then removed, washed in sucrose-free buffer, and air-dried. The chromosomes can then be stained by Giemsa or fluorescent stains for microscopic examination. Figure 1 shows a series of fractions from a gradient of separated chicken chromosomes that were centrifuged onto coverslips, stained with 50 μg/ml ethidium bromide, and photographed by their fluorescence excited at about 510 nm (green) light. The chromosomes fluoresce a bright red-orange color because of the dye which is concentrated by intercalation into the DNA. Visual identification of which chromosomes are present is still difficult, but such preparations have been stained to give banded chromosomes (*13*), which are easier to identify.

An alternate procedure for concentrating the chromosomes from a fraction is to filter them onto Millipore or Nuclepore filters. Such preparations are viewed best when stained by a fluorescent dye and examined by vertical illumination.

B.   Flow Microfluorometry

In contrast to the qualitative data which visual examination provides, precise quantitative data are readily obtained by the particle analysis tech-

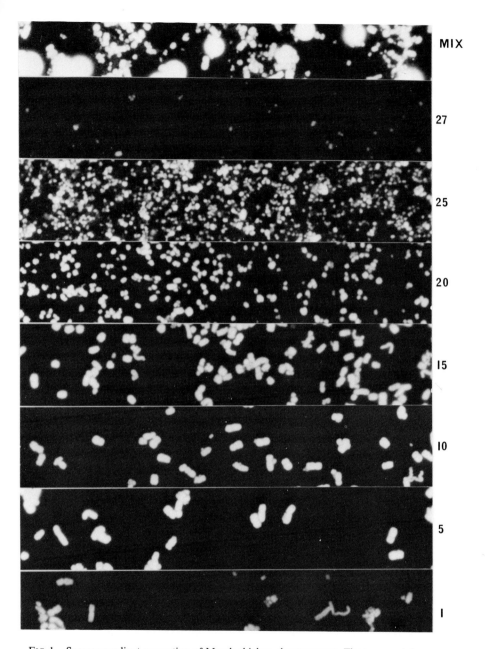

MIX

27

25

20

I5

I0

5

I

Fɪɢ. 1.   Sucrose gradient separation of Marek chicken chromosomes. The top panel shows the mixture of chromosomes and nuclei applied to the top of the gradient. The remaining panels show the various fractions at the conclusion of the centrifugation. The nuclei sediment to the rotor wall; the bottom fractions contain chromosome aggregates. The chromosomes were sedimented from the various samples onto coverslips and stained with ethidium bromide for fluorescence photomicroscopy.

nique known as flow microfluorometry (FMF) (*10*). For such an analysis an aliquot of a gradient fraction is mixed with an equal volume of 50 μg/ml ethidium bromide solution in 1.15% sodium citrate. After a 10-minute wait to allow the dye to intercalate into the chromosomal DNA, the sample is placed in a flow system that carries the chromosomes in a narrow stream through the beam of a laser tuned to 514.5 nm (green). As each chromosome passes through the laser beam it fluoresces with an intensity that is proportional to its DNA content. The pulse of fluorescent light is detected by a photomultiplier tube, and the intensity is measured by a multichannel analyzer. This analysis can be done at rates between 100 and 1000 particles (chromosomes) per second. The accumulated pulses in each channel combine to give a curve which represents the distribution of different sizes of

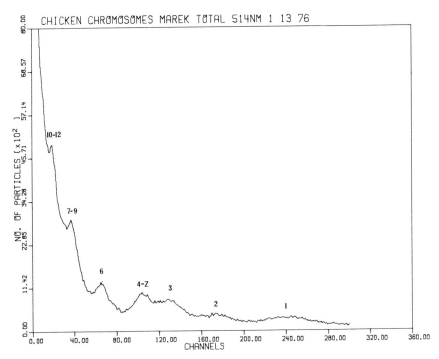

FIG. 2. FMF analysis of isolated chicken chromosomes. Only the first dozen macrochromosomes are large enough to resolve by this technique. Compare peak positions with the chromosome sizes shown in Fig. 3.

FIG. 3. Karyotype of the MSB-1 Marek chicken lymphoma cell line. Beyond chromosome ▶ 12 the microchromosomes cannot be paired reliably, nor can they be resolved by FMF analysis. The microchromosomes do add considerable noise to the system, however, and reduce the resolution of the macrochromosome peaks.

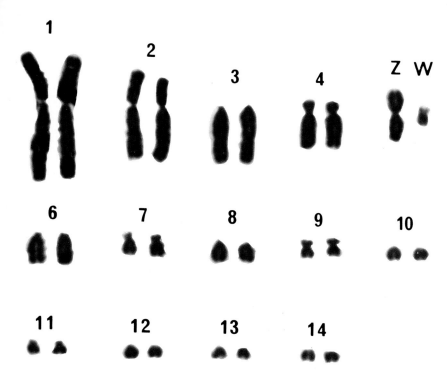

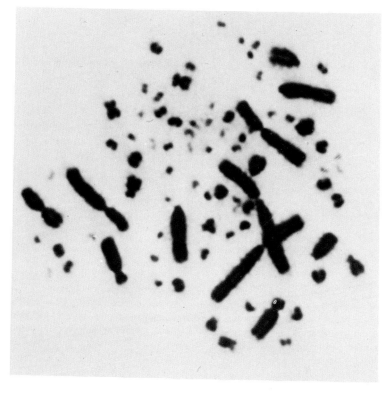

chromosomes in the sample. Such a curve for the Marek chicken chromosomes is shown in Fig. 2. Comparing this curve with the karyotype of the Marek cells (Fig. 3), we can assign specific chromosomes to each of the peaks for the largest macrochromosomes of the chicken. The large number of microchromosomes in chicken interfere with the resolution of the peaks since they are more abundant than the macrochromosomes and are too small to be resolved. The coefficients of variation are about three times the values that we see with chromosomes from other species, such as Chinese hamster where no microchromosomes are present. However, peaks for chromosomes 1, 2, 3, 4 and Z, and 6 are easily located. Their positions on the abscissa are proportional to their relative lengths.

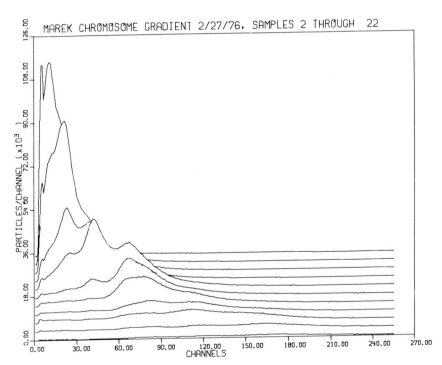

FIG. 4. Multiple plot of the even-numbered fractions of Marek chromosomes separated according to size by sucrose gradient velocity sedimentation.

FIG 5. (pp. 109–110). Curve fitting of Gaussian peaks to fractions 4, 8, 12, and 16. The areas under the peaks are proportional to the number of chromosomes of that type in the fraction. Peaks near the origin do not fit well because of microchromosome noise which greatly increases the coefficient of variation for the smaller chromosomes. However, the percentage contribution to the total sample is not changed appreciably by improving the fit to the experimental curve.

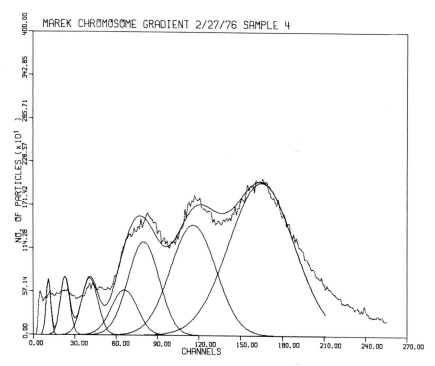

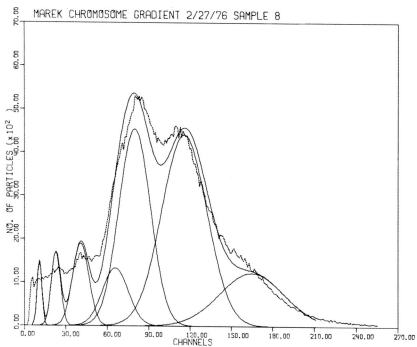

FIG. 5. Fractions 4 and 8. Caption appears on facing page.

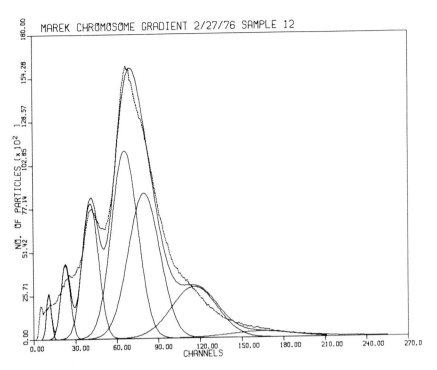

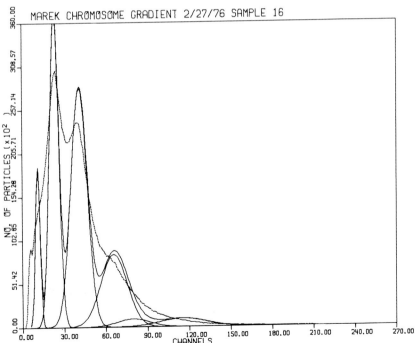

FIG. 5. Fractions 12 and 16. Caption appears on page 108.

Fractions from an experiment like that shown in Fig. 1 were analyzed for their chromosome composition by FMF analysis. In Fig. 4 we have plotted the profiles of every other fraction through the gradient. The reduced numbers of microchromosomes among the larger chromosomes improves the resolution somewhat. It is clear that most fractions contain a mixture of about three chromosomes.

The data shown in Fig. 4 were collected in a way that allows us to quantitate the relative abundance of chromosomes from fraction to fraction. The counts were normalized to a constant volume by adding a known concentration of a suspension of fluorescent microspheres (larger than the chromosomes) to each sample and counting these in a separate channel as the chromosomes were counted. Chromosome data were accumulated until 300 microspheres had been observed in each sample. The bottom fractions contain chromosomes in more dilute concentrations than those fractions nearer the center of the rotor.

V. Computer Analysis

Analysis of the numbers from which these curves were generated is important, but it is almost overwhelming unless computer assistance is available. Knowing where the chromosome peaks should be placed, we fit Gaussian curves to the profiles in such a way that the sum of the peaks best matched the observed data. In this case we first made the fit by hand on a DuPont Curve Resolver (model 310) and found an average coefficient of variation of 14% for the individual peaks. This constant coefficient of variation was then used for the computer data fit, using a program called CONVEX (14), on a CDC Cyber 73 computer. Examples from fractions 4, 8, 12, and 16 are shown in Fig. 5.

Since the areas under the peaks correspond to the abundance of a particular chromosome in a fraction, and the peak position corresponds to the relative DNA content, we can readily calculate the contribution of each chromosome to the total DNA in a fraction of the gradient. These data are given in Tables I and II for this gradient. Recall that in this experiment we were interested in the annealing of the DNA from the different fractions to specific nucleic acid probes. By comparing the degree of hybridization to the DNA in each fraction contributed by the various chromosomes of that fraction and by following the variations in these parameters through the fractions, we can readily decide which chromosome is contributing the DNA involved in the annealing if it comes from only one chromosome.

ELTON STUBBLEFIELD *et al.*

TABLE I

CHROMOSOME COMPOSITION OF MAREK CHROMOSOME FRACTIONS FROM A SUCROSE GRADIENT

Fraction number	Total chromosomes (per 300 microspheres)	Estimated percentage area under chromosome peaks						
		Chromosome number						
		1	2	3	4–7	6	7–9	10–12
2	1.53×10^5	50.5	23.5	14.5	3.6	4.2	2.5	1.3
4	2.70	47.5	24.3	14.2	5.7	4.5	2.6	1.3
6	5.43	28.2	37.3	14.9	9.2	5.6	3.2	1.6
8	4.80	15.5	39.2	28.0	6.8	6.1	3.1	1.4
10	7.62	4.9	26.0	36.4	20.8	7.1	3.5	1.4
12	8.51	2.5	15.3	29.8	32.0	14.5	4.5	1.4
14	14.61	0.8	7.8	12.4	31.2	33.9	11.2	2.7
16	12.48	0.1	3.7	2.4	18.7	38.6	29.0	7.4
18	19.16	0.0	1.7	0.7	9.4	26.4	43.5	18.3
20	20.38	0.0	0.9	0.0	5.7	18.7	40.6	34.1
Mean peak position (channel number)		166	117	81	67	42	24	12

TABLE II

CALCULATED DNA COMPOSITION OF MAREK CHROMOSOME FRACTIONS

Fraction number	Estimated DNA contribution to total DNA in fraction						
	Chromosome number						
	1	2	3	4–7	6	7–9	10–12
2	65.5	21.5	9.2	1.9	1.4	0.5	0.0
4	63.0	22.7	9.2	3.0	1.5	0.5	0.1
6	41.8	39.0	10.7	5.5	2.1	0.7	0.2
8	25.1	44.9	22.2	4.5	2.5	0.7	0.2
10	9.6	35.4	34.2	16.2	3.5	1.0	0.2
12	5.6	23.9	32.2	28.6	8.1	1.5	0.2
14	2.1	15.6	17.1	35.7	24.3	4.6	0.6
16	0.5	10.1	4.6	29.1	37.6	16.1	2.1
18	0.1	6.2	1.7	19.3	34.0	32.0	6.7
20	0.0	3.9	0.1	14.5	29.5	36.6	15.4

Since the proteins associated with the chromosomes may in some cases also be unique (*1*) to a certain chromosome, the identical approach can be used to localize a specific protein in the karyotype. If the larger chromosomes contribute more protein to the total in a fraction, as they almost surely do, then the relative abundance of a certain protein as seen in acrylamide gel separations can be compared with the relative contribution of a specific chromosome to the total proteins of a gradient fraction; a match in the abundance profiles through the gradient fractions would then locate the chromosomal source of the protein. Of course, very pure fractions of single chromosomes would give unequivocal answers to such questions, but current technology is already adequate to attack these problems without further development.

ACKNOWLEDGMENTS

The authors wish to thank Mrs. Dolores Scott for technical assistance and Miss Dorothy Holitzke for help in preparing the manuscript. We also thank Dr. Stephen Bloom for permission to use his previously unpublished karyotype of the Marek cell line (Fig. 3). This work was supported in part by Grants BMS-75-05622 and BMS-75-16778 from the National Science Foundation and Grant CA-11430 from the National Cancer Institute.

REFERENCES

1. Stubblefield, E., and Wray, W., *Cold Spring Harbor Symp. Quant. Biol.* **38**, 835 (1974).
2. Padgett, T. G., Stubblefield, E., and Varmus, H. *Cell* **10**, 649 (1977).
3. Somers, C. E., Cole, A., and Hsu, T. C., *Exp. Cell Res. Suppl.* **9**, 220 (1963).
4. Cantor, K. P., and Hearst, J. E., *Proc. Natl. Acad. Sci. U.S.A.* **55**, 642 (1966).
5. Huberman, J. A., and Attardi, G., *J. Mol. Biol.* **29**, 487 (1967).
6. Salzman, N. P., Moore, D. E., and Mendelsohn, J., *Proc. Natl. Acad. Sci. U.S.A.* **56**, 1449 (1966).
7. Maio, J. J., and Schildkraut, C. L., *J. Mol. Biol.* **24**, 29 (1969).
8. Wray, W., and Stubblefield, E., *Exp. Cell Res.* **59**, 469 (1970).
9. Wray, W., Stubblefield, E., and Humphrey, R., *Nature (London) New Biol.* **238**, 237 (1972).
10. Stubblefield, E., Cram, S., and Deaven, L., *Exp. Cell Res.* **94**, 464 (1975).
11. Akiyama, Y., and Kato, S., *Biken J.* **17**, 105 (1974).
12. Mendelsohn, J., Moore, D. E., and Salzman, N. P., *J. Mol. Biol.* **32**, 101 (1968).
13. Wray, W., and Stefos, K. *Cytologia* **41**, 729 (1976).
14. Hartley, H. O., and Hicking, R. R., *Management Sci.* **9**, 600 (1963).

Chapter 11

Methods and Mechanisms of Chromosome Banding

Department of Medical Genetics,
City of Hope National Medical Center,
Duarte, California

I. Introduction

A report by Caspersson and his colleagues in 1968 (*1*) that chromosomes stained with quinacrine mustard and examined by fluorescent microscopy showed a distinct banding pattern opened the field of chromosome banding. There had been some previous suggestions—in the form of the introduction of cold heterochromatin by Darlington and LaCour (*2*), differential staining of heterochromatin by Levan (*3*), and induction of distinct bands by excess treatment with Colcemid by Stubblefield (*4*)—that some differential staining along chromosomes could be produced. These studies by Caspersson, however, were the first to provide a readily usable and reproducible technique to develop bands all along the chromosomes. This technique was soon followed by a procedure to show centromeric-type constitutive heterochromatin (*5*) and a Giemsa banding procedure producing bands which mimic those seen with quinacrine (*6*). By international convention these three procedures have been termed Q-, C-, and G-banding, respectively (*7*). A modification of the G-banding procedure resulted in a pattern opposite to G-banding and this was termed R-banding (*8*). Since then a whole series of additional modifications of G-banding and numerous other types of chromosome banding have been reported. These are all outlined in Table I. It is not the purpose of this paper to detail all these different techniques; the interested reader can obtain them from the respective papers. Here, I will present some standardized methods for Q-banding, C-banding, R-banding, two techniques for G-banding, and some

TABLE I
Types of Chromosome Banding

Types	Treatment	References	Stain	Characteristics
Q	None	(19,20)	Quinacrine	Q-bands, some centromeric heterochromatin[3+], other ± or −
G	a. G_2 treatments			
	Actinomycin D	(21,22)		
	Rifampicin	(D. A. Shafer, personal communication)		
	Tetracycline	(L. F. Meisner, personal communication)		
	Azure B	(22)		
	b. Treatment of fixed chromosomes		Giemsa	G-bands, some centromeric heterochromatin[3+], other ± or −
	2 × SSC, 60°, 1 hour	(2)		
	Variations	(6,23–31)		
	Proteolytic enzymes	(32–38)		
	Urea	(27,39–42)		
	Detergents	(39,43)		
	Oxidizing agents	(44–46)		
	Anti-Antibodies	(47)		
	c. Treatment in the stain	(48–52)	Giemsa pH 9 Dilute Giemsa	
R	20mM PO$_4$, pH 615, 87°	(8,53)	Giemsa, AO	Opposite of G-banding
	Earle's salts, pH 6.5, 87°	(26,54–60)		
	Variations			
	Cold Treatment	(61)		
	Anti-C Antibodies	(62)		
T	Earle's salts, pH 5.1, 87°	(54)		Stains telomere regions (extension of R-banding)
	Variations	(63)		

BUdR	a. Exposure to high conc. of BUdR during latter part of the S period	(64)	Giemsa	Detection of regions of early and late DNA replication
	b. Exposure to BUdR during one and part of 2nd S period	(65)	Hoechst 33258	Detection of regions of early and late DNA replication
C	0.07 N NaOH 2–6 × SSC / Variations / DNase / Methyl-cytosine antibodies	(5,) (3, 28, 66–72) (72) (73)	Giemsa	Stains centromeric heterochromatin
Giemsa 11	None	(74, 75)	Giemsa pH 11 37°, 10–20 minutes	In human stains 9h and some other centromeric heterochromatin
A1	0.9% NaCl, 100°, 5 minutes, then 0.2 M CsCl 65° 10 minutes	(28)	Leishman's	In human stains 1h and 16h
Cd	Earle's salts pH 8.5–9.0, 85°, 45 minutes after fixation of chromosomes	(76)	Geimsa	Stains kinetochores
N	5% TCA, 30 minutes 85–90°; then 0.1 N HCl 30–45 minutes, 60° / Silver stain	(77,78) (79)	Giemsa / Silver stain	Stains nucleolus organizer region / Stains nucleolus organizer region
Hoechst 33258	None	(80–82)	Hoechst 33258	Stains some centromeric heterochromatin intensely
BG(FPG)	Incorporation of BudR for two complete S periods	(65, 83, 84)	Hoechst 33258 AO, Giemsa	Detects semiconservative segregation of DNA strands

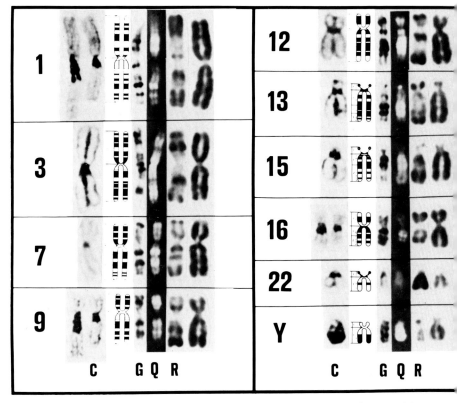

FIG. 1. Comparison of the C-, G-, Q-, and R-banding techniques for several human chromosomes. The chromosomes to the left of C-banding for 1, 9, and 16 show variants containing increased amounts of centromeric heterochromatin. The diagrammed banding represents the Paris Conference Standardization (7). The right-most chromosome is unbanded and stained with Giemsa. The C-banded chromosomes were supplied by Dr. O. Alfi, the Q-banded by Drs. Lin and Uchida, and the remaining chromosomes by Helen Lawce.

brief comments on the probable mechanism of these different techniques. A comparison of these techniques for several human chromosomes is shown in Fig. 1.

II. Slide Preparation

None of the banding techniques will work if the chromosome preparations are poor to begin with. The ideal chromosomes are those which are contracted just enough so that there is little problem with overlapping.

Chromosomes which are overcontracted, due to too long exposure to anti-mitotic agents, give poor resolution of the bands. Cells may be treated with Colcemid, colchicine, or vinblastine for 1 to 2 hours to produce metaphase arrest. Various final concentrations of these drugs ranging from 1 to 0.01 μg/ml or less should be tried to determine the minimum concentration needed to obtain good elongated chromosomes.

Exposure of the cells to 0.075 M KCl at 37°C for 5–15 minutes is a widely used hypotonic treatment, but simple dilution of one part of the media with three parts of water may also be satisfactory. The cells are then gently pelleted and resuspended in freshly made fixative consisting of three parts methanol with one part glacial acetic acid. The fixative is changed three times, by centrifugation at 500 g for 5 minutes. The cells should be left in the fixative at room temperature for at least 10 minutes with each change. On the final resuspension, add a volume of fixative which will give a somewhat opalescent suspension, but not too thick. Take acid-cleaned, dry glass slides, flood one with distilled water, and drain off excess water. Then add one or two drops of suspension from a distance of a few inches above the slide and let the cells spread out on the slide. Allow the slides to air-dry. The drying may be facilitated by gently blowing on the slide. A slide can then be examined under phase contrast to determine if the chromosomes are well spread and the cell concentration is appropriate. If the chromosomes are not adequately spread, increasing the height of dropping onto the slide or diluting the chromosomes may correct this.

III. Q-Banding

A. Method

Staining solution: Quinacrine mustard, 0.005% (50 μg/ml) or quinacrine dihydrochloride, 0.2% (2 mg/ml) in distilled water, pH 5 to 7. At these concentrations quinacrine mustard and quinacrine dihydrochloride are essentially interchangeable.

Staining time: Thirty minutes.

Wash: McIlvaine's buffer (0.2 M disodium phosphate–0.1 M citric acid), pH 5 to 7. One to three 30-second washes and mount in same buffer. Distilled water may also be used for washing and mounting.

Microscope: Any good fluorescence microscope is satisfactory. The two most often used are Leitz Ortholux and Zeiss Photomicroscope. The elaborate Zeiss Photomicroscope is not necessary. A far less expensive

Zeiss standard microscope, RA, can be fitted with a mercury vapor lamp and either epi- or transmitted illumination.

Optical system: Either a transmitted light with a bright field condenser or epi-illumination may be used; preferences vary from investigator to investigator. If one is purchasing a new fluorescent microscope it is wise to have a trial period where two or more brands of microscope can be compared under laboratory conditions by examining quinacrine-stained chromosomes and comparing the duration of exposure necessary to produce a good photograph. Systems which require only a short exposure time will allow the use of slower, fine-grain films and thus produce better photographs.

Filters: When quinacrine is bound to DNA it shows two optical density peaks, one at 460 nm and a second at 435 nm. The latter corresponds to one of the mercury lamp emission peaks. Quinacrine fluorescence shows a single peak at 500 nm with a gradual broad, decreasing slope to 600 nm (9). Given these characteristics, the excitation filter most often used is the BG-12, utilizing 2 to 4 mm of filter. This passes light of 350 to 475 nm. The Schott 436 excitor filter may also be used. The barrier filters may be 500 nm, 510 nm, or 530 nm, allowing emission of light at 500 nm and greater.

Lamp: The most universally used is a high-pressure mercury lamp, HBO 200 W/4, with a dc power supply.

Objective lens: A 100/1.3 oil-immersion objective lens with an iris is commonly used. An alternative is a 40/1.0 oil-immersion lens. Its lower magnification allows transmission of more light and shorter exposures.

Film: Kodak Tri-X Panatomic (ASA 400), Kodak Panatomic-X (ASA 32), High-Contrast Copy Film (ASA 64), or Kodak 50–410 photomicroscopy monochrome film may be used. Because quinacrine fluorescence fades, the most important objective is to attempt to get as much illumination as possible. If less light reaches the camera, faster but grainier films will have to be used; if more light reaches the camera, finer-grain films can be used which tolerate magnification much more readily. There are several things which can be done to increase the amount of light entering the camera. Perhaps the most important is to decrease the distance the light has to travel. In the old Zeiss Photomicroscopes the light was diverted in so many directions that it was difficult to use for quinacrine fluorescence studies. The best arrangement is to allow light to come directly from the mercury lamp to an epi-illuminator, or mirror for transmitted light, and then travel directly from the slide straight up to a top-mounted camera.

B. Mechanism

Very AT-rich DNA markedly enhances quinacrine fluorescence, while DNA containing GC-bases causes quenching of quinacrine fluorescence

(9–15). The intensity of the fluorescence varies with the fourth power of the base composition (12,16). A change in base composition of 6% is adequate to result in a 50% change in the intensity of quinacrine fluorescence, and this is sufficient to account for most Q-banding (16). Proteins can also inhibit the binding of quinacrine to DNA (15,17). The degree to which such protein–DNA interactions play a role in Q-banding is unknown.

IV. G-Banding

Methods

G-banding has the following advantages over Q-banding. (1) It does not require a fluorescence microscope. (2) The stain is permanent and does not fade as does quinacrine fluorescence. (3) The bands are easier to photograph since regular transmitted light can be used. The exposure is rapid; an automatic photoscope and high-contrast film can be utilized. (4) The resolution of the bands is usually better. (5) The slides can be stored and reexamined without restaining. (6) Some structural polymorphisms are seen that would be missed with Q-banding.

A major advantage of Q-banding is that some polymorphisms are seen that do not show up with G- or C-banding (18).

Many different techniques can be used to give G-banded chromosomes (Table I). They can be divided into: (a) treatment of the living cells during the G_2 phase; (b) treatment of fixed chromosomes followed by Giemsa staining; and (c) treatment in the stain. The treatments of the fixed chromosomes include the use of hot salt solutions, proteolytic enzymes, urea, and detergents. The major feature these treatments have in common is their ability to denature proteins. The possible relevance of this is discussed under mechanisms of G-banding. Only two of the most widely used techniques, the ASG and trypsin banding, will be described here.

1. ASG TECHNIQUE (ACETIC/SALINE/GIEMSA)

1. Incubate slides in 2 × SSC (0.3 M NaCl, 0.03 M trisodium citrate, pH 7.0) at 60°C for 1 hour.
2. Rinse with distilled water.
3. Stain in Giemsa (see below).

Some of the variables involved in banding with salts have been examined (27). The pH is especially critical and should be 7.0 or greater (Fig. 2).

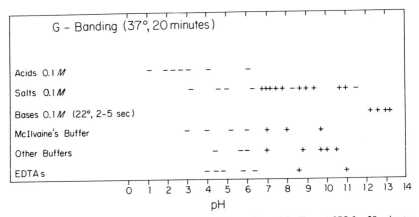

FIG. 2.   Effect of pH on G-banding using various salts and buffers at 37° for 20 minutes.
Data from Kato and Moriwaki (27).

2.   TRYPSIN TECHNIQUE

1. Prepare a solution of 0.025% trypsin in a balanced salt solution such as Hanks', without Ca^{2+} or Mg^{2+}, pH 7 to 8. Immerse three slides containing fixed air-dried chromosomes for 1, 2, and 4 minutes at room temperature.

2. Wash once for a few seconds in Hanks' without trypsin, a few seconds in 70% ethanol, then 95% ethanol, and let air-dry. Then stain for 10 to 30 minutes in Giemsa (see below). Wash several times in distilled water and air-dry. The chromosomes can be examined directly under oil or mounted under a coverslip using Permount or Eukitt. Choose the best of the three slides and use this time for subsequent G-banding.

Frequently good G-banding may not be obtained after the first try. The single most important feature to obtain successful G-banding is to examine a series of variables. First, run a series of slides using variations in the time of exposure to trypsin from a few seconds to minutes and pick the times which give the best results. Too little incubation gives no banding; too much incubation gives swollen chromosomes with the narrowed G-bands or crude C-bands (56,85). If over-trypsinization persists, decrease the concentration of trypsin to as little as 0.005% and repeat the series. An additional variable is the age of the slides. Optimally, they should be 5 to 7 days old. They can be made ready for immediate staining by thorough drying in a 60° oven for an hour. If one is in a humid area they should be stored in a moisture-proof box with a drying agent. Seabright (86) has suggested aging slides by pouring H_2O_2 on them for 5 to 10 minutes, then washing it off with isotonic saline. This also allowed old slides to be banded. A combination of the ASG and trypsin banding may sometimes work when either alone does not (37).

A final variable which should be evaluated is Giemsa stains from different sources [Gurr; National Biological Stains and Reagents Department (formerly National Aniline Division), Allied Chemical, Morristown, New Jersey; Halleco; Matheson, Coleman and Bell Giemsa Blood Stain (liquid stock)]. These differ in composition (87,88).

It has been my experience that a procedure that works in one laboratory may not work in another, and a procedure that works on the chromosomes of one species may not work on another. However, by testing the above variables a satisfactory procedure for a given laboratory and a given species can be obtained.

GIEMSA STAINING

a. Method. Liquid stocks of Giemsa can be purchased or made by dissolving 4 g of a Giemsa powder in 250 ml of absolute methanol plus 250 ml of glycerol. This should be stirred for at least 24 hours and filtered. A 1 to 4% solution of Giemsa is then made by adding 1 to 4 ml of this stock to 96 to 99 ml of a 0.003 to 0.01 M phosphate buffer (pH 6.5). The slides are first wetted in phosphate buffer alone (not essential) and then stained in Giemsa for 30 minutes, washed twice with distilled water, and dried. They can be examined as is under oil or mounted in Permount.

b. Mechanisms. One of the most important features of G-banding is that it corresponds exactly to the chromomers of meiotic pachytene chromosomes (89,90). Since pachytene chromosomes are several-fold longer than mitotic chromosomes, the chromomeres can be seen by simple Giemsa staining of the untreated chromosomes. During mitosis the chromosomes are much more condensed, the chromomeres become obscured, and the mitotic chromosomes stain uniformly. The chromomeres, however, are still there, and this is why G-banding can sometimes be observed by whole-mount electron microscopy (91,92), UV or phase microscope (93), staining with Feulgen (94,95), or dilute Giemsa (49). The centromeric heterochromatin or C-bands, intercalary heterochromatin or G-bands, and euchromatin or R-bands each contain chromatin with distinct characteristics (96–98). Although the centromeric heterochromatin is genetically inactive and corresponds to the classical heterochromatin, the precise nature of the intercalary heterochromatin is not well known. It tends to be condensed during interphase, relatively AT-rich, undermethylated, and late replicating, although this correlation may not be precise (99).

Studies of the interaction of the thiazin dyes in Giemsa with DNA and chromatin indicate the positively charged thiazin dyes bind to DNA by side stacking on the phosphate groups (88, 100). The dye does not bind significantly to proteins. Thus, any time the chromosome regions stain well with Giemsa this means there are a number of free, uncovered phosphate groups available for dye binding. Conversely, there are several possible explanations for the pale staining interband regions in G-banded chromosomes: (1) the DNA has been extracted; (2) the DNA has been redistributed; and (3) the DNA phosphate groups have been covered by protein and are unavailable for binding thiazin dyes.

Since after G-banding, even with trypsin, little protein or DNA is lost (56), the first explanation is unlikely. Electron microscopy of trypsin-treated chromosomes shows good G-bands (101, 102). Since these chromosomes are unstained the banding cannot be explained by the presence of proteins covering the phosphate groups. Thus it is quite likely that some redistribution of the chromatin revealing the underlying chromomere pattern is taking place. Still the maximum contrast of bands versus interbands is seen when the chromosomes are stained with Giemsa (56). Equilibrium dialysis studies (where the amount of DNA is held constant) using unfixed and fixed chromatin and fixed chromatin treated by the ASG technique indicate that an additional factor may be that the G-banding treatments result in an increased ability of certain chromosomal proteins to cover the phosphate groups, presumably in the interband regions (100). Since G-bands can be produced whether histones are present or not (103, 104), it is probably the nonhistone proteins denatured by the G-banding agents which are masking the DNA in the interband euchromatic regions.

V. C-Banding

The C-banding procedure specifically stains the centromeric type of heterochromatin which is usually located around the centromeres and usually contains highly repetitious satellite DNA. In humans and other species polymorphisms consisting of variations in the amount of C-banding heterochromatin are frequently seen (Fig. 1). C-banding has the advantage that it stains a specific type of heterochromatin that is not always well visualized by G- or Q-banding. Since it is not necessarily present on all chromosomes and is localized to only a small portion of the chromosome, it is not as useful for the specific identification of individual chromosomes as Q- or G-banding.

A.   Method (5, 105)

1. Treat slides with 0.2 N HCl at room temperature for 15 minutes and rinse 3 times with distilled water.
2. Treat slides with 0.07 N NaOH for 2 minutes. Rinse with 70% ethanol; then rinse with 95% ethanol 3 times for 5 minutes each. Let slides air-dry.
3. Place the slides in a moist chamber and cover with 2 × SSC [0.3 M NaCl–0.03 M Sodium citrate (pH 7.5)] and place a coverglass over the SSC solution. Incubate in the moist chamber at 60° for 16–20 hours.
4. Rinse in 2 × SSC and three times, 5 minutes each, in 70% ethanol, 95% ethanol. Let air-dry.
5. Stain with Giemsa.
If problems occur, consult some of the articles that have examined the technical details and mechanism of the technique (26,56,70,105).

B.   Mechanism

The treatment of chromosomes with sodium hydroxide and salts results in extraction of up to 80% of the DNA from the chromosome (56,106). The DNA is preferentially extracted from the non-C-band regions resulting in poor staining of the arms and intense staining of the centromeric heterochromatin. We originally suggested the centromeric heterochromatin was protected because it was bound to some nonhistone proteins not present in the euchromatin. However, some recent studies which we have done on isolated Drosophila virilis heterochromatin indicate it is markedly deficient in nonhistone proteins compared to euchromatin (106a). This is consistent with other observations which indicate that generally active chromatin is much more enriched in nonhistone proteins than genetically inactive chromatin. This suggests that chromatin which is coupled only with histone is much more highly compacted than chromatin which contains a significant amount of nonhistone proteins. It may be this tight compaction which protects the centromeric heterochromatin from destruction by sodium hydroxide and salts and allows C-banding to occur.

VI. R-Banding

R-banding is one of the most intriguing of the banding procedures. It was first described by Dutrillaux and Lejeune (8) who observed that if chromosomes were treated with 0.02 M phosphate buffer (pH 6.5) at 87° for 10 minutes, stained with Giemsa, and examined under phase microscopy, they showed a staining pattern that was the exact opposite of G-banding.

R-banding has the advantage that it stains telomeres well. A modification of R-banding results in the staining of telomeres to the exclusion of anything else to give T-bands (63).

A. Method

Those interested in R-banding should read Dutrillaux and Covic's paper (54) detailing a wide range of conditions which result in R-banding. Figures 3 and 4 show some of these data. The following conditions are used for R-banding. Earle's balanced salt solution adjusted to pH 6.5 (1 ×, ionic strength 0.14) or 0.02 to 0.07 M phosphate buffer (pH 6.5). One- to three-day-old slides are used and incubated at 87° for 30 minutes. After being rinsed in buffer the slides may be stained with Giemsa or acridine-orange. With Giemsa stain the chromosomes may be somewhat lighter than with G-banding and the contrast may have to be enhanced by phase

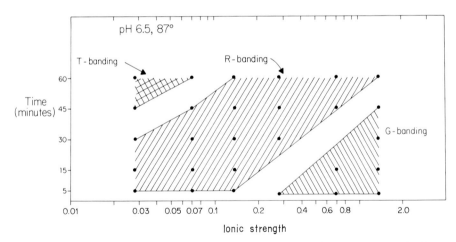

FIG. 3. Effect of ionic strength (Earle's salts) and duration of treatment on production or T-, R-, and G-banding at pH 6.5, 87°. Data from Dutrillaux and Covic (54), excluding overlap regions.

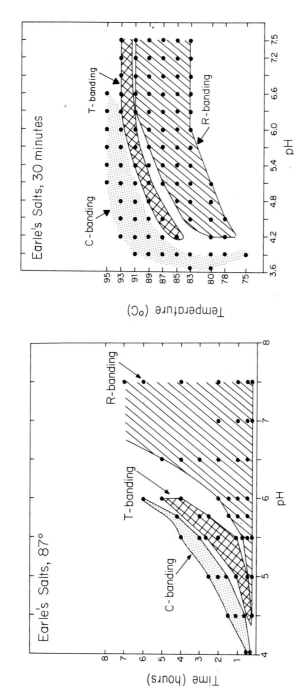

FIG. 4. Left: Effect of pH and duration of treatment on C-, T-, and R-banding with Earle's salts at 87°. Right: Effect of pH and temperature on C-, T-, and R-banding with Earle's salts for 30 minutes. Data from Dutrillaux and Covic (54), excluding overlap regions.

microscopy. Striking R-banded pictures can be obtained by staining the chromosomes with acridine-orange. The slides are immersed in 0.01% acridine-orange in 0.07 M phosphate buffer (pH 6.5) for 4–6 minutes, rinsed in the same buffer for 1.5 to 3 minutes, and mounted in phosphate buffer (59). Overstaining results in orange-red fluorescence which can be corrected by additional rinsing. If the chromosomes are still orange-red the duration of heat incubation should be decreased. Understaining results in green fluorescence without banding. This is corrected by longer staining.

The best photographic results with acridine-orange-stained slides are obtained by photographing with Kodachrome-X (ASA 64) followed by printing the transparencies in black and white (59). This gives a white background with dark telomers like those in Fig. 1. An advantage of the Giemsa staining is that the chromosomes may be photographed directly on black and white film and printed to give a white background.

When chromosomes are stained with acridine-orange the AT-rich G-bands are red and the GC-rich R-bands are green. If the centromeric heterochromatin contains AT-rich DNA, it stains red (or pale green) (56–58, 97,107). If it contains GC-rich DNA, it stains green (108). Verma and Lubs (59) have found that polymorphisms may be observed in the human chromosomes based upon variations in red-green color.

B.   Mechanisms

When acridine-orange is used for stain, the mechanism of R-banding seems straightforward in that AT-rich DNA, which has a lower DNA T_m, stains red (side stacking of acridine-orange on single-stranded DNA), while the GC-rich DNA, which has a higher melting point, stains green (intercalation of acridine-orange in double-stranded DNA). However, the fact that R-banding can also be seen with Giemsa, which does not differentiate well between single and double-stranded DNA (100), suggests that proteins might also be involved. It is most likely, however, that G-banding represents a denaturation of proteins and R-banding represents denaturation of DNA.

VII.   BG-Banding

A final important form of banding should be mentioned. In 1973 Latt *et al.* (9) observed that if cells were exposed to bromodeoxyuridine (BudR) for two cycles of DNA replication such that one chromatid contained DNA substituted with BudR in both strands (BB) while the other chromatid contained DNA which was substituted in only one strand (BT) and then stained with Hoechst 33258, the chromatid with the BT DNA fluoresced much more brightly than the chromatid containing the BB DNA. This was due to the fact that the DNA doubly substituted BudR was much more effective in quenching the fluorescence of Hoechst than a singly substituted DNA. This provided a powerful new technique for examining the distribution of replicating DNA, which gave much better resolution than autoradiography. If the BudR was incorporated for only one cell cycle and just part of the next S period, the distribution of early versus late replicating DNA could be observed on the chromosome, again with much greater resolution than with autoradiography. Chromosomes stained with acridine-orange instead of Hoechst gave the same result (*109, 110*).

A modification of this technique allowed the chromosomes to be stained with Giemsa to give intense staining of the chromatid containing BT DNA and poor staining of the chromatid containing BB DNA. One step in the procedure involved the exposure of the chromosomes to light followed by subsequent exposure to hot salt solutions. The most likely mechanism is that the BB DNA, which is very sensitive to light, is broken down and washed off the slide and thus the chromatid stains very poorly (*111*). Although the term FPG (fluorescent plus Giemsa) has been suggested for this type of banding (*112*), it was found that Hoechst staining was not necessary (*84*). For this reason we prefer the term BUdR-Giemsa, or BG-banding (*113*).

This technique is very useful for examining sister chromatid exchange. Some chemicals which cause chromosome breakage cause a much higher frequency of sister chromatid exchange (*114*). For this reason the BG-banding will find great utility in the evaluation of mutagens and clastogens.

ACKNOWLEDGEMENT

This work was supported by NIH grant GM 15886.

REFERENCES

1. Caspersson, T., Farber, S., Foley, G. E., Kudynowski, J., Modest, E. J., Simonsson, E., Wagh, U. and Zech, L., *Exp. Cell Res.* **49**, 212 (1968).
2. Darlington, C. D., and LaCour, L., *J. Genet.* **40**, 185 (1940).
3. Levan, A., *Hereditas* **32**, 449 (1946).
4. Stubblefield, E., *in* "Cytogenetics of Cells in Culture" (J. R. C. Harris, ed.), Vol. III, p. 223. Academic Press, New York, 1964.
5. Arrighi, F. E., and Hsu, T. C., *Cytogenetics* **10**, 81 (1971).
6. Sumner, A. T., Evans, H. J., and Buckland, R. A., *Nature (London) New Biol.* **232**, 31 (1971).

7. *Stand. Human Cytogenet. Birth Defects, Paris, 1971.*
8. Dutrillaux, B., and Lejeune, J., *C. R. Hebd. Seances Acad. Sci. Ser. C.* **272**, 2638 (1971).
9. Latt, S. A., Brodie, S., and Munroe, S. H., *Chromosoma* **49**, 17 (1974).
10. Weisblum, B., and deHaseth, P. L., *Proc. Natl. Acad. Sci. U.S.A.* **69**, 629 (1972).
11. Weisblum, B., and deHaseth, P. L., *in* "Chromosomes Today" (C. D. Darlington and K. R. Lewis, eds.), Vol. IV, p. 35. Plenum, New York, 1973.
12. Pachmann, U., and Rigler, R., *Exp. Cell Res.* **72**, 602 (1973).
13. Michelson, A. M., Monny, C., and Kovoor, A., *Biochimie* **54**, 1129 (1972).
14. Selander, R. K., and de la Chapelle, A., *Nature (London) New Biol.* **245**, 240 (1973).
15. Comings, D. E., Kovacs, B. W., Avelino, E., and Harris, D. C., *Chromosoma* **50**, 111 (1975).
16. Comings, D. E., and Drets, M. E., *Chromosoma* **58**, 199 (1976).
17. Gottesfeld, J. M., Bonner, J., Radda, G. K., and Walker, I. O., *Biochemistry* **13**, 2937 (1974).
18. McKenzie, W. H., and Lubs, H. A., *Cytogenet. Cell Genet.* **14**, 97 (1975).
19. Caspersson, T., Zech, L., Johansson, C., and Modest, E. J., *Chromosoma* **30**, 215 (1970).
20. Caspersson, T., Lomakka, G., and Zech, L., *Hereditas* **67**, 89 (1971).
21. Shafer, D. A., *Lancet* **1**, 828 (1973).
22. Hsu, T. C., Pathak, S., and Shafer, D. A., *Exp. Cell Res.* **79**, 484 (1973).
23. Drets, M. E., and Shaw, M. W., *Proc. Natl. Acad. Sci. U.S.A.* **68**, 2073 (1971).
24. Crossen, P. E., *Clin. Genet.* **3**, 169 (1972).
25. Crossen, P. E., *Histochemie* **35**, 51 (1973).
26. Eiberg, H., *Clin. Genet.* **4**, 556 (1973).
27. Kato, H., and Moriwaki, K., *Chromosoma* **38**, 105 (1972).
28. Geraedts, J., and Pearson, P., *Humangenetik* **20**, 171 (1973).
29. Utakoji, T., and Matsukuma, S., *Exp. Cell Res.* **87**, 111 (1974).
30. Schnedl, W., *Chromosoma* **34**, 448 (1971).
31. Meisner, L. F., Chuprevich, T. W., Johnson, C. B., Inhorn, S. L., and Carter, J. J., *Lancet* **1**, 100 (1973).
32. Seabright, M., *Lancet* **2**, 971 (1971).
33. Dutrillaux, B., de Grouchy, J., Finaz, C., and Lejeune, J., *C. R. Hebd. Seances Acad. Sci.* **273**, 587 (1971).
34. Dutrillauz, B., Finaz, C., de Grouchy, J., and Lejeune, J., *Cytogenetics* **11**, 113 (1972).
35. Finaz, C., and de Grouchy, J., *Ann. Genet.* **14**, 309 (1971).
36. Wang, H. C., and Federoff, S., *Nature (London) New Biol.* **235**, 52 (1972).
37. Gallimore, P. H., and Richardson, C. R., *Chromosoma* **41**, 259 (1973).
38. Howard, P. N., Stoddard, G. R., and Seely, J. R., *Clin. Genet.* **4**, 162 (1973).
39. Kato, H., and Yosida, T. H., *Chromosoma* **36**, 272 (1972).
40. Shiraishim, Y., and Yosida, T. H., *Chromosoma* **37**, 75 (1972).
41. Hansen-Melander, E., Melander, Y., and Olin, M. L., *Hereditas* **76**, 35 (1974).
42. Berger, R., *Exp. Cell Res.* **75**, 298 (1972).
43. Lee, C. L. Y., Welch, J. P., and Lee, S. H. S., *Nature (London) New Biol.* **241**, 142 (1973).
44. Utakoji, T., *Nature (London)* **239**, 168 (1972).
45. Utakoji, T., *in* "Chromosomes Today" (J. Wahrman, ed.), Vol. IV, p. 53. Plenum, New York, 1973.
46. Sumner, A. T., *Exp. Cell Res.* **83**, 438 (1973).
47. Dev, V. G., Warburton, D., Miller, O. J., Miller, D. A., Erlanger, B. F., and Beiser, S. M., *Exp. Cell Res.* **74**, 288 (1972).
48. Patil, S. R., Merrick, S., and Lubs, H. A., *Science* **173**, 821 (1971).

49. Sanchez, O., Escobar, J. I., and Yunis, J. J., *Lancet* 2, 269 (1973).
50. Nombela, J. J., and Murcia, C. R., *Chromosoma* 37, 63 (1972).
51. Takayama, S., *Jpn. J. Genet.* 49, 189 (1974).
52. Yunis, J. J., and Sanchez, O., *Chromosoma* 44, 15 (1973).
53. Carpentier, S., Dutrillaux, B., and Lejeune, J., *Ann. Genet.* 15, 203 (1972).
54. Dutrillaux, B., and Covic, M., *Exp. Cell Res.* 85, 143 (1974).
55. Eiberg, H., *Lancet* 2, 836 (1974).
56. Comings, D. E., Avelino, E., Okada, T. A., and Wyandt, H. E., *Exp. Cell Res.* 77, 469 (1973).
57. Bobrow, M., and Madan, K., *Cytogenet. Cell Genet.* 12, 145 (1973).
58. De la Chapelle, A., Schroder, J., Selander, R.-K., *Chromosoma* 42, 365 (1973).
59. Verma, R. S., and Lubs, H. A., *Am. J. Human Genet.* 27, 110 (1975).
60. Schested, J., *Humangenetik* 21, 55 (1974).
61. Shiraishi, Y., *Jpn. J. Genet.* 45, 429 (1970).
62. Schreck, R. R., Warburton, D., Miller, O. J., Beiser, S. M., and Erlanger, B. F., *Proc. Natl. Acad. Sci. U.S.A.* 70, 804 (1973).
63. Dutrillaux, B., *Chromosoma* 41, 395 (1973).
64. Zahkarov, A. F., and Egolina, N. A., *Chromosoma* 38, 341 (1972).
65. Latt, S. A., *Proc. Natl. Acad. Sci, U.S.A.* 70, 3395 (1973).
66. Alfi, O. S., and Menon, R., *J. Lab. Clin. Med.* 82, 692 (1973).
67. Yunis, J. J., Roldan, L., Yasmineh, W. G., and Lee, J. C., *Nature (London)* 231, 532 (1971).
68. Sumner, A. T., *Exp. Cell Res.* 75, 304 (1972).
69. Chuprivich, T. W., Meisner, L. F., Inhorn, S. L., and Indriksons, A., *Lancet* 1, 1453 (1973).
70. McKenzie, W. H., and Lubs, H. A., *Chromosoma* 41, 175 (1973).
71. Scheres, J. M. J. C., *Humangenetik* 23, 311 (1974).
72. Alfi, O. S., Donnell, G. N., and Derencsenyi, A., *Lancet* 2, 505 (1973).
73. Miller, O. J., Schnedl, W., Allen, J., and Erlanger, B. F., *Nature (London)* 251, 636 (1974).
74. Bobrow, M., Madan, K., and Pearson, P. L., *Nature (London) New Biol.* 238, 122 (1972).
75. Gagne, R., and Laberge, C., *Exp. Cell Res.* 73, 239 (1972).
76. Eiberg, H., *Nature (London)* 248, 55 (1974).
77. Matsui, S., and Sasaki, M., *Nature (London)* 246, 148 (1973).
78. Matsui, S., *Exp. Cell Res.* 88, 88 (1974).
79. Goodpasture, C., and Bloom, S. E., *J. Cell Biol.* 67, 140a (1975).
80. Hilwig, I., and Gropp, A., *Exp. Cell Res.* 75, 122 (1972).
81. Gropp, A., Hilwig, I., and Seth, P. K., *in* "Chromosome Identification—Techniques and Applications in Biology and Medicine" (T. Caspersson and L. Zech, eds.), Nobel Symp. No. 23, p. 300. Academic Press, New York, 1973.
82. Raposa, T., and Natarajan, A. T., *Humangenetik* 21, 221 (1974).
83. Perry, P., and Wolff, S., *Nature (London)* 251, 156 (1974).
84. Kornberg, J. R., and Freedlender, E. F., *Chromosoma* 48, 355 (1974).
85. Merrick, S., Ledley, R. S., and Lubs, H. A., *Pediat. Res.* 7, 39 (1973).
86. Seabright, M., *Lancet* 1, 1249 (1973).
87. Lillie, R. D., "H. J. Conn's Biological Stains." Williams & Wilkins, New York, 1969.
88. Comings, D. E., *Chromosoma* 50, 89 (1975).
89. Ferguson-Smith, M. A., and Page, B. M., *J. Med. Genet.* 10, 282 (1973).
90. Okada, T. A., and Comings, D. E., *Chromosoma* 48, 65 (1974).
91. Bahr, G. F., Mikel, U., and Engler, W. F., *in* "Chromosome Identification—Technique

132 DAVID E. COMINGS

and Applications in Biology and Medicine" (T. Caspersson and L. Zech, eds.), Nobel Symp. No. 23, p. 280. Academic Press, New York, 1973.
92. Comings, D. E., and Okada, T. A., *Exp. Call Res.* **93**, 267 (1975).
93. McKay, R. D. G., *Chromosoma* **44**, 1 (1973).
94. Rodman, T. C., *Science* **184**, 171 (1974).
95. Rodman, T. C., and Tahiliani, S., *Chromosoma* **42**, 37 (1973).
96. Comings, D. E., *in* "Advances in Human Genetics" (H. Harris and K. Hirschhorn, eds.), Vol. III, p. 237. Plenum, New York, 1972.
97. Comings, D. E., *in* "Chromosome Identification—Techniques and Applications in Biology and Medicine" (T. Caspersson and L. Zech, eds.), Nobel Symp. No. 23, p. 293. Academic Press, New York, 1973.
98. Comings, D. E., *Excerpta Med. Int. Congr. Ser.*, 44 (1974).
99. Stubblefield, E., *Chromosoma* **53**, 209 (1975).
100. Comings, D. E., and Avelino, E., *Chromosoma* **51**, 365 (1975).
101. Burkholder, G. D., *Nature (London)* **247**, 292 (1974).
102. Burkholder, G. D., *Exp. Cell Res.* **90**, 269 (1975).
103. Comings, A. E., and Avelino, E., *Exp. Cell Res.* **86**, 202 (1974).
104. Sivak, A., and Wolman, S. R., *Histochemistry* **42**, 345 (1974).
105. Arrighi, S. E., and Hsu, T. C., *in* "Human Chromosome Methodology" (J. J. Yunis, ed.), p. 59. Academic Press, New York, 1974.
106. Pathak, S., and Arrighi, F. E., *Cytogenet. Cell Genet.* **12**, 414 (1973).
106a. Comings, D. E., Harris, D. C., Okada, T. A. and Holmquist, G., *Exp. Cell Res.* **105**, 349 (1977).
107. Stockert, J. C., and Lisanti, J. A., *Chromosoma* **37**, 117 (1972).
108. Comings, D. E., and Wyandt, H. E., *Exp. Cell Res.* **99**, 183 (1976).
109. Dutrillaux, B., Laurent, C., Couturier, J., and Lejeune, J., *C. R. Hebd. Seances Acad. Sci., Ser. D.* **276**, 3179 (1973).
110. Dutrillaux, B., Fosse, A. M., Prieur, M., and Lejeune, J., *Chromosoma* **48**, 327 (1974).
111. Goto, K., Akematsu, T., Shimazu, H., and Sugiyama, T., *Chromosoma* **53**, 223 (1975).
112. Wolff, S., and Perry, P., *Chromosoma* **48**, 341 (1974).
113. Holmquist, G. P., and Comings, D. E., *Chromosoma* **52**, 245 (1975).
114. Latt, S. A., *Proc. Natl. Acad. Sci. U.S.A.* **71**, 3162 (1974).

Chapter 12

Cytological Analysis of Drosophila Polytene Chromosomes

J. DERKSEN

Department of Genetics,
Katholieke Universiteit,
Nijmegen, Holland

I. Introduction

Polytene chromosomes as they occur in *Drosophila* larvas offer the possibility for studying local genome activities in interphase cells on the cytological level. Polytene chromosomes represent bundles of paired interphase chromosomes in which the chromomeres are visible as bands, and the interchromomeres as interbands. If regions are highly active in transcription, then they can be distinguished cytologically as puffs. The hypothesis that one band forms one puff is still valid (*1*).

A variety of techniques are available for the study of this system on the cytological and ultrastructural level. Patterns of local genome activities (i.e., puffs) can be studied by examining chromosome squash preparations from various tissues. Usually the salivary glands are used because of the high degree of polyteny in that tissue. Preparations commonly are stained with aceto-orcein for this purpose, but the band structure of the chromosomes can also be seen in nonstained preparations in a phase-contrast microscope. Nonhistone proteins can be detected qualitatively by staining with fast green FCF at low pH (*2*); for a quantitative analysis staining with naphtol-yellow S is required (*3*). Since the degree of polyteny differs from cell to cell within any one tissue, the amount of dye bound to the chromosome should be related, for a quantitative estimation, to the amount of DNA. Since Feulgen staining is stochiometric and specific for DNA and its absorbtion spectrum has very little overlap with that of naphtol-yellow S, this stain should be used as a counterstain for naphtol-yellow S (*4,5*) in such studies.

For studies at well-defined loci on the ultrastructural level the squash and selection technique must be used (*6,7*). In this technique chromosome parts are identified in squashed material before sectioning. Thus longitudinal sections through well-defined and long parts of the chromosomes can be obtained. The formation of puffs as well as ribonucleoprotein (RNP) formation within puffs can then be studied directly (8). Using various staining procedures differences in protein composition of such RNP particles can be detected (9).

II.   Staining with Aceto-orcein and Fast Green FCF

A.   Preparation of the Stains

1.   ACETO-ORCEIN

Dissolve 3 gm of orcein (natural, BDH Chemicals Ltd., Poole, England) in 100 ml of 70% acetic acid by boiling for 3 hours in the presence of glass beads and by using a water-cooled reflux column. The solution is allowed to cool down to room temperature and then is filtered. This solution is stable for several months when stored in a closed bottle at room temperature; if necessary it can be refiltered before use.

2.   FAST GREEN FCF

Dissolve 3 gm of the dye (Chroma, Stuttgart) in 100 ml of 50% acetic acid. The solution should have an intense green color; if the solution has a bluish appearance it will not give the expected results. The solution is stable if stored in a closed bottle.

3.   FAST GREEN–ORCEIN MIXTURE

For simultaneous staining of the tissue with orcein and fast green. Mix equal volumes of the orcein and the fast green solutions and add 5–10% glacial acetic acid to the mixture. This mixture is stable for not more than 2 weeks; preferably fresh solutions should be used.

B.   Fixation and Staining of the Tissue

Transfer the hand-isolated tissue to, preferably, freshly made ethanol–acetic acid mixture (3:1 v/v) for 2–10 minutes. Place the tissue on a grooved slide, remove with a piece of filter paper any excess of the fixative, and add a drop of the staining solution. Stain for at least 3 minutes. Rinse the preparation with 45% acetic acid, cover with a large coverslip (24 × 24 or 32 mm), and squash by tapping the preparation several times with the back end of pair of forceps. Freeze the preparation on dry ice (a few minutes), flip off the coverslip with a sharp knife or a razor blade, and transfer immediately to absolute ethanol.

After 1 minute transfer to fresh absolute ethanol. After 10 minutes in ethanol and two further changes in xylene for 10 minutes each, mount the preparation in a proper medium, for example caedax (Merck, Darmstadt), cover with a coverslip, and blot off excess mounting medium with a filter paper. Dry the preparations for 1 hour at 60°C in an oven. Preparations stained with orcein alone can also be mounted, after the last dehydration step in ethanol, in euparal (Chroma); the fast green staining is not stable in euparal and thus different mounting media are required.

C.   Appearance in the Microscope

1.   ORCEIN

Orcein is a stain for deoxyribonucleoprotein (DNP); bands are darkly stained and have a dark reddish to brown color.

2.   FAST GREEN FCF

Fast green is a stain for proteins. Stained are cytoplasm, nucleoli, puffs, bands, and some interbands. Bands and stained interbands often cannot be distinguished. The color of the stained structures is bright green.

3.   SIMULTANEOUS STAINING WITH ORCEIN AND FAST GREEN FCF

Bands stain brown, somewhat less reddish than with orcein alone; other structures stain as described under 2. Stained interbands can be distinguished now from bands.

III.   Staining with Naphtol-Yellow S (NYS)

A.   Preparation of the Stains

1.   NYS

Dissolve 0.25 gm NYS (Fluka, Buchs, Switzerland) in 100 ml of 2.5% acetic acid. Fresh solutions should be used.

2.   SCHIFFS REAGENT

Dissolve 1 gm pararosaniline (Chroma; acridin free) in 200 ml of distilled water at 100°C. The solution is allowed to cool down to 58°C, filtered and

20 ml of 1 N HCL are added. After cooling down to room temperature, 1 gm of $K_2S_2O_5$ and 0.5 gm of active charcoal are added. Shake the mixture and keep it in the dark for 24 hours in a closed bottle. The now-colorless solution is filtered and can be stored for long periods in a closed flask in the dark.

B.   Fixation and Staining of the Tissue

Since ethanol–acetic acid fixation followed by squashing in acetic acid extracts some protein, a different fixation is required, namely in 7.5% formaldehyde at pH 7.2 in, for example, 0.01 M phosphate buffer. The tissue is squashed in 45% acetic acid, but does not squash as well as after fixation in ethanol–acetic acid. After freezing on Dry Ice the coverslip is removed and the preparation is postfixed for 1 hour in a mixture of formalin (37%; commercially available) and absolute methanol (1:9 v/v). The preparations are passed through a methanol–water series (90, 70, 50, and 30%, 10 minutes each) to distilled water and rinsed thoroughly several times. Hydrolyze the preparations for 10 minutes in 1 N HCl at 60°C; stain for 90 minutes in Schiff's reagent and wash 3 times for 10 minutes each in 1% $K_2S_2O_5$ acidified with 5% 1 N HCl. The preparations are thoroughly washed in distilled water, rinsed in 2.5% acetic acid, and stained for 1 hour in the NYS staining solution. After staining the preparations are shortly rinsed in 1% acetic acid and dehydrated in three changes of tertiary butyl alcohol. The butyl alcohol should be kept slightly above its freezing point. Finally the preparations are mounted in caedax and covered with a coverslip. An excess of caedax can be blotted off with a piece of filter paper. Dry the preparations for 1 hour at 60°C in an oven.

Except for the colors, the stained image is identical to that obtained after staining with orcein–fast green. The preparations can be scanned in a cyto-spectrofotometer at 430 and 550 nm for NYS and Feulgen, respectively (3–5).

IV.   The Squash and Selection Technique

The use of this technique is restricted to tissues with a very high degree of polyteny, i.e., the salivary glands. In *D. hydei* sometimes good results can be obtained with midgut cells, but chromosomes from tissues with a low degree of polyteny will generally not spread out properly during squashing and are too thin to permit a proper sectioning on the ultratome.

A.   Preparation of the Stains

1.   HEMALUM (MAYER)

Dissolve 0.1 gm of hematoxylin (Merck, Darmstadt) in 100 ml of distilled water and add 0.02 gm of $NaJO_3$ and 5 gm of $AlK(SO_4) \cdot 12 \ H_2O$. The solution has a blue to violet color. Leave the solution overnight at room temperature; then add 0.1 gm citric acid. The solution is stable for a long time and has a red to violet color.

2.   TOLUIDINE BLUE

Dissolve 0.1 gm of toluidine blue (Fluka) in 100 ml of distilled water.

3.   URANYL-ACETATE (Ur-ac)

Dissolve 3 gm of Ur-ac (the low radioactive Ur-ac from BDH is preferred) in 100 ml of 70% ethanol. Stir until the Ur-ac has dissolved completely. Use freshly prepared solutions.

4.   PHOSPHOTUNGSTIC ACID (PTA), ALCOHOLIC

Dissolve 3 gm of PTA in 100 ml of absolute ethanol (reagent grade) and cool to 0°C. Use freshly prepared solutions.

5.   PHOSPHOTUNGSTIC ACID (PTA), AQUEOUS AT pH 1

Dissolve 5 gm of PTA in 100 ml of distilled water and bring to pH 1 with concentrated HCl. Use freshly prepared solutions.

B.   Fixation and Squashing of the Tissue

The hand-isolated tissue is transferred to 1 ml of 3% glutaric aldehyde (Merck, Darmstadt) in 0.1 M cacodylate buffer (BDH), pH 7.2 at 0°C; after 15 minutes the fixation is continued at room temperature for another 15 minutes.

The tissue is transferred to a siliconized object slide and the remaining fixative is carefully removed with a piece of filter paper. A drop of not more than 40 μl of 50% acetic acid is added and the tissue is left in it for 3 minutes. If the squashing of the tissue is difficult, the tissue can be kept longer than 3 minutes in the acetic acid and may even be heated up to 35°C. This treatment, however, should be as short as possible since a prolonged treatment can give a fluffy appearance to the chromosome. Squash under a siliconized coverslip by making a gliding movement over it with the blunt end of a forceps or a preparation needle. Siliconization is required since both coverslip and object slide have to be removed from the tissue during the preparation [to siliconize dip in a 1% Siliclad (Clay and Adams, New York) solution].

Blot off the preparation with filter paper, freeze on dry ice (a few minutes), remove the coverslip, and transfer the preparation to methanol-formalin mixture (9:1 v/v) (common Coplin jars can be used).

The fixative used here, glutaric aldehyde, has given the best results so far. Osmium tetroxide (OsO_4), another often-used fixative in electron microscopy, is completely unsuitable as a fixative in this technique. The most common fixative for light microscopy, ethanol–acetic acid, can be used, but this fixative does not prevent the extraction of proteins during the procedure and, together with the strong stretching of the chromosome after ethanol–acetic acid fixation, this may be responsible for the lower diameter (from 105 to 75 Å in *D. hydei*) of DNP fibrils. Furthermore, RNP structures are insufficiently preserved. Also, especially in *D. hydei*, chromosomes often collapse during the procedure after ethanol–acetic acid fixation. On the other hand, squashing after glutaric aldehyde fixation is more difficult and, especially in *D. melanogaster*, chromosomes fragment frequently.

C.　Staining and Embedding of the Tissue

After freezing on dry ice and removal of the coverslip, the preparation is kept for 10 minutes (possibly longer) in methanol-formalin mixture (9:1 v/v) and hydrated in a methanol–water series (80, 50, and 30%; 10 minutes each). Rinse carefully in distilled water and leave for another 10 minutes in fresh distilled water. If Ur-ac is to be used as a stain the preparation can be postfixed by using the commercially available 37% formalin (Merck, Darmstadt) instead of distilled water in the hydration series.

Stain the preparations, for light microscopical examination, for 1 minute in hemalum, rinse in distilled water, and leave the preparations in 0.5% K_2CO_3 in distilled water until the color of the stain has turned blue.

Dehydrate the preparation by passing it through an ethanol–water series (30, 50, and 70%; 10 minutes each). Stain for 1 hour in Ur-ac. Sometimes precipitation occurs in the Ur-ac; this precipitation does not affect the staining, but do not use the solution again. The dehydration is continued for 10 minutes in 96% ethanol and further by two changes in absolute ethanol for 30 minutes each. Eventually the preparations can be stored in the last dehydration step.

For staining with alcoholic PTA the staining is as follows. Stain, after hydration, for 3 minutes in toluidine blue; rinse in 30% ethanol and dehydrate for $\frac{1}{2}$ minute in 50% ethanol, for 1 minute in 70% ethanol, and for 5 minutes in 96% ethanol. Complete the dehydration in two changes of absolute ethanol for 30 minutes each. Stain the preparations in alcoholic

PTA at 0 °C overnight. Then rinse the preparations in absolute ethanol and wash again for 10 minutes in fresh absolute ethanol. Do not leave the preparations in ethanol!

For staining with aqueous PTA the procedure is as follows. Stain, after hydration, for 5 hours in aqueous PTA, pH 1, at room temperature; rinse in 0.01 N HCl and wash for 5 minutes in distilled water. Stain and dehydrate as described for alcoholic PTA.

When the dehydration is completed, leave the preparations for 15 minutes in an ethanol–propylene oxide mixture (1:1 v/v), transfer to a epon propylene oxide mixture (1:1 v/v) and keep the preparations in it for at least 3 hours, possibly overnight at 4°C. Then the preparations are transferred to epon for 1–3 hours. Let the epon drip off the object slide and clean the back of the object slide with paper wetted with ethanol. Cover the preparations with fresh epon. The epon layer should not be thicker than 1 mm. Place the preparations strictly horizontal (for light microscopical examination an equal thick epon layer is required) in an oven, for 12 hours at 40°C, 12 hours at 50°C, and finally for 12 hours at 60°C to completely polymerize the epon [the epon is prepared according to Luft *(10)*]. The preparations can now be examined in the light microscope.

Select the desired chromosome or chromosome part and place a mark on it under the microscope. A gelatin capsule (Lilly and Co.) is filled with epon and placed on the marked spot. After polymerization overnight at 60°C in an oven, the epon is split off from the object slide with a sharp knife. Cut out the gelatin capsule which now carries the epon layer with the wanted genome part and place it in a holder. [This holder consists of a metal plate the size of an object slide with an opening in the middle. The diameter should be slightly larger than the diameter of the capsule. On this opening an 8-mm-high funnel with the same (inner) diameter as the opening is fixed. With a screw through a side-opening it is possible to hold an object tightly in the funnel.] The capsule with the chromosome on top is fixed in the holder and can be examined in the microscope. Mark the desired part again by encircling it with a needle.

Trimming and sectioning occur following standard techniques to allow examination in the electron microscope.

REFERENCES

1. Beermann, W., *in* "Results and Problems in Cell Differentiation" (W. Beerman, ed.), Vol. 4, p. 1. Springer-Verlag, Berlin and New York, 1972.

2. Alfert, M., and Geschwind, J. J., *Proc. Natl. Acad. Sci. U.S.A.* **39**, 991 (1953).

3. Holt, T. K. H., *Chromosoma* **32**, 64 (1970).

4. Deitch, A. D., *Lab. Invest.* **4**, 324 (1955).

5. Deitch, A. D., *in* "Introduction to Quantitative Cytochemistry" (G. L. Wied, ed.), p. 451. Academic Press, New York, 1966.

6. Sorsa, M., and Sorsa, V., *Chromosoma* **22**, 32 (1967).
7. Berendes, H. D., and Meyer, G. F., *Chromosoma* **25**, 184 (1968).
8. Berendes, H. D., Alonso, C., Helmsing, P. J., Leenders, H. J., and Derksen, J., *Cold Spring Harbor Symp. Quant. Biol.* **38**, 645 (1973).
9. Derksen, J., and Willart, E., *Chromosoma* **55**, 57 (1976).
10. Luft, J. H., *J. Biophys. Biochem. Cytol.* **9**, 409 (1961).

Chapter 13

Isolation and Purification of Nucleoli and Nucleolar Chromatin from Mammalian Cells

MASAMI MURAMATSU AND TOSHIO ONISHI

Department of Biochemistry,
Tokushima University School of Medicine,
Tokushima, Japan

I. Introduction

The nucleolus is known to be the site of ribosomal RNA (rRNA) synthesis and ribosome assembly (*1*). Although a large body of information concerning these functions has been accumulated for two decades, the detailed process and their precise mechanisms are yet to be uncovered. Since the nucleolus contains rRNA genes or the DNA coding for rRNA together with DNA-dependent RNA polymerase which is engaged in the transcription of rRNA precursor, this organelle is thought to be an excellent model system whereby regulation of RNA transcription may be studied in eukaryotic cells. Association of histones and nonhistone proteins with the nucleolar DNA provides an unique opportunity to study the regulatory function of these proteins in a DNA–protein complex, designated as chromatin, which is actively synthesizing one species of RNA.

Besides the transcription of rRNA, its processing as well as the assembly into ribosomes are other major points of interest (*2,3*). The presence of both specific methylases and processing enzymes for rRNA precursors has been suggested (*4–6*) but they have not yet been studied in detail. The assembly process of the ribosome is not scrutinized in eukaryotes as compared with that in *E. coli*.

In addition to these functionally established processes, other attributes whose functional relationship is hardly known are also present in the nucleolus including heterochromatin (*7*), nucleolus-specific low-molecular-weight RNA (*8*), certain enzymes such as nicotinamide adenine dinucleotide (NAD) pyrophosphorylase (*9*), and so forth.

Isolation of nucleoli from various eukaryotic cells has been utilized for some time to solve many of the above problems (*10,11*). However, as research progresses the demand for a better preparation of nucleoli with higher purity and intactness is increasing. Also, application to wider spectrum of cells is needed in many instances.

In this chapter, we will describe the newest version of our technique of isolation of nucleoli from various cells in some detail. Special attention was paid to the intactness of the macromolecular structure of the nucleolus in order to retain *in vivo* function in isolated nucleoli. For a more general discussion readers are referred to a previous review (*12*).

II.   Isolation of Nucleoli from Various Mammalian Cells

A.   General Principles

Theoretically, isolation of nucleoli consists of three steps: (1) isolation of nuclei from whole homogenate; (2) disruption of nuclei; and (3) purification of nucleoli from the disintegrated nuclear material.

It is possible to destroy nuclei in whole homogenate without isolating nuclei (*13*). However, the possibility of contact with cytoplasmic or lysosomal degradating enzymes must be considerably greater under such conditions. Therefore, purified nuclei are an apparently more preferred starting material. Isolation of cytoplasmic-free nuclei is known to be relatively easy for liver and for kidney cells since the advent of Chauveau's high-density sucroce procedure (*14*). However, this procedure cannot be used for most tumor cells or for such somatic cells as spleen and fetal liver because the usual homogenization procedure with even the tightest pestle clearance cannot strip off cytoplasmic tags adhering to the nuclear membranes of these cells. Since this is especially so when divalent cations such as Ca^{2+} or Mg^{2+} are included in the homogenization medium, and the presence of these cations is essential to strengthen nucleoli against the sonication or pressure that disrupts nuclei (*15*), more drastic procedures are required to remove perinuclear cytoplasm from these cells. For this purpose, a combination of hypotonic shock with various detergents is usually employed together with a tightly fitting Dounce or Potter–Elvehjem type homogenizer. In special cases, a Chaikoff homogenizer, which provides a very narrow defined clearance, gives better results (*16*). The optimal conditions of hypotonicity and detergents vary from one cell species to another, so that they should be determined for each material (see below).

There are, in principle, three major techniques for disruption of nuclei (*12*).

One is sonication; the other two are compression–decompression and chemical fractionation procedures. The sonication procedure was first introduced by Monty et al. (17). It was after adaptation of this technique by Maggio et al. (18) and Muramatsu et al. (15) to a form which could produce nucleoli sufficiently intact for biochemical analyses that this method was widely used as a standard technique for isolation of nucleoli. Later a compression–decompression technique was developed by Desjardins et al. (13) as an alternative, but it has not been used as frequently as sonication, probably partly because it requires more experience for the proper operation of the pressure apparatus. The underlying principle of the sonication procedure is that an appropriate concentration of either Ca^{2+} or Mg^{2+} can confer nucleoli with enough rigidity so that they are not destroyed when the nuclei are disintegrated completely by sonic oscillation. The concentration of Ca^{2+} or Mg^{2+} in the initial homogenizing medium and in the sonicate, when necessary, is critical to obtain complete destruction of nuclei with minimum damage to nucleoli. Interestingly, for a wide variety of cells, the optimal concentrations of these cations are very similar if other conditions are carefully controlled. But the concentrations of these cations do vary with the concentration of the homogenate (tissue weight/homogenizing medium). The more concentrated the homogenate, the higher the concentration of divalent cations required. It should also be noted that several times higher molarity is required with Mg^{2+} than with Ca^{2+}. Actual concentrations are described in the next section.

Chemical fractionation of nuclei as developed by Penman et al. (19) makes use of DNase with a high salt buffer (0.5 M NaCl–0.05 M $MgCl_2$–0.01 M Tris-HCl, pH 7.4) to solubilize nuclear material, nucleoli being sedimented through a layer of sucrose solution. This procedure has been used successfully for the study of processing of preribosomal RNA and of preribosomal particles (11). However, it cannot be used to study chromatin and RNA polymerase in the nucleolus since DNase destroys chromatin structure and knocks off RNA synthetic activity of the nucleolus completely (20).

For studying nucleolar functions in vitro, a procedure which preserves fine structure of nucleolar chromatin and nucleolar RNA polymerase bound to it is required. At present, our sonication procedure appears most satisfactory for these purposes.

Purification of nucleoli from the sonicate or pressate (in the case of the compression–decompression procedure) can be done by differential centrifugation through a layer of sucrose solution. It is not difficult to sediment nucleoli from the destroyed chromatin material since the latter becomes very fine particles after sonication. In the case of pressate, the chromatin material tends to become rather dense fibers which more easily contaminate the nucleolar pellet. Therefore, a higher concentration of sucrose solution is usually required to separate nucleoli from this material (*13*).

B.   Procedures

1.   METHODS FOR RAT LIVER

The following procedures are conducted at 2–4°C (*21*).

1. Perfuse the liver thoroughly from the portal vein with 10–20 ml of ice-cold 0.25 M sucrose–3.3 mM $CaCl_2$ or 0.25 M sucrose–10 mM $MgCl_2$ to get rid of the blood and also to cool down the tissue quickly.

2. Mince the liver with scissors or pass it through a tissue press (or Harvard press) to remove the bulk connective tissue.

3. Homogenize the tissue with approximately 2 volumes of 2.3 M sucrose–10mM $MgCl_2$ (in the Ca^{2+} procedure, replace $MgCl_2$ with 3.3mM $CaCl_2$) by means of a loosely fitting Potter–Elvehjem type homogenizer with a Teflon pestle. A few strokes are usually enough to destroy liver cells and release most of the nuclei.

4. Filter through four layers of cheesecloth or gauze to remove residual connective tissue. Wash the homogenate remaining on the cheesecloth with some of the same homogenization medium and dilute the homogenate to 8–10 volumes (12–15 volumes for Ca^{2+} procedure) (v/w) of the original liver. In a large-scale preparation, the homogenate may be mixed with a more concentrated sucrose (e.g. 2.5 M) to a density of 1.28, which can be checked easily by a refractometer. In this procedure, however, a higher concentration of Mg^{2+} or Ca^{2+} is always required because the tissue:medium ratio becomes quite high. For example, we routinely use 2.5 M sucrose–20 mM $MgCl_2$ for homogenization of 500 gm of liver, adjusting the homogenate to 1.5 liters with the same sucrose solution. Although the nuclear preparation thus prepared is slightly contaminated with cytoplasmic components, it may be used quite well for isolation of nucleoli. The homogenate must be stirred well before centrifugation.

5. Centrifuge the homogenate at 40,000 g for 1 hour. The pellet contains purified nuclei. For extra-pure nuclei, the Blobel–Potter (*22*) type two-layer system may be used. However, it requires a swing rotor and therefore the amount of tissue which can be treated is limited. In usual cases, centrifugation in an angle rotor without layering over is sufficient.

6. Suspend the nuclear pellet in 0.5–1 ml of 0.34 M sucrose per gram of original tissue containing 0.05 mM $MgCl_2$ (in the Ca^{2+} procedure, omit any divalent cations) using a loosely fitting Potter–Elvehjem type homogenizer with a Teflon pestle.

7. Sonicate the suspension with a sonicator in 20 ± 5 ml batches for 30–60 seconds (Either a 10 or 20 kH apparatus will do; 200–250 W work better than 150 W). Usually, a drop of the sonicate is examined under a light microscope after mixing with a drop of 0.1% azure C or toluidine blue (dissolved in physiological saline) after the first 30 seconds of sonication. Most of the nuclei are broken by this time, but a few unbroken nuclei may still be seen per one medium magnification field. In most cases, an additional 15 seconds sonication will break up completely the remaining nuclei. Sometimes 15–30 seconds more sonication is necessary, but when a total of 90 seconds sonication does not destroy all the nuclei, other errors such as the presence of excess divalent cations, etc., should be suspected. In these cases prolonged sonication may not work, and rather serious damage may occur to the nucleolar macromolecular components. In our experience the nuclei of regenerating rat liver are more easily broken down than those of normal liver; 20–30 seconds sonication is usually enough. An insufficient amount of divalent cations as well as a high pH of either the homogenization or sonication medium tend to weaken the nucleoli and sometimes most of them are broken into pieces as the nuclei are disrupted.

8. Pour each 20 ± 5 ml of the sonicate into a 50-ml centrifuge tube and underlay an equal volume of 0.88 M sucrose containing 0.05 M $MgCl_2$.

9. Centrifuge at 2000 g (3000–3500 rpm) for 20 minutes. The pellet contains purified nucleoli.

2. METHODS FOR TUMOR AND TISSUE CULTURE CELLS

For some slow-growing tumors such as Morris hepatoma 5123D, the original high-density sucrose procedure for isolation of nuclei may be applied with considerable success (23), but in most cases the use of detergents together with hypotonic shock is necessary if one wants to start with cytoplasm-free nuclei. Crude "nuclei" can be obtained by homogenizing tumor cells in 1.8–2.0 M sucrose using a relatively tight glass homogenizer. A higher concentration of sucrose cannot be used since crude nuclei with cytoplasmic tags will float up during centrifugation in the high-density solution. Although these "nuclear preparations" are heavily contaminated with adhering cytoplasm, they may be used for isolation of nucleoli since cytoplasmic tags are eventually broken up during sonication and remain in the supernatant when nucleoli are spun down. The following paragraphs describe a rapid and reproducible procedure for isolation of nucleoli from various tumor and tissue culture cells starting from purified nuclei (24).

1. Harvest the cells with a phosphate-buffered saline [0.14 M NaCl–2.7 mM KCl–8.1 mM Na_2HPO_4–1.5 mM KH_2PO_4(pH 7.2)] and collect by centrifugation at 500 g (1500 rpm) for 5 minutes.

2. Suspend the pellet in 20 volumes of reticulocyte standard buffer

TABLE I

OPTIMAL CONCENTRATION OF DETERGENTS FOR COMPLETE
REMOVAL OF PERINUCLEAR CYTOPLASM[a]

| Cell species | Final concentration (%) | |
	Nonidet P-40	Deoxycholate
L cell	0.3	0
HeLa cell	0.3	0.2
Ehrlich ascites		
tumor	0.3	0.2
MH134	0.3	0.3
C3H2K	0.2	0

[a]The indicated concentrations of both detergents were added
to the cell suspension made in the hypotonic solution as described in the text.

[RSB, 0.01 M Tris–HCl–0.01 M NaCl–1.5 mM MgCl$_2$(pH 7.4)] and stand
it in an ice bath for 10 minutes.

3. Centrifuge at 600 g (2000 rpm) for 5 minutes to collect swollen cells.
Resuspend the pellet in the same volume (as employed in the first hypotonic
shock) of RSB.

5. Add a 10% solution of Nonidet P-40 (Shell Chemical Co) to a final
concentration of 0.2–0.3% according to the cell species (Table I). Further,
if necessary, add a freshly prepared 10% solution of sodium deoxycholate to
the final concentration determined for each cell type (Table I).

6. Homogenize the suspension in a tightly fitting (0.4 mm clearance)
Potter–Elvehjem type homogenizer with a Teflon pestle giving 10 up-and-
down strokes.

7. Centrifuge at 1200 g (2500 rpm) for 5 minutes to sediment crude nuclei.

8. Homogenize the pellet in 10 volumes (of the original cell volume) of
0.25 M sucrose–10 mM MgCl$_2$ (in the Ca^{2+} procedure, replace MgCl$_2$ with
3.3 mM CaCl$_2$).

9. Underlay an equal volume of 0.88 M sucrose–0.05 mM MgCl$_2$ and
centrifuge at 1200 g (2500 rpm) for 10 minutes. The pellet contains purified
nuclei.

10. Suspend the pellet in 10 volumes (of the initial cell volume) of 0.34 M
sucrose–0.05 mM MgCl$_2$ (in the Ca^{2+} procedure, no divalent cations are
added) and sonicate for 45–60 seconds to destroy all the nuclei. The method
of sonication and of checking the sonicate is the same as described for rat
liver nuclei (see above).

11. Underlay an equal volume of 0.88 M sucrose–0.05 mM MgCl$_2$ and
centrifuge at 2000 g (3000–3500 rpm) for 15–20 minutes. The pellet contains purified nucleoli.

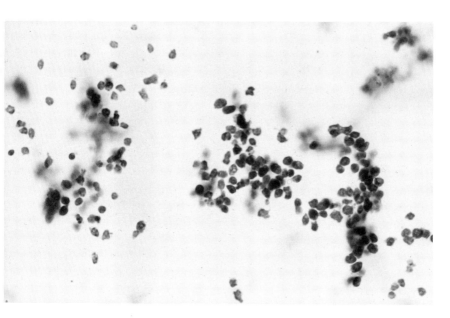

FIG. 1. HeLa cell nucleoli isolated by the procedure described in II,B,2, stained with azure C. Blurred spots are not the contaminants but nucleoli that are out of focus. × 785.

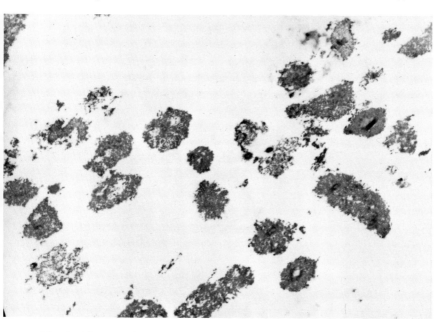

FIG. 2. Electron micrograph of a HeLa cell nucleolar preparation. × 11,5000,

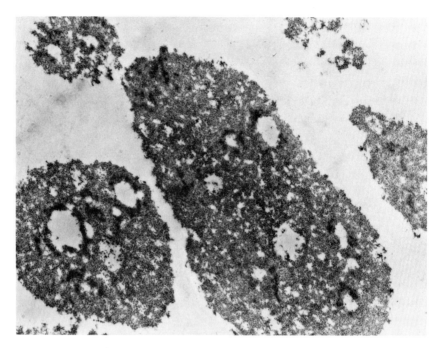

FIG. 3. Higher magnification of isolated nucleoli of the HeLa cell. × 34,500. Structure of nucleolonema with granular and fibrillar components is clearly seen as *in situ.*

The morphology of isolated nucleoli is shown in Figs. 1–3. The preparation is over 99.9% pure in terms of particle counts, and the ultrastructure is well preserved as demonstrated by electron micrograph.

C.   Biochemical Properties of Isolated Nucleoli

1.   NUCLEIC ACID

The size of DNA in isolated nucleoli was determined by directly dissolving them in alkali and centrifuging in an alkaline sucrose density gradient. Nucleolar DNA sedimented with a rather broad peak with a modal S value of 27, which corresponded to a molecular weight of 5.7×10^6 (Fig. 4). This size, approximately 1.7×10^4 nucleotides, is sufficient to code for one stretch of 45 S ribosomal RNA precursor (1.4×10^4 nucleotides), although much smaller fragments were also present. When the nucleolar DNA was extracted by the method of Quagliarotti *et al.* (25), the mean molecular weight decreased to about 5.3×10^6. When nucleoli were washed thoroughly to prepare a nucleolar chromatin (see next section), the DNA became even smaller–presumably due to nicks caused by both endogenous DNase activity and mechanical shear (Fig. 4).

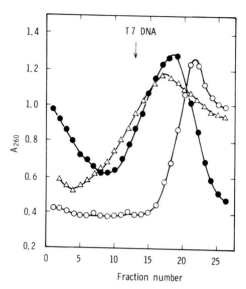

FIG. 4. Alkaline sucrose density gradient profiles of nucleolar DNA from rat liver. Isolated nucleoli or nucleolar chromatin containing 50 µg DNA were incubated in 0.2 ml of 0.5 N KOH at 37°C for 1 hour and layered on 5–20% alkaline sucrose gradient. Centrifugation was performed at 16,000 rpm for 28 hours at 20°C. ($\triangle$) isolated nucleoli; ($\bigcirc$) nucleolar chromatin; ($\bullet$) DNA purified from isolated nucleoli by phenol–chloroform procedure. The arrow indicates the peak of T7 DNA (37 S).

Isolated nucleoli contain a series of ribosomal RNA precursors and certain species of low-molecular-weight nuclear RNA. In order to obtain intact RNA species precautions must be taken to prevent their degradation by endogenous and exogenous RNases. Three important factors for this are the low temperature during the whole procedure, the rapidity of the separation of nucleoli from other nuclear material after sonication, and the purity of nuclear and final nucleolar preparation. Typical sucrose density gradient patterns of nucleolar RNA from rat liver and mouse hepatoma, MH134, cells are presented in Fig. 5. Note that the relative pool size of various precursors differs between these cells.

When the cells are pulse-labeled with either labeled orotic acid or uridine, or methyl-labeled methionine, typical kinetics of labeling pattern could be demonstrated; i.e., only the 45 S RNA peak is labeled for the first 10 minutes, the 32 S RNA peak being labeled thereafter (19, 26–29). Early label appears on the 20–23 S peak which seems to be a precursor to 18 S RNA (30,31). Nucleolar low-molecular-weight RNAs could be resolved by acrylamide gel into bands including the U₃ RNA which is unique to the nucleolus (8,32,33). 5 S RNA, a part of which is actually associated with a maturing

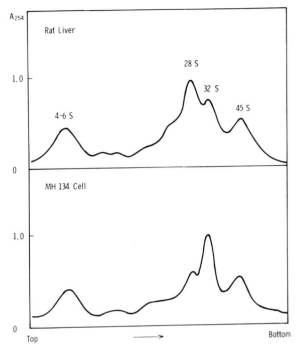

FIG. 5.   Sucrose density gradient profiles of nucleolar RNA. RNA was extracted from nucleolar pellet with sodium dodecyl sulphate (SDS)–hot phenol procedure (59) and centrifuged through a linear 10–40% sucrose gradient buffered with 0.1 M sodium acetate (pH 5.1) for 17 hours at 30,000 rpm at 4 °C. Absorbance at 254 nm was monitored with an ISCO apparatus.

larger ribosomal subunit, is also present. When nucleolar RNA is extracted at a high temperature, 5.8 S RNA which is originally hydrogen-bonded with 28 S RNA is found in the low-molecular-weight RNA fraction. The significance of other low-molecular-weight RNAs is not apparent, although they are known to be very stable metabolically.

2.   PROTEINS

More than 80% of the dry nucleolar mass represents protein (10). It includes proteins of preribosomal particles, histones, and other nuclear proteins. Preribosomal proteins do not necessarily comprise all the ribosomal proteins, but contain other protein components which cannot be found in mature ribosomes (34–40). Most of these basic proteins could be extracted with either 0.25 N HCl or 67% acetic acid, leaving acid-insoluble proteins behind. Nonribosomal nucleolar proteins may be divided into three categories: saline–soluble, acid–soluble, and acid-insoluble proteins. When isolated nucleoli are extracted with dilute salt solutions, so-called globular proteins are extracted together with a part of preribosomal particles. Usually, relatively mature preribosomal particles, such as "60 S" particles containing 28 S, 30 S, and 32 S RNA, are more easily extracted than those containing 45 S RNA (39). These premature particles are in fact hard to extract and usually require DNase and some detergent to be liberated (41). Acid extraction of nucleoli with cold 0.25 N HCl solubilizes histones together with preribosomal proteins if the latter is remained. No difference has ever been reported in the relative amounts of histone fractions between the nucleus and the nucleolus, although a more elaborate determination has to be made before any conclusion is drawn. Modifications of histones in the nucleolus also, remain to be studied in detail. The residue, after acid extraction, contains so-called nonhistone proteins or acidic proteins of the nucleolus. They are known to be very heterogeneous and metabolically active. A high rate of phosphorylation has also been reported (42–44). Recently we have also demonstrated that the SDS polyacrylamide gel electrophoresis pattern of nonhistone proteins is distinctly different between nucleolar and extranucleolar chromatin (44a). The species and function of nucleolar nonhistone proteins should be an exciting target of future studies in relation to gene regulation.

3. RNA Polymerase

Isolated nucleoli contain a high concentration of RNA polymerase I (or A) which is engaged in the transcription of rRNA. This enzyme could be solubilized either by sonicating the nucleoli in a high-salt solution (45) or by incubating at 37°C in an appropriate buffer (46) and purifying free from DNA. We have recently purified RNA polymerase I from rat liver nucleoli to a homogeneity and determined all the subunit structure (47,47a). Similar purification was accomplished by Schwartz et al. (48) from a mouse myeloma.

4. In Vitro RNA Synthetic Activity

Isolated nucleoli, when incubated under appropriate conditions, can incorporate labeled nucleoside triphosphate into acid-insoluble material in the presence of three other nucleoside triphosphates. That this represents a real RNA synthesis is indicated by the following criteria; i.e., besides the product becoming acid-soluble by RNase treatment, the incorporation is (a) dependent on all four nucleoside triphosphates, (b) inhibited strongly by a low concentration of actinomycin D, and (c) knocked off completely by a small amount of DNase I (20). The real proof came from a nearest-neighbor analysis in which all four nucleoside 2′,3′-monophosphates were found to be labeled after alkaline hydrolysis of the product when only one of the

$[\alpha\text{-}^{32}P]$-labeled nucleoside triphosphates was used as the precursor (49). Recently, Grummt et al. (50,51) found that under appropriate conditions isolated rat liver nucleoli could synthesize 45 S RNA. We have confirmed their experiments and defined detailed conditions for successful incubation (51a). It must be added that methylation of newly synthesized 45 S RNA may be accomplished in this system by including S-adenosyl methionine in the incubation medium (4,5,51). The details of the techniques of 45 S RNA synthesis in vitro will be described in another chapter (51b).

III.   Purification of Nucleolar Chromatin

A.   General Principles

In order to study the regulation mechanisms of transcription, it is desirable to have an in vitro system in which isolated template can be reacted with purified enzyme and other regulating factors. Since isolated nucleoli are the organelles containing both DNA–protein complexes and RNA polymerase, attempts have been made to separate these components, namely, nucleolar chromatin and RNA polymerase.

The isolation of the latter has already been referred to in Section II,C. The rationale of isolation of nucleolar chromatin comes from a finding that purified rDNA of Xenopus laevis could not be transcribed faithfully with either E. coli or homologous RNA polymerase (52,53). Since whole nuclear chromatin is known to have tissue as well as species specificity (54) and to serve as a template in which similar sequences as in vivo mRNA could be transcribed by exogenously added E. coli RNA polymerase (55–57), it may well be anticipated that chromatin prepared from isolated nucleoli could support specifically transcription of rRNA genes in the nucleolus by exogenously added enzymes. And if the template requires a specific RNA polymerase, homologous RNA polymerase I must be the one that can transcribe rRNA genes in high fidelity. With these concepts in mind, we have purified chromatin from isolated nucleoli (47,57a). The principle of purification of chromatin from isolated nucleoli is almost the same as that from nuclei. However, nucleoli contain preribosomal particles which are rather hard to extract completely. Nor is it easy to remove RNA polymerase I activity completely from isolated nucleoli without extensive extraction which tends to cause nicks on the DNA in the nucleolus. In the following, we shall describe three alternative procedures which are useful to prepare nucleolar templates which depend on exogenously added RNA polymerase.

B. Procedures

1. Nucleolar Chromatin Prepared with Extensive Salt Extraction

To achieve complete removal of associated RNA polymerase I from isolated nucleoli, the nucleolar preparation as isolated by the procedure described in the previous section was extracted successively with dilute salt solutions as described by Spelsberg and Hnilica (58).

1. Suspend the nucleoli from 80 gm of rat liver in 20 ml of 80 mM NaCl–20 mM ethylenediaminetetraacetic acid (EDTA)(pH 6.3) and homogenize in a Potter–Elvehjem homogenizer with a Teflon pestle by a gentle grinding by hand.

2. Centrifuge at 5000 g for 10 minutes. Discard the supernatant and repeat steps 1 and 2.

3. Homogenize the pellet in 20 ml of 0.15 M NaCl and centrifuge as above. Repeat.

4. Homogenize the pellet in 20 ml of 1.5 mM NaCl–0.15 mM sodium citrate and centrifuge as above. Repeat.

5. Homogenize the pellet in 20 ml of 0.35 M NaCl–0.01 M EDTA and centrifuge as above. Repeat. This step is included to remove RNase activity from the nucleoli which tends to degrade RNA synthesized *in vitro.*

6. Rinse the final pellet with 10 mM Tris-HCl (pH 7.4)–0.1 mM EDTA, suspend in the same buffer by a gentle homogenization, and store in ice until use.

2. Polymerase-Depleted Nucleoli Prepared with Mild Extraction of RNA Polymerase I at 0°C

This procedure was designed to obtain nucleolar template with least damage on nucleolar DNA. Nucleoli from 40 gm of liver were suspended in 20 ml of 10 mM Tris-HCl(pH 7.1)–0.15 M KCl and kept in ice for 16–18 hours. The nucleoli were recovered by a low-speed centrifugation and washed once with the same buffer solution. The final pellet was usually suspended in an appropriate amount of 0.34 M sucrose–1 mM MgCl$_2$ to deliver to incubation tubes.

3. Polymerase-Depleted Nucleoli Prepared with Incubation at 37°C

In this procedure nucleoli from 40 gm of liver were suspended in 20 ml of 50 mM Tris-HCl (pH 7.9)–25 mM KCl–0.5 mM dithiothreitol and incubated at 37°C for 30 minutes. After centrifugation the pellet was washed once with the same buffer solution. In method 1, residual RNA polymerase activity was hardly detected. But in methods 2 and 3, approximately 20–25% of the original activity remained in the preparation.

C.   Properties of Isolated Nucleolar Chromatin

Although the nucleolar templates prepared by methods 2 and 3 respond well to added RNA polymerases and show certain specific properties (57b), they contain some endogenous RNA polymerase activity and are not yet fully characterized. Therefore, the properties of nucleolar chromatin isolated by method 1 will be described here in some detail.

1.   Chemical Composition

Nucleic acids and protein composition of isolated nuclear and nucleolar chromatin are presented in Table II. It can be seen that nucleolar chromatin contained higher amounts of both RNA and protein. RNA must consist of preribosomal RNA including 45 S RNA which is known to be bound tenaciously to nucleoli. Acid-soluble proteins consist of histones and certain preribosomal proteins. The high content of nonhistone (residual) proteins is interesting but needs further elaboration to establish what kind of proteins are really increased in nucleoli (see below).

2.   Nucleic Acids and Proteins

The size of DNA in the nucleolar chromatin was examined by directly dissolving the chromatin in alkali and centrifuging in an alkaline sucrose gradient. The DNA sedimented with an average molecular weight of 1.6×10^6, which was similar to the DNA of nuclear chromatin prepared in the same way (Fig. 4). As compared with DNA in isolated nucleoli, the DNA must have suffered from two to three nicks on the average during the extensive washing. The thermal denaturation profile of nucleolar chromatin was compared with those of nuclear chromatin and nucleolar DNA. As shown in Fig. 6, although the DNA in chromatin was much more stabilized than the free DNA, there was no gross difference between chromatin preparations prepared from nucleoli and nuclei. Whether or not the small

TABLE II

CHEMICAL ANALYSIS OF NUCLEAR AND NUCLEOLAR CHROMATIN[a]

	mg/mg DTA	
	Nuclear	Nucleolar
DNA	1.00	1.00
RNA	0.11	0.45
Acid-soluble protein	1.27	2.03
Alkaline-soluble protein	0.95	1.77

[a]Nuclear and nucleolar chromatin were prepared by method 1 in the text. Acid-soluble proteins were extracted with 0.25 N HCl.

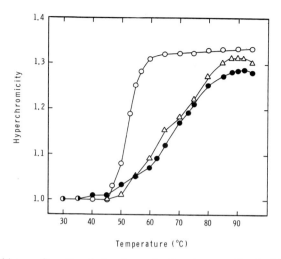

FIG. 6. Melting profiles of nucleolar chromatin. Samples were dissolved in 4 *M* urea and absorbance at 260 nm was determined at 5 °C intervals with a Gilford spectrophotometer. Hyperchromicity was normalized with respect to the absorbance at 30 °C. (O) nucleolar DNA; (●) nucleolar chromatin; (△) nuclear chromatin.

differences detected have some structural or functional significance remains to be determined.

The RNA which is always present in the nucleolar chromatin preparation appears almost ribosomal in nature. This is exemplified by the hybridization study described in the next paragraph. However, the size of the remaining RNA is invariably small, presumably due to RNase attack during preparation of chromatin.

The proteins of nucleolar chromatin have been analyzed with acid extraction and subsequent SDS solubilization followed by polyacrylamide gel electrophoresis. The data showed that although histone species were very similar, nonhistone proteins were very much different in nucleolar chromatin than in extranucleolar nuclear chromatin (*44a*). Metabolic as well as functional aspects of these proteins with special reference to regulation of nucleolar genes are now under study.

3. TEMPLATE PROPERTIES OF NUCLEOLAR CHROMATIN

While nucleolar chromatin was a good template for exogenously added RNA polymerase, it reacted differentially with different RNA polymerases. In Fig. 7 are shown template activities of nucleolar chromatin against various RNA polymerases. RNA polymerase I purified from rat liver nucleoli transcribed nucleolar chromatin approximately 50% better than RNA polymerase II. However, *E. coli* RNA polymerase (holoenzyme) could

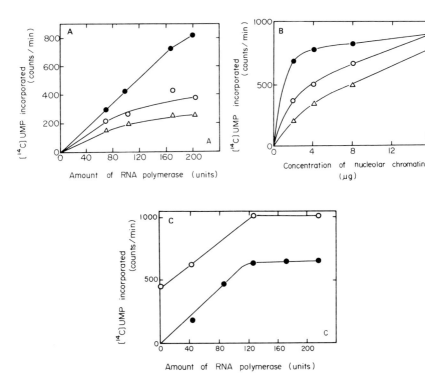

FIG. 7. Transcription of nucleolar chromatin with different RNA polymerases. Nucleolar chromatin was prepared from isolated rat liver nucleoli by the method described in the text. RNA polymerase I and II were purified from either rat liver nucleoli or nucleoplasmic fraction by our procedure (*47a*). *E. coli* RNA polymerase was purified according to Chamberlin and Berg (*60*) with minor modifications. Reaction mixture contained in a final volume of 0.15 ml: 5 μmol of Tris-HCl, pH 7.9 (at 37°C); 0.05 μmol of dithiothreitol; 0.24 μmol of MnCl$_2$; 0.1 μmol each of adenosine, guanosine, and cytidine 5'-triphosphate (ATP, GTP, and CTP); 0.1 μCi of [^{14}C]uridine 5'-triphosphate (UTP); a defined amount of nucleolar chromatin; and a defined amount of specified RNA polymerase. Ammonium sulfate concentration was adjusted to 0.05 M and 0.15 M for RNA polymerase I and RNA polymerase II, respectively. For *E. coli* RNA polymerase, 0.15 M ammonium sulfate was used. After incubation at 37°C for 10 minutes, acid-insoluble radioactivity was determined. A unit of RNA polymerase is defined as that amount causing incorporation of 1 pmol of [^{14}C]uridine 5'-monophosphate (UMP) into acid-insoluble material in 10 minutes at 37°C using native calf thymus DNA as the template. (A) Saturation of nucleolar chromatin with different RNA polymerases. Four μg of DNA equivalent nucleolar chromatin were incubated with various amounts of RNA polymerase I (O) II ($\triangle$), or *E. coli* RNA polymerase (●). (B) Saturation of a fixed amount of different RNA polymerase with nucleolar chormatin. Various concentrations of nucleolar chromatin were incubated with 100 units of RNA polymerase I (O), II ($\triangle$), or *E. coli* RNA polymerase (●). (C) Additive effects of *E. coli* RNA polymerase and RNA polymerase I on nucleolar chormatin transcription. Four μg of DNA equivalent nucleolar chromatin were incubated with various amounts of RNA polymerase I in the presence (O) and absence (●) of 100 units of *E. coli* RNA polymerase.

transcribe it much better (Fig. 7A). Although this seems rather paradoxical, it may indeed suggest that *E. coli* RNA polymerase can initiate transcription more randomly than eukaryotic RNA polymerase. That it is so may be demonstrated in a reciprocal experiment shown in Fig. 7B. Clearly, *E. coli* RNA polymerase was saturated with much smaller amount of nucleolar chromatin, suggesting more binding sites were available for this enzyme. This situation could be confirmed by another experiment in which both *E. coli* and nucleolar RNA polymerases were used on the same chromatin template. As seen in Fig. 7C, when a certain amount of *E. coli* RNA poly-

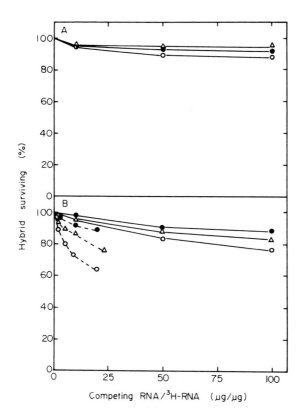

FIG. 8. Hybridization–competition of nuclear DNA and nuclear chromatin transcripts with ribosomal RNA and 45 S RNA. Millipore filters containing 0.2 μg (for nucleolar DNA transcripts) or 2 μg (for nucleolar chormatin transcripts) nucleolar DNA were incubated with 0.1 μg nucleolar DNA transcripts (A) or 0.5 μg nucleolar chromatin transcripts (B) in the presence of various amounts of ribosomal (solid lines) or nucleolar 45 S (broken lines) RNA. Hybridized counts are expressed as a percentage of those without a competitor. Products of RNA polymerase I (○); RNA polymerase II (△); and *E. coli* RNA polymerase (●).

merase was added to an increasing amount of RNA polymerase I, the total incorporation increased by exactly the amount caused by the former enzyme alone, even after a saturation was reached with respect to the latter enzyme. This strongly supports the previous idea that *E. coli* enzyme could initiate transcription at the sites where RNA polymerase I could not initiate.

In order to clarify this point further, *in vitro* products have been analyzed with respect to rRNA sequence; i.e., the *in vitro* labeled RNA was extracted with phenol and hybridized with nucleolar DNA and *competed* with cold rRNA. The results shown in Fig. 8A, indicate that when purified nucleolar DNA was used as the template, there was little difference, if any, between the products made *in vitro*, no matter what RNA polymerase was employed as the transcribing agent. In fact, slightly more competition was always found when RNA polymerase I was used to prepare labeled RNA, although the competition leveled off at around 10%. However, when nucleolar chromatin was used as the template, a distinct difference was noted as shown in Fig. 8B. The product made with RNA polymerase I was competed to the greatest extent, followed by those made with RNA polymerase II and *E. coli* polymerase. Although the maximum competition was only slightly more than 20%, this may be at least partly due to the presence of cold rRNA precursors in the nucleolar chromatin. It should be recalled that 18 S plus 28 S RNA sequences comprise only 50% or slightly more of the total 45 S RNA stretch. Accordingly the maximum competition that could be expected with rRNA is about 50% even if labeled products contained an identical sequence with 45 S RNA. Therefore, 45 S RNA was prepared from isolated nucleoli and used for competition experiments. If *in vitro* products contain nonconserved regions or transcribed spacers of 45 S RNA, then more competition could be expected with this competitor than with mature rRNA. The data in Fig. 8B show that 45 S RNA does compete with *in vitro* synthesized RNA much better than rRNA does, reaching approximately 40%—twice that of rRNA. It is also clear that the competition was most remarkable when the labeled product was made with RNA polymerase I. Competition was less with the product made with RNA polymerase II and least for that made with *E. coli* RNA polymerase with 45 S RNA.

These results strongly support the following conclusions. (1) The structure of chromatin is required for ribosomal RNA genes in the nucleolus to be transcribed in high fidelity. (2) RNA polymerase I is a better agent for the selective transcription of ribosomal RNA genes in the nucleolar chromatin than either RNA polymerase II or *E. coli* RNA polymerase.

The molecular mechanisms of this specificity and selectiveness will be the subject of future studies.

IV. Commentary

In this chapter, procedures for isolation of nucleoli from various cell types and those for purification of chromatin from isolated nucleoli were described in some detail. Stress was placed on the intactness of the nucleoli and chromatin as the RNA synthesizing organelles. Therefore, more drastic procedures utilizing DNase and high salts were not included here. The technique of isolation of nucleoli appears to have reached a point of sophistication where not much improvement can be expected in the near future. However, the method of purification of chromatin from isolated nucleoli has just started, so that improvements aiming at less endogenous RNA polymerase activity, less DNA damage, less contamination with preribosomal particles, and possibly more template specificity will emerge in future elaborations. Combined with advanced procedures for purification of eukaryotic RNA polymerases from various sources and with the improved techniques of analysis of histone and nonhistone proteins, isolated nucleoli and nucleolar chromatin will constitute a good system for the study of gene regulation in eukaryotic cells.

ACKNOWLEDGMENTS

The authors would like to thank all the colleagues who cooperated in the studies described here. The research described in this chapter was supported by Grants-in-aid from the Ministry of Education, Science and Culture.

REFERENCES

1. Busch, H., and Smetana, K., "The Nucleolus." Academic Press, New York, 1970.
2. Perry, R. P., *Ann. Rev. Biochem.* **45**, 605 (1976).
3. Warner, J. R. *in* "Ribosomes" (M. Nomura, A. Tissiéres, and P. Lengyel, eds.), p. 461. Cold Spring Harbor Lab., Cold Spring Harbor, New York, 1974.
4. Liau, M. C., and Hurlbert, R. B., *Biochemistry* **14**, 127 (1975).
5. Liau, M. C., and Hurlbert, R. B., *J. Mol. Biol.* **98**, 321 (1975).
6. Mirault, M. E., and Scherrer, K., *FEBS Lett.* **20**, 233 (1972).
7. Mattoccia, E., and Comings, D. E., *Nature (London) New Biol.* **229**, 175 (1971).
8. Prestayko, A. W., Tonato, M., Lewis, B. C., and Busch, H., *J. Biol. Chem.* **246**, 182 (1971).
9. Siebert, G., Villalobos, J., Jr., Ro, T. S., Steele, W. J., Lindenmayer, G., Adams, H., and Busch, H., *J. Biol. Chem.* **241**, 71 (1966).
10. Muramatsu, M., and Busch, H., *Methods Cancer Res.* **2**, 303 (1967).
11. Darnell, J. E., *Bacteriol. Rev.* **32**, 262 (1968).
12. Muramatsu, M., *Methods Cell Physiol.* **4**, 195 (1970).
13. Desjardins, R., Smetana, K., Steele, W. J., and Busch, H., *Cancer Res.* **23**, 1819 (1963).
14. Chauveau, J., Moulé, Y., and Rouiller, C., *Exp. Cell Res.* **11**, 317 (1956).

15. Muramatsu, M., Smetana, K., and Busch, H., *Cancer Res.* **23**, 510 (1963).
16. Fukuda, T., Akino, T., Amano, M., and Izawa, M., *Cancer Res.* **30**, 1 (1970).
17. Monty, K. J., Litt, M., Kay, E. R., and Dounce, A. L., *J. Biophys. Biochem. Cytol.* **2**, 127 (1956).
18. Maggio, R., Siekevitz, P., and Palade, G. E., *J. Cell Biol.* **18**, 293 (1963).
19. Penman, S., Smith, I., and Holtzman, E., *Science* **154**, 786 (1966).
20. Ro, T. S., Muramatsu, M., and Busch, H., *Biochem. Biophys. Res. Commun.* **14**, 149 (1964).
21. Higashinakagawa, T., Muramatsu, M., and Sugano, H., *Exp. Cell Res.* **71**, 65 (1972).
22. Blobel, G., and Potter, V. R., *Science* **154**, 1662 (1967).
23. Muramatsu, M., Higashinakagawa, T., Ono, T., and Sugano, H., *Cancer Res.* **28**, 1126 (1968).
24. Muramatsu, M., Hayashi, Y., Onishi, T., Sakai, M., Takai, K., and Kashiyama, T., *Exp. Cell Res.* **88**, 345 (1974).
25. Quagliarotti, G., Hidvegi, E., Wikman, J., and Busch, H., *J. Biol. Chem.* **245**, 1962 (1970).
26. Muramatsu, M., Hodnett, J. L., Steele, W. J., and Busch, H., *Biochim. Biophys. Acta* **123**, 116 (1966).
27. Greenberg, H., and Penman, S., *J. Mol. Biol.* **21**, 527 (1966).
28. Zimmerman, E. F., and Holler, B. W., *J. Mol. Biol.* **23**, 149 (1966).
29. Muramatsu, M., and Fujisawa, T., *Biochim. Biophys. Acta* **157**, 476 (1968).
30. Egawa, K., Choi, Y. C., and Busch, H., *J. Mol. Biol.* **56**, 565 (1971).
31. Wellauer, P. K., and Dawid, I. B., *Proc. Natl. Acad. Sci. U.S.A.* **70**, 2827 (1973).
32. Prestayko, A. W., Tonato, M., and Busch, H., *J. Mol. Biol.* **47**, 505 (1970).
33. Weinberg, R. A., and Penman, S., *Biochim. Biophys. Acta* **190**, 10 (1969).
34. Shepherd, J., and Maden, B. E. H., *Nature (London)* **236**, 211 (1972).
35. Kumar, A., and Warner, J. R., *J. Mol. Biol.* **63**, 233 (1972).
36. Kumar, A., and Subramanian, A. R., *J. Mol. Biol.* **94**, 409 (1975).
37. Tsurugi, K., Morita, T., and Ogata, K., *Eur. J. Biochem.* **29**, 585 (1972).
38. Tsurugi, K., Morita, T., and Ogata, K., *Eur. J. Biochem.* **32**, 555 (1973).
39. Higashinakagawa, T., and Muramatsu, M., *Eur. J. Biochem.* **42**, 245 (1974).
40. Prestayko, A. W., Klomp, R. G., Schmoll, D. J., and Busch, H., *Biochemistry* **13**, 1945 (1974).
41. Matsuura, S., Morimoto, T., Tashiro, Y., Higashinakagawa, T., and Muramatsu, M., *J. Cell Biol.* **63**, 629 (1967).
42. Grummt, I., *FEBS Lett.* **39**, 125 (1974).
43. Olson, M. O., Ezrailson, E. G., Guetzow, K., and Busch, H., *J. Mol. Biol.* **97**, 611 (1975).
44. Ezrailson, E. G., Olson, M. O. J., Guetzow, K. A., and Busch, H., *FEBS Lett.* **62**, 69 (1976).
44a. Tokugawa, S., Onishi, T., and Muramatsu, M., manuscript in preparation.
45. Roeder, R. G., and Rutter, W. J., *Nature (London)* **224**, 234 (1969).
46. Jacob, S. T., Sajdel, E. M., and Munro, H. N., *Biochem. Biophys. Res. Commun.* **32**, 831 (1968).
47. Muramatsu, M., Onishi, T., Matsui, T., Kawabata, C., and Tokugawa, S., *Fed. Eur. Biochem. Soc. Meet., Proc. 1974*, **33**, 325 (1975).
47a. Matsui, T., Onishi, T., and Muramatsu, M., *Eur. J. Biochem.* **71**, 351 (1976).
48. Schwartz, L. B., Sklar, V. E. F., Jaehning, J. A., Weinmann, R., and Roeder, R. G., *J. Biol. Chem.* **249**, 5889 (1974).
49. Ro, T. S., and Busch, H., *Cancer Res.* **24**, 1630 (1964).
50. Grummt, I., and Lindigkeit, R., *Eur. J. Biochem.* **36**, 244 (1973).

51. Grummt, I., Loening, U. E., and Slack, M. W., *Eur. J. Biochem.* **59**, 313 (1975).
51a. Onishi, T., and Muramatsu, M., manuscript in preparation.
51b. Onishi, T., and Muramatsu, M., *Meth. Cell. Biol.* **18**, in press.
52. Reeder, R. H., and Brown, D. D., *J. Mol. Biol.* **51**, 361 (1970).
53. Roeder, R. G., Reeder, R. H., and Brown, D. D., *Cold Spring Harbor Symp. Quant. Biol.* **35**, 727 (1970).
54. Bonner, J., Dahmus, M. E., Fambrough, D., Huang, R. C. C., Marushige, K., Tuan, D. Y. H., *Science* **159**, 47 (1968).
55. Paul, J., and Gilmour, R. S., *J. Mol. Biol.* **34**, 305 (1968).
56. Smith, K. D., Church, R. B., and McCarthy, B. J., *Biochemistry* **8**, 4271 (1969).
57. Hnilica, L. S., "The Structure and Biological Function of Histones" CRC Press, Cleveland, Ohio, 1972.
57a. Onishi, T., Matsui, T., and Muramatsu, M., in preparation.
57b. Onishi, T., Matsui, T., and Muramatsu, M., *J. Biochem.*, in press.
58. Spelsberg, T. C., and Hnilica, L. S., *Biochim. Biophys. Acta* **228**, 202 (1971).
59. Muramatsu, M., *Methods Cell Biol.* **7**, 23 (1973).
60. Chamberlin, M., and Berg, P., *Proc. Natl. Acad. Sci. U.S.A.* **48**, 81 (1962).

Chapter 14

Nucleolar Proteins

MARK O. J. OLSON AND HARRIS BUSCH

Department of Pharmacology,
Baylor College of Medicine,
Houston, Texas

I. Introduction

The nucleolus contains a variety of proteins or protein subunits (Table I), of which more than 100 have been separated (*1, 2*). Many of these proteins, such as preribosomal particle proteins, RNA polymerase I, and certain non-

TABLE I

PROTEINS OF THE NUCLEOLUS

I. Localized to the Nucleolus
   A. Enzymes
      1. RNA polymerase I (A, B)
      2. Methylases for preribosomal RNA
      3. Cleavage enzymes for ribosomal RNA (rRNA) synthesis
      4. ATPase A
      5. Nicotinamide adenine dinucleotide (NAD) synthetase (diphosphopyridine nucleotide pyrophosphorylase)
   B. Phosphoproteins
      1. Chromatin protein C18
      2. Acid-soluble protein C23
   C. Special structural proteins
      1. Proteins of preribosomal nucleolar (RNP) ribonucleoprotein particles
      2. Proteins associated with NOR (nucleolus organizer region)
      3. Proteins associated with low-molecular-weight RNA U-3
   D. Special nucleolar antigens
II. Found in other cellular sites
   A. Ribosomal proteins
   B. Histones
   C. Protein A24
   D. Nonhistone proteins common to nucleolar and extranucleolar chromatin.

histone proteins, can be considered uniquely nucleolar and are not found in other parts of the nucleus. However, other nucleolar proteins such as the histones and many nonhistone proteins are components of chromatin which are distributed throughout the nucleus in both nucleolar and extranucleolar fractions. At present, no clear-cut means of extraction is available for selective separation of the various functional classes of nucleolar proteins. Therefore, this chapter presents general methods for nucleolar protein extraction and specific methods for assay and isolation of nucleolar enzymes.

Methods for isolation of nucleoli and their products have been presented in a special monograph (3) and in previous reviews (4, 5). The techniques for the study of the chromatin components of the nucleus, which are also applicable to the nucleolar fraction of chromatin, were recently reviewed in detail (6). Although advances in methodology in the past 5 years have allowed more precise definition of the numbers of nucleolar proteins, the detailed chemistry and precise functions of these proteins remain largely unknown. It is hoped that the methods presented in this chapter will aid in the elucidation of these problems in the future.

II.   Direct Extraction of Nucleolar Proteins

The following procedures may be applied directly to isolated nucleoli prior to fractionation of nucleolar subcomponents. Many of the methods are primarily of historical interest but may be applied to obtain general classes of proteins.

A.   Sodium Chloride–Hydrochloric Acid Extraction of Proteins

Early studies by Birnstiel *et al.* (7) and Chakravorty and Busch (8) showed the similarities in amino acid composition between nucleolar and microsomal proteins. Subsequent fractionations of nucleolar proteins were performed by using differential solubility in NaCl–HCl solutions (Table II) (9, 10); these studies were reviewed by Busch and Mauritzen (5) and Busch and Smetana (3). Amino acid analysis of the various fractions showed that isolated nucleoli contain histones, acidic proteins, and ribosomelike proteins.

B.   Chloroethanol Extraction

More recently, various methods used to extract ribosomal proteins were employed in our laboratory to extract nucleolar proteins. One of these

TABLE II
PROCEDURE FOR NUCLEOLAR PROTEIN FRACTIONS[a]

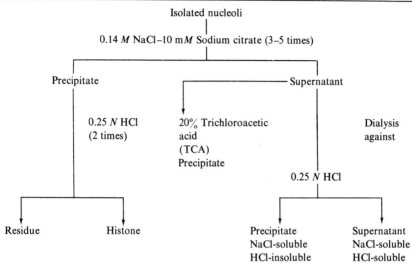

[a] From Grogan et al. (10).

methods, that used by Fogel and Sypherd (11), utilizes 2-chloroethanol. Novikoff hepatoma or rat liver nucleoli are suspended in buffer [50 mM NaCl–10 mM Tris-HCl (pH 8.0)–5 mM ethylenediaminetetraacetic acid (EDTA)–1 mM dithiothreitol]. The solution is adjusted to 0.6 N HCl and 5 volumes of 2-chloroethanol are added. The sample is kept at 0–4°C for 4–8 hours and centrifuged at 27,000 g for 15 minutes. The supernatant is dialyzed against water and lyophilized, and the protein is suspended in 5% sodium dodecyl sulfate (SDS). The sample is diluted to 1.0–0.5% SDS for electrophoresis on SDS–polyacrylamide gels according to Shapiro et al. (12).

The one-dimensional elecrophoresis patterns of chloroethanol-extracted proteins from rat liver nucleoli and isolated ribonucleoprotein particles showed that many bands in the RNP extract had mobilities similar to the bands from whole nucleoli.

C. Urea Extraction Procedure

High concentrations of urea have been shown to be effective in extraction of nonhistone chromatin proteins. Gronow and Thackrah (13) employed 8 M urea–1 mM N-ethyl maleimide–50 mM sodium phosphate (pH 7.6) to extract whole nuclei or nuclear chromatin. This extraction solubilizes approximately 70% of the total nuclear proteins without removing histones. A similar solvent containing 8 M urea–10 mM Tris–10 mM Mg acetate–1 mM EDTA–1 mM 2-mercaptoethanol was employed in this laboratory to extract a group of nucleolar proteins (3).

D. Acid Extraction

Extraction of nucleolar proteins by 0.4 N H$_2$SO$_4$ has been employed extensively in this laboratory for preparative as well as analytical studies (1, 14–18). Nucleoli are extracted for 2 hours at 4°C with 0.4 N H$_2$SO$_4$, centrifuged at 39,000 g for 20 minutes, and the protein is precipitated from the supernatant with 4 volumes of ethanol at −20°C. The proteins are pelleted at 35,000 g for 10 minutes, washed twice with cold absolute ethanol, and dried over CaCl$_2$. This procedure removes approximately 60% of the nucleolar proteins (18). The proteins are then dissolved in an appropriate buffer and applied to a system of separation. When acid-soluble proteins from Novikoff hepatoma are analyzed by two-dimensional polyacrylamide gel electrophoresis (1), approximately 100 distinct protein spots are observed (Fig. 1).

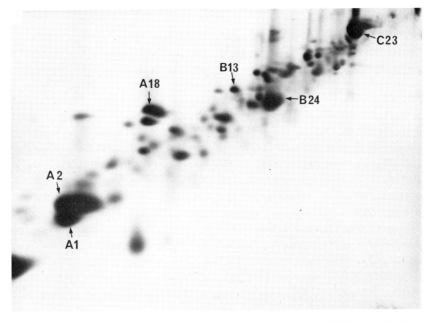

FIG. 1A.   Two-dimensional polyacrylamide gel electrophoresis of 250 μg of Novikoff hepatoma nucleolar proteins. Samples were first loaded on tube gels of 10% acrylamide–6 M urea and run in the first dimension for 5 hours at 120 V constant voltage. For the second dimension, a 12% acrylamide–0.1% SDS slab gel was run for 14 hours at 50 mA constant amperage and then stained with Coomassie blue. From Orrick et al. (1).

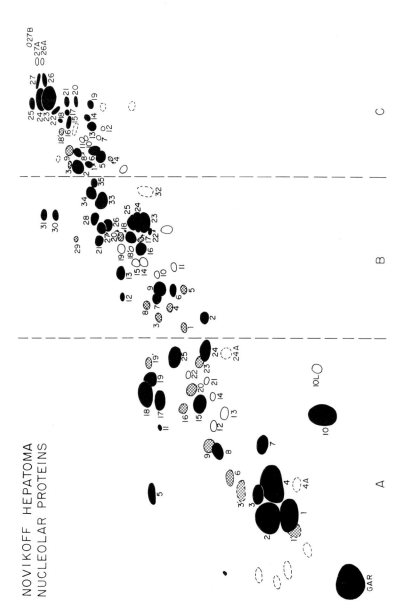

FIG. 1B.   Diagram of Fig. 1A. From Orrick *et al.* (*1*).

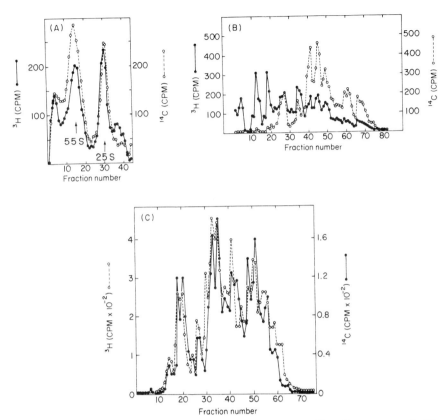

FIG. 2.   Proteins from nascent nucleolar particles. (A) 2.5 × 10⁸ cells were grown for 20 hours in Eagle's minimal essential medium (MEM) with leucine at one-half the normal concentration and [³H]leucine added to 1 μCi/ml. The cells were then collected by centrifugation and resuspended in 1/10 volume of fresh MEM containing no leucine and pulsed with 1 μCi/ml of [¹⁴C]leucine for 30 minutes. Following the pulse-label unlabeled leucine was added to 10 times the normal concentration. Three minutes later the cells were harvested, nucleoli purified, and the particles extracted in the presence of unlabeled 18 S rRNA. The resulting RNP particles were resolved in 36 ml of 15–30% sucrose gradient made in NEB buffer, [10 mM NaCl–10 mM EDTA–10 mM Tris (pH 7.4)] by centrifugation for 16 hours at 26,000 rpm on a Spinco SW-27 rotor. Fractions (25 μl) of the sucrose gradient were used for acid-insoluble radioactivity. (B). The 55 S regions of the sucrose gradient (fractions 14 and 15) shown in (A) were pooled, and the RNP particles concentrated by precipitation with 10% TCA. Proteins were solubilized in urea, SDS, and phosphate buffer and analyzed in a 20 cm, 10% SDS–polyacrylamide gel as described earlier (19, 21). Crushed gel samples were counted in a Triton X-100–toluene scintillator (19). (C) Nucleolar 55 S with cytoplasmic 50 S proteins. 1.75 × 10⁸ cells were concentrated 10-fold in fresh MEM containing no leucine, and labeled with 1 μCi/ml [¹⁴C]leucine for 30 minutes followed by a 2-minute chase with excess unlabeled leucine. Pulse-labeled proteins from the 55 S nucleolar ribosomal precursor particles were prepared and mixed with 20-hour [³H]leucine-labeled cytoplasmic 50 S proteins and analyzed in SDS–acrylamide as above. From Kumar and Warner (20). Reproduced with permission.

E.   SDS Extraction

Extraction of nucleolar proteins with the detergent SDS (*19, 20*) has proved to be more efficient than the previous method of Birnstiel *et al.* (*7*) using sodium deoxycholate (DOC). Nucleolar samples are suspended in 10 mM sodium phosphate buffer (ph 7.0)–1% SDS–1% β-mercaptoethanol–0.5 M urea. The sample is immersed in boiling water for 90 seconds and then dialyzed for 2 hours against 0.5% SDS–10 mM sodium phosphate buffer (pH 7.4)–0.5 M urea–0.1% β-mercaptoethanol. Electrophoresis on 10% polyacrylamide gels is performed according to the procedure of Maizel (*21*). Figure 2 shows the gel electrophoresis pattern of HeLa cell nucleolar proteins uniformly labeled with [^{3}H]leucine and pulse-labeled with [^{14}C]-leucine. The proteins that are rapidly labeled correspond to cytoplasmic ribosomal proteins, while the slower moving proteins which are labeled only after long periods appear to be localized in the nucleolus and have been designated nucleolar "stable" proteins (*20*).

III.   Proteins of Nucleolar Subcomponents

For functional studies it may be advantageous to fractionate the nucleolus into "native" subcomponents prior to protein isolation. Several procedures are available for isolation of substructures such as nucleolar chromatin, preribosomal RNP particles, and fibrillar elements.

A.   Nucleolar Chromatin

Chromatin, a term originally used by Flemming (*22*) to describe the basic staining material of the nucleus, is now considered to be the interphase state of chromosomes. A precise definition of chromatin is elusive; however, it is generally regarded as the nuclear DNA and its associated histones, non-histone proteins, RNA, and small molecules. Isolated chromatin is operationally defined by the procedure used to obtain it (*6*). Although the nucleolar component of chromatin is a relatively small fraction of total nuclear chromatin, it would be expected to fit into the above general definition. Two representative methods of chromatin preparation are detailed below.

1. ISOLATION OF NUCLEOLAR CHROMATIN

Recent studies in this laboratory compared the nonhistone chromatin proteins from whole nuclei, nucleoli, and the extranucleolar fraction (*2*). Chromatin was prepared by the saline–EDTA and dilute Tris washing method

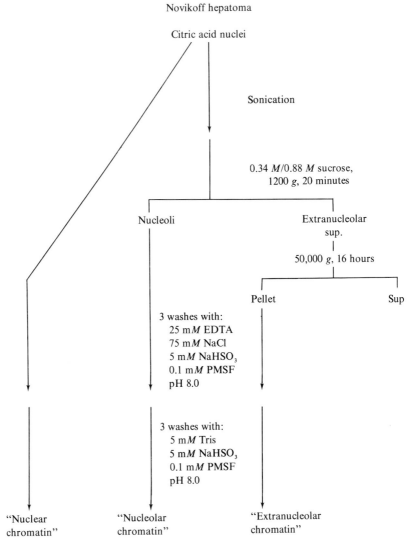

FIG. 3.   Flow diagram for chromatin fractions.

(Fig. 3) similar to the procedure of Marushige and Bonner (23). Nucleoli are isolated by the citric acid method (24). The nucleoli are then washed 3 times with 10 volumes of 25 mM EDTA–75 mM NaCl–5 mM NaHSO$_3$–10 mM KF–0.1 mM phenylmethylsulfonyl fluoride (PMSF) (pH 8.0) by 10 strokes in a tight-fitting Teflon–glass homogenizer. After each wash, the chromatin is pelleted by centrifugation for 10 minutes at 12,000 g. This is followed by three washes in the same manner with 10 volumes of 5 mM Tris-HCl–5 mM KF (pH 8.0). For each of the Tris washes, the chromatin is pelleted by centrifugation at 27,000 g for 10 minutes. The final pellet is defined as nucleolar chromatin. For comparative studies, whole nuclear and extra-nucleolar chromatin are prepared in a similar manner (Fig. 3).

An earlier study comparing nucleolar and extranucleolar chromatin proteins employed pre-extraction with 0.3 M NaCl (25). Nuclei are prepared from normal rat livers by the method of Chauveau et al. (26). After isolation of nucleoli by the sonication procedure, chromatin is prepared by a modification of the method described by Spelsberg and Hnilica (27). Nucleoli are homogenized in 50 volumes of 80 mM NaCl–20 mM EDTA (pH 6.3) and then centrifuged at 5000 g for 10 minutes. This homogenization and centrifugation are repeated twice. The pellet is then homogenized twice as above in 5 volumes of 1.5 mM NaCl–0.15 mM sodium citrate (pH 7.0). In the two final washes, chromatin is sedimented by centrifugation at 20,000 g for 10 minutes. The chromatin is then extracted with 0.3 M NaCl, as suggested by Johns and Forrester (28), to remove nonspecifically adsorbed cytoplasmic "contaminants."

2. ISOLATION OF NUCLEOLAR CHROMATIN PROTEINS

In the past decade, numerous methods of chromatin dissociation have been developed (6). Two representative methods of nucleolar chromatin protein isolation primarily for analytical purposes are described below. A third method which may be applied to preparative procedures is also described.

a. Nuclease digestion after acid extraction. A procedure currently in use in this laboratory (2) is based on the method of Wilson and Spelsberg (29). Chromatin is extracted twice with 10 volumes of 0.4 N H$_2$SO$_4$ at 4° by stirring for 4 hours. This extract, referred to as "chromatin fraction I," contains the histones and other acid-soluble proteins (30). After acid extraction, the pellet is homogenized in 2 mM CaCl$_2$–2 mM MgCl$_2$–0.1 M Tris-HCl (pH 7.5) at a concentration of 2 mg/ml. The DNA is digested by addition of 25 μg of deoxyribonuclease I per milligram of pellet at 37° for 30 minutes. The proteins are precipitated by the addition of perchloric acid to a concentration of 0.4 N. For electrophoretic analysis the protein pellet is dissolved in 0.9 N acetic acid–10 M urea–1% β-mercaptoethanol and dialyzed against two changes of 500 volumes of the same solution. The proteins soluble after this treatment are designated "chromatin fraction II" (30). When this fraction is separated on two-dimensional polyacrylamide gel electrophoresis, at least 50 protein spots are seen (Figs. 4 and 5). One of these, C18, is highly phosphorylated (Fig. 6).

b. Dissociation by high-molarity salt and urea, then SDS. An alternative

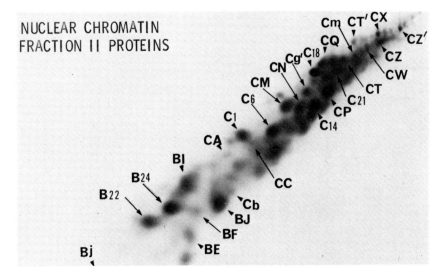

FIG. 4.   Two-dimensional polyacrylamide gel electrophoretic patterns of 250 μg of nuclear chromatin fraction II proteins. Samples were run in the first dimension on disc gels of 6% polyacrylamide–6 *M* urea–0.9 *N* acetic acid at 120 V for 6 hours. For the second dimension, an 8% polyacrylamide–0.1% SDS slab gel was run for 14 hours at 50 mA/slab. Gels were stained with Coomassie brilliant blue R.

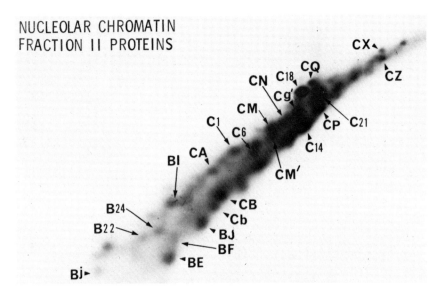

FIG. 5.   Nucleolar chromatin fraction II proteins.

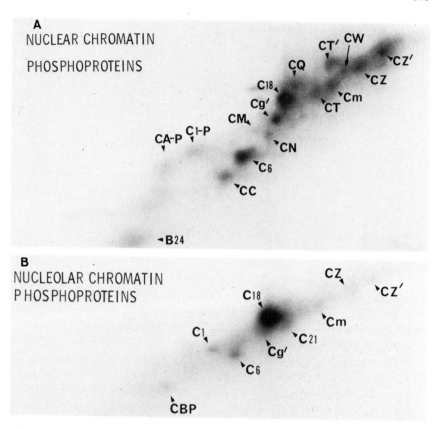

FIG. 6. Autoradiograms on X-omat X-ray film of ^{32}P-labeled chromatin fraction II proteins subjected to two-dimensional electrophoresis as described in the legend to Fig. 4. Spots on autoradiograms were matched with stained spots and numbered as in Fig. 4. Numbers followed by P indicate radioactive spots that do not comigrate with stained spots. (A) Whole nuclear chromatin fraction II proteins. (B) Nucleolar chromatin fraction II proteins.

method of nucleolar chromatin protein isolation is that of Wilhelm *et al.* (*25*). This procedure is primarily for analytical studies since the final dissociating agent is SDS.

The nucleolar chromatin is gently homogenized in 5 M urea–2 M NaCl–10 mM sodium acetate (pH 6.0). Particulate material is removed by centrifugation at 15,000 g for 10 minutes. After adjusting the concentration to 200–300 µg DNA/ml, the dissociated chromatin is centrifuged for 3.5 × 10^6 g hours in a Spinco 50.1 rotor. The supernatant which contains all of the histones and about 15% of the nonhistone proteins is dialyzed against 20 mM HCl, then deionized water, and appropriately concentrated for analytical studies. The pellet which contains the DNA–nonhistone protein complex is dissolved in 1% SDS–1% β-mercaptoethanol–50 mM Tris-HCl (pH 8.0). After adjusting to a concentration of 200–300 μg DNA/ml, the solution is centrifuged at 4.5 × 10⁶ g × hour. The upper ⅔ of the supernatant of each tube is removed and dialyzed against 5 volumes of deionized water and lyophilized. The samples from both fractions may then be dissolved in an appropriate buffer and subjected to polyacrylamide gel electrophoresis in the presence of SDS.

B. Proteins of Nucleolar Ribonucleoprotein Particles

1. ISOLATION OF NUCLEOLAR RNP PARTICLES

a. EDTA–dithiothreitol procedure. This procedure was introduced by Warner and Soeiro (*31*) to isolate RNP particles from HeLa cell nucleoli. They demonstrated by analysis of RNA and protein that these nucleolar particles were precursors of ribosomes.

The procedure involves suspension of nucleoli prepared according to Penman (*32*) in NEB buffer ⌊10 mM NaCl–10 mM EDTA–10 mM Tris (pH 7.4)⌋ containing 10 mM dithiothreitol. The suspension is pipetted vigorously at 0 °C until free of visible particles and gently stirred at room temperature for 15 minutes. After centrifugation at 12,000 g for 10 minutes, the supernatant is layered on 15–30% sucrose gradients prepared in NEB buffer, and the nucleolar RNP particles are separated by centrifugation for 16 hours at 22,000 rpm in a Spinco SW-27 rotor.

An initial mild DNAse treatment of the nucleoli is an optional step which is used to increase yield of the RNP particles (*33*). Nucleoli are suspended at a concentration of approximately 1 mg/ml wet weight in a buffer containing 10 mM NaCl–5 mM MgCl₂–10 mM Tris-HCl (ph 7.4). Deoxyribonuclease I is added to a concentration of 10 μg/mg of nucleoli. After slowly stirring the mixture for 15 minutes at 0°, it is centrifuged at 20,000 g for 10 minutes. The pellet is then suspended in the NEB buffer as above and carried through the isolation procedure.

Further characterization of the nucleolar RNP particles was performed by Kumar and Warner (*20*), who used this procedure. They showed that the buoyant densities of the nucleolar 80 and 55 S particles and the cytoplasmic 50 S ribosomal subunits are 1.51 gm/cm³, 1.55 gm/cm³, and 1.57 gm/cm³, respectively. These lower buoyant densities suggest that there is an excess of proteins associated with the nucleolar particles which have a higher protein/ RNA ratio than the ribosomes.

b. Polyvinyl sulfate–Mg²⁺–dithiothreitol procedure. This procedure as developed by Liau and Perry (*34*) involves suspension of L-cell nucleoli in a buffer containing 0.25 M sucrose–2 mM MgCl₂–50 mM KCl–10 mM sodium acetate (pH 6.0). The suspension is adjusted to a concentration of 3.3 A_{260} units/ml and made 40 μg/ml in polyvinyl sulfate (PVS). After remaining in an ice bath for 10 minutes with occasional pipetting for mixing, the sample is sedimented at 20,000 g for 10 minutes. The gelatinous pellet is dispersed in a Dounce homogenizer in a buffer containing 10 mM Tris-HCl (pH 7.4)–10 mM KCL–0.5 mM $MgCl_2$–20 mM dithiothreitol, and kept at room temperature for 15 minutes. After centrifugation at 20,000 g for 15 minutes, the RNP particles in the supernatant are fractionated on 15–30% sucrose gradients containing 0.5 mM $MgCl_2$ –10 mM KCl–10 mM dithiothreitol–10 mM triethanolamine (pH 7.4). As in the previous procedure, the DNase treatment may be used as an initial step (33).

The buoyant density of these particles is as low as 1.46–1.49 gm/cm^3. These studies suggested that some of the nucleolar RNP particles are very large, containing a high protein/RNA ratio. Studies from our laboratory (35) and that of Shinozawa et al. (36) have suggested that this procedure, which utilizes PVS, may produce aggregates of nucleolar RNP particles and perhaps nonspecific adsorption of proteins. However, by morphological criteria the RNP particles prepared by the PVS method are more satisfactory than those prepared by the EDTA procedure (37).

   c. *Heparin–Brij 35 procedure.* A method for the isolation of nucleoli and nucleolar RNP particles that avoids EDTA and high-salt buffers was recently developed by Mirault and Scherrer (38). HeLa cell nuclei are prepared using a Tween–DOC mixture as described by Penman (32) with the following modification of the method: final concentration 1% Tween–0.33% DOC; stirring 30 seconds on a Vortex-type mixer. Nucleoli are prepared from these nuclei by the heparin method (38). To extract the preribosomal particles, the nucleolar pellet is resuspended at 0°C with a small Dounce homogenizer in a buffer containing 10 mM triethanolamine (pH 7.4)–10 mM KCl–0.1 mM $MgCl_2$–10 mM dithiothreitol–0.1% Brij 35. After gentle homogenization at 25°C for 10 minutes, the nucleolar suspension is centrifuged at 20,000 g for 20 minutes. The supernatant is used for further purification of RNP particles by sucrose density gradient centrifugation and polyacrylamide gel electrophoresis.

The buoyant densities of the two main classes of preribosomal particles are 1.52 and 1.54 gm/cm^3; these contained 45 S and 32 S RNA, respectively. These data are compatible with results from Koshiba et al. (39) which indicate that the RNP particles with a large sedimentation coefficient had a lower protein/RNA content than the slower-sedimenting particles. The buoyant densities of the preribosomal particles obtained by the heparin–Brij procedure are quite similar to those obtained by Warner and Soeiro (31) using the EDTA–dithiothreitol procedure.

2. EXTRACTION OF PROTEINS FROM RNP PARTICLES

Proteins may be extracted from the preribosomal particles by several techniques. An effective method is extraction by 0.4 N H_2SO_4 (Section II,D) which has been applied to preribosomal particles of Novikoff hepatoma nucleoli (*15*). The chloroethanol procedure (Section II,B) has also been employed. For analytical studies, proteins may be extracted directly with SDS (*31*).

Acetic acid extraction. A widely used solvent for extraction of ribosomal proteins as well as preribosomal RNP proteins is 66% acetic acid. The method based on that of Tsurugi *et al.* (*40*) has been applied to nucleolar preribosomal proteins in our laboratory (*33*). Nucleolar particles are extracted with 40 volumes of 66% acetic acid–0.1 M magnesium acetate–5 mM dithiothreitol for 8 hours at 4° with continuous stirring. The insoluble RNA precipitate is removed by centrifugation at 27,000 g for 15 minutes. The extraction is repeated and the supernatants are combined and precipitated overnight with 4 volumes of 100% ethanol at −30°C. The precipitate is collected by centrifugation, washed twice with 100% ethanol, air-dried, and stored as a dry powder. As an alternative to ethanol precipitation, the sample may be dialyzed against an appropriate buffer and concentrated for further studies.

3. CHARACTERISTICS OF NUCLEOLAR RIBONUCLEOPROTEIN PARTICLES (GRANULAR COMPONENTS)

Early studies by Warner and Soeiro (*31*) showed the similarity of many proteins of the nucleolar 55 S RNP particles isolated by the EDTA procedure and of proteins of the cytoplasmic 50 S ribosomal subunit. Subsequently, Kumar and Warner (*20*) and Soeiro and Basile (*40*) showed that the nucleolar 55 S particle contained "nonribosomal" proteins as well as ribosomal proteins. The ribosomal proteins were rapidly labeled whereas the nonribosomal proteins were slowly labeled and were therefore designated nucleolar "stable" proteins (Fig. 2).

A recent study in this laboratory employed two-dimensional electrophoresis to show that proteins from both ribosomal subunits are present in 80 S ribosomal precursor particles (*33*). Electron micrographs indicate that the isolated particles are 200–250 Å in diameter (Fig. 7). Of the 60 protein spots seen in the nucleolar RNP particles, 21 had mobilities identical with large ribosomal subunit spots and 10 were identical in mobility with small subunit proteins (Figs. 8–10; Table III). The nonribosomal proteins tended to be higher in molecular weight than the ribosomal proteins.

In other studies, 60 S nucleolar particles which apparently are precursors to the large subunits have been isolated. Tsurugi *et al.* (*39*) found all but three of the major large subunit proteins and at least four additional major pro-

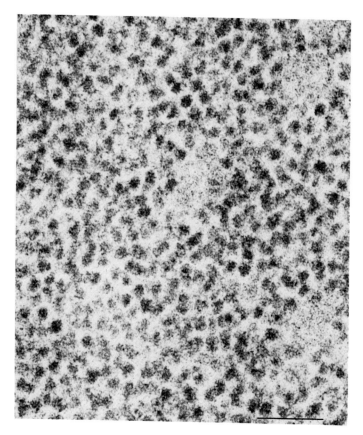

Fɪɢ. 7. Electron micrograph of nucleolar 78 S RNP particles from Novikoff hepatoma. Samples were fixed in 1% glutaraldehyde, postfixed in 2% osmium tetroxide, and stained with uranyl acetate and lead citrate. Note the RNP particles containing electron-dense filamentous substructures. × 152, 500. This micrograph was provided by Dr. K. Koshiba.

teins in the 60 S precursor particle. In a similar study, Higashinakagawa and Muramatsu (42) concluded that the 60 S particles discard at least two proteins and pick up at least 10 proteins to become large ribosomal subunits.

The functions of the nonribosomal or "stable" nucleolar RNP proteins are presently unknown but it has been suggested that they are involved in maturation and processing of the particles (43). A diagrammatic representation of the process is presented in Fig. 11. One of the nonribosomal proteins may be an endoribonuclease associated with the nucleolar RNP particles (35). Another nonribosomal protein of the preribosomal particles designated as protein C23 in highly phosphorylated (15).

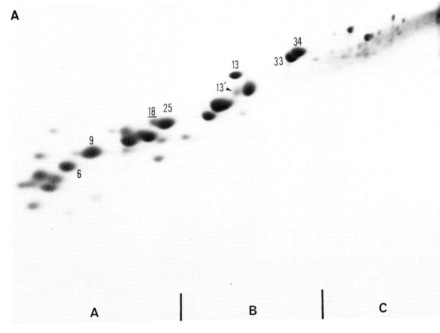

FIG. 8A.   Two-dimensional gel electrophoresis of 250 μg of Novikoff hepatoma ribosomal large (60 S) subunit proteins. Samples were run in the first dimension on disc gels of 10% acrylamide–6 M urea–0.9 N acetic acid (pH 2.5) at 120 V for 5.0 hours. For the second dimension, a 12% acrylamide–0.1% SDS slab gel was run for 14 hours at 50 mA/slab. Gels were stained with Coomassie brilliant blue R.

C.   Proteins of the Nucleolar Fibrillar Elements

A number of studies indicate that precursor rRNA is synthesized in the fibrillar portion of the nucleolus from which the RNA moves to the granular

TABLE III

PROTEINS FROM NUCLEOLAR PRERIBOSOMAL PARTICLES[a]
WHICH ARE PRESENT IN RIBOSOMAL SUBUNITS

Region	Small ribosomal subunit (40 S)	Large ribosomal subunit (60 S)
A	4, 7, 12, 13, 15, 21, 22	2, 3, 4, 6, 8, 15, 16, 20, 23, 24A, 25
B	1′, 3, 6	2, 4, 7, 9, 11, 13, 15, 33, 34
C		23

[a]These comparisons were made with preribosomal particles prepared by the EDTA method.

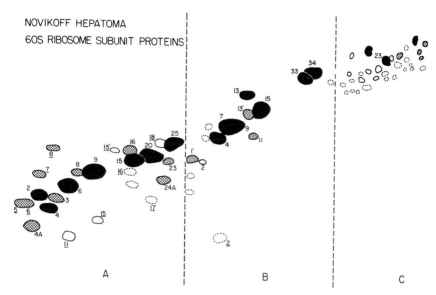

FIG. 8B.   Diagrammatic representation of the electrophoretic pattern of Fig. 8A. The most dense spots are black, the less dense spots are cross-hatched, the even less dense spots are open circles, and minor or faint spots are broken circles. The gel was arbitrarily divided into A, B, and C regions (1) on the basis of mobility of spots A24 and B34. The ribosomal spots with underlined numbers were not found in nucleoli. The spots with prime numbers were found in nucleolar preribosomal proteins but were not present in sufficient concentration to be detected in whole nucleolar acid extracts. From Prestayko *et al.* (*33*). Reproduced with permission.

portion on the RNP particles (*3, 44–46*). Isolation of fibrillar elements was recently achieved in our laboratory (*37*). The kinetics of isotope incorporation indicated that the ^{32}P accumulated in the granular elements subsequent to the fibrillar fraction, thereby confirming the above *in situ* studies.

1. ISOLATION OF FIBRILLAR COMPONENTS AND ASSOCIATED PROTEINS

The polyvinyl sulfate–Mg^{2+} method was found to be superior and is therefore described in preference to the EDTA procedure. Nucleoli are suspended at a concentration of 1 mg/ml (wet weight) in a buffer containing 25 mM sucrose–2 mM MgCl$_2$–50 mM KCL–10 mM sodium acetate (pH 6.0). The mixture is centrifuged at 20,000 *g* for 10 minutes, and the pellet is resuspended in the above buffer to which is added 40 µg/ml polyvinyl sulfate. After standing at 0° for 10 minutes, the mixture is centrifuged at 20,000 *g* for 10 minutes. The pellet is resuspended in 10 mM KCl–0.5 mM MgCl$_2$–10 mM Tris-HCl–20 mM dithiothreitol (pH 7.4) at a concentration of 3 mg/ml

A

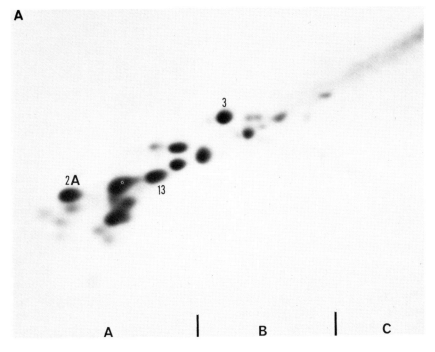

Fig. 9A. Two-dimensional polyacrylamide gel electrophoresis of 250 μg of Novikoff hepatoma ribosomal small (40 S) subunit proteins. See Fig. 8A for conditions.

of the starting nucleolar preparation. The mixture is centrifuged at 20,000 g for 15 minutes. Aliquots (2–3 ml) of the supernatant are layered onto 15–55% sucrose gradients (30 ml) containing 0.5 mM MgCl$_2$–10 mM KCl–1 mM dithiothreitol–10 mM triethanolamine HCl (pH 7.4) and centrifuged for 16 hours at 85,000 g in a Spinco SW-27 rotor. The gradients are monitored at 254 nm and divided into 1-ml fractions. A peak of approximately 12 S contains the fibrillar component whereas the granular component sediments at 80 S (Fig. 12). The fractions from the fibrillar peak are pooled and rerun on an identical gradient for further purification. An electron micrograph of this isolated fibrillar component is shown in Fig. 13.

For isolation of proteins, the fibrillar fraction is first precipitated with trichloroacetic acid. The proteins are extracted with 66% acetic acid as described in Section III, B, 2.

2. CHARACTERISTICS OF PROTEINS OF ISOLATED NUCLEOLAR FIBRILLAR ELEMENTS

Two-dimensional gel electrophoresis of the proteins associated with the isolated fibrillar component of nucleoli revealed the presence of 11 major

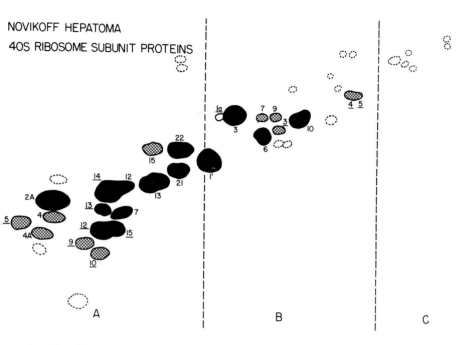

NOVIKOFF HEPATOMA

40S RIBOSOME SUBUNIT PROTEINS

FIG. 9B. Diagrammatic representation of the electrophoretic pattern of Fig. 9A. See Fig. 8B for legend. From Prestayko *et al.* (*33*). Reproduced with permission.

and 18 minor protein spots (Fig. 14). Comparison with proteins (Table IV) of the granular (RNP) component and with ribosomal proteins indicated that 6 proteins were unique to the fibrillar fraction, 5 more were also found in the granular component as nonribosomal proteins, and an additional 13 were ribosomal proteins found in both granular and fibrillar components. These data suggest that certain ribosomal and nonribosomal proteins associate with the preribosomal RNA almost immediately after synthesis.

IV. Analyses and Fractionation of Nucleolar Proteins

A. Two-Dimensional Polyacrylamide Gel Electrophoresis

For analyses of the many protein components of the nucleolus, one-dimensional polyacrylamide gel electrophoresis is inadequate. Therefore, two-dimensional systems are currently used for protein separation.

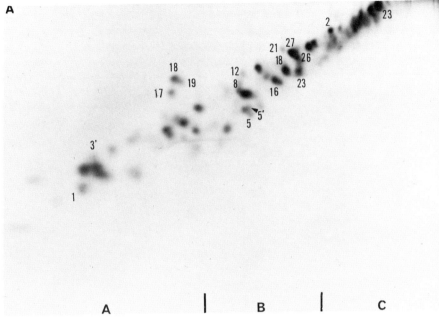

FIG. 10A. Two-dimensional polyacrylamide gel electrophoresis of 250 μg of proteins from Novikoff hepatoma nucleolar preribosomal (nucleolar RNP) particles prepared by the EDTA method. See Fig. 8A for conditions.

A two-dimensional system was developed by Kaltschmidt and Wittmann (47) and was subsequently used by several groups (48–52) for separation of ribosomal proteins. The system has also been applied to the study of nucleolar 60 S preribosomal RNP particle proteins (39).

The principle of the method involves placing the proteins in the middle of an acrylamide gel (8% acrylamide, pH 9.6) polymerized in a glass tube (180 × 5 mm). In this first dimension the proteins migrate certain distances toward the anode or cathode according to the strength of their charge. After the first run the gel is removed from the tube, dialyzed for a period of 3 hours against the starting buffer of the next run, and used as horizontal starting gel for the second dimension (gel slab, 200 × 200 mm; polymerized with 18% acrylamide, pH 4.6). After the run in the second dimension is terminated, the gel slabs are taken off the chamber and the proteins are made visible by means of Amido black staining.

Because of the low solubility of whole nucleolar proteins in nondenaturing solvents, a two-dimensional polyacrylamide gel system for separation of nucleolar proteins was developed in our laboratory (1). For whole nucleolar acid-soluble proteins the samples are dissolved in 0.9 N acetic acid–10

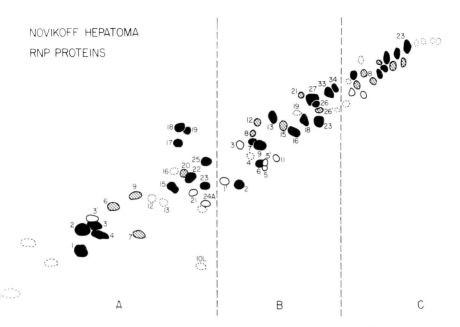

FIG. 10B. Diagrammatic representation of the electrophoretic pattern of Fig. 10A. See Fig. 8B for legend. From Prestayko *et al.* (*33*). Reproduced with permission.

M urea–1% β-mercaptoethanol at a concentration of 10 mg/ml. Samples containing 500 μg of protein are applied to the first dimension (9.5 × 0.6 cm) gels containing 10% acrylamide–0.035% bisacrylamide–0.9 N acetic acid–4.5 M urea. The gels are run at 120 V for 5 hours using 0.9 N acetic acid–4.5 M urea as a tank buffer. After extrusion, gels are bisected longitudinally and adapted for 30 minutes at 45° in each of the following series of solutions: (1) 2% SDS–0.1 M sodium phosphate–6 M urea–1% β-mercaptoethanol (pH 7.1); (2) 1% SDS–10 mM sodium phosphate–6 M urea–1% β-mercapto-ethanol (pH 7.1); (3) 0.1% SDS–10mM sodium phosphate–6 M urea–1% β-mercaptoethanol (pH 7.1).

The second-dimensional gels (100 × 95 × 3 mm) formed in an Ortec apparatus consist of 12% acrylamide–0.31% bisacrylamide–0.1% SDS–6 M urea–0.1 M phosphate buffer (pH 7.1) with 0.05% N, N, N', N'-Tetramethylethyle-nediamine (TEMED)–0.075% ammonium persulfate to facilitate polymeri-zation.

The adapted portion of the bisected first-dimensional gel is placed on top of the second-dimensional slab and polymerized into a gel cap identical to

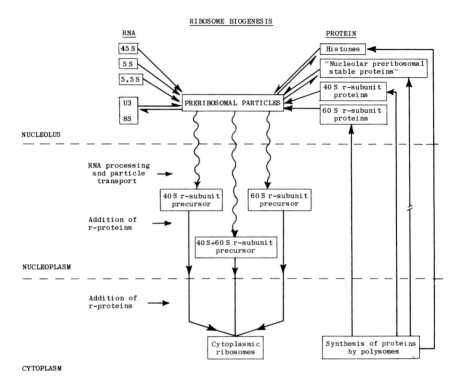

FIG. 11.   Schematic representation of events involved in the synthesis and transport of ribosomes in eukaryotic cells. From Prestayko *et al.* (*33*). Reproduced with permission.

the slab gel except that the phosphate concentration is 10 mM. The second dimension is then run for 16 hours at 50 mA per slab with upper and lower electrolytes consisting of 0.1 M phosphate–0.1% SDS (pH 7.1).

After removal from the cells the gels are stained with Coomassie brilliant blue R (0.25%) in an aqueous solution of 4.5% acetic acid–22.7% methanol. After staining for 5–6 hours with occasional agitation, the gels are rinsed in deionized water and destained in 10% acetic acid–5% methanol.

This method applied to nucleolar acid-soluble proteins is capable of resolving approximately 100 polypeptide spots (Fig. 1). For analyses of non-histone chromatin proteins the system is modified to use 6% and 8% acrylamide gels in the first and second dimensions, respectively (*53*). Figure 5 shows that nucleolar nonhistone chromatin proteins are resolved into approximately 50 spots by use of this technique (*2*).

Recently much greater resolution of nonhistone chromatin proteins was achieved by employing isoelectric focusing in the first dimension and SDS gel electrophoresis in the second dimention (*54, 55*). More than 450 compo-

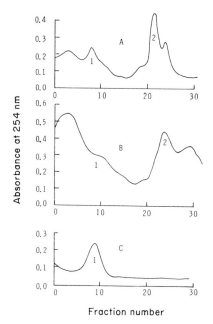

Fraction number

Fig. 12. Sucrose gradient analysis of nucleolar extracts obtained by the PVS–Mg^{2+} isolation procedure (A) and the EDTA procedure (B). Extracts obtained by either method were layered on 15–55% sucrose (A) and 10–40% sucrose (B) gradients containing the appropriate buffers and centrifuged for 16 hours at 25,000 rpm in the Spinco SW-27 rotor at 4°C. Peak 1 and peak 2 were used for the electron microscopic study. The peak sedimenting faster than peak 2 was shown earlier to also contain granular elements. (C) Purification of peak 1 from (A). From Daskal et al. (37). Reproduced with permission.

nents were detected in the HeLa cell nonhistone fraction. This method may also be generally applicable to nucleolar chromatin proteins.

B. Fractionation and Purification of Nucleolar Proteins

Aside from certain nucleolar enzymes (see Section V), very few nucleolar proteins have been purified to homogeneity. The methods of fractionation which have been applied to chromatin proteins (6) are generally applicable to nucleolar proteins. Some selected methods are described below.

1. PREPARATIVE POLYACRYLAMIDE GEL ELECTROPHORESIS

For isolation of small quantities of nucleolar proteins for chemical studies, preparative gel electrophoresis has been found to be extremely useful in our laboratory. A nucleolus-specific protein, band 15, was first purified to homogeneity (Fig. 15; Table V) using this technique (14).

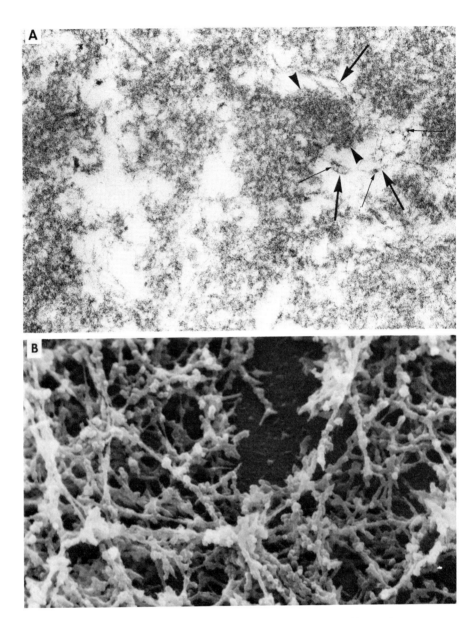

Fig. 13. (A) Electron micrograph of structures of peak 1. "Long filaments" are seen to interconnect fibrous clusters (large arrows). Electron-dense "chromatinlike tufts" are in close association with the fibrous clusters (pointer). Granules of variable size are seen connected to and dispersed in the fibrillar mass and along the interconnecting filaments (small arrows). ×137,500. These figures were kindly supplied by Dr. I. Daskal. (B) Scanning photomicrograph of nucleolar fibrillar fraction obtained as described above.

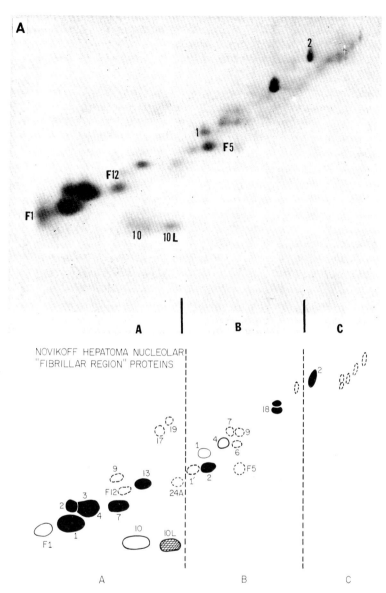

FIG. 14. (A) Two-dimensional polyacrylamide gel electrophoresis of Novikoff hepatoma nucleolar "fibrillar region" proteins extracted from the isolated fibrillar region (peak 1). Samples were run in the first dimension on disc gels of 10% acrylamide–6 M urea–0.9 N acetic acid at 120 V for 5–6 hours. For the second-dimension electrophoresis, a 12% acrylamide–0.1% SDS slab gel was run for 14 hours at 50 mA/slab. Gels were stained with Coomassie brilliant blue R. From Daskal et al. (37). Reproduced with permission. (B) Diagrammatic representation of the electrophoretic pattern of (A). The most dense spots are black, the less dense spots are cross-hatched, even less dense spots are open circles, and minor or faint spots are broken-line circles. From Daskal et al. (37). Reproduced with permission.

TABLE IV

Proteins of the Fibrillar and Granular Components of the Nucleolus[a]

Region	Proteins unique to the fibrillar fraction	Proteins common to both granular and fibrillar fractions	
		Nonribosomal proteins	Ribosomal proteins[b]
A	F1, 10, 10L, F12	1, 17, 19	2, 3, 4, 7, 9, 13, 24A
B	1, F5	18	1', 2, 4, 6, 7, 9
C	—	2	—

[a] From Prestayko et al. (33).
[b] These nucleolar proteins were previously shown to be present in cytoplasmic ribosomes (33).

Recently the technique was modified to include preparative electrophoresis in two dimensions. Nucleolar extracts are dissolved in sample buffer ($0.9 N$ acetic acid–10 M urea–1% β-mercaptoethanol) at a concentration of 10 mg/ml. The samples are applied to slabs ($9 \times 10 \times 0.3$ cm) containing 4.5% acrylamide–0.9 N acetic acid–4.5 M urea. After electrophoresis for 5 hours at 20 V (constant voltage) per slab, the slab gels are removed, stained with Buffalo black, and destained. Strips containing the desired bands are cut from the gels and adapted for the second dimension (see above, Section IV,A). The adapted strips are annealed to the second dimension which in this case contains 8% acrylamide–4.5 M urea–0.1% SDS–0.1 M sodium phosphate (pH 7.1). This dimension is run overnight at 50 mA/slab, constant current. The desired bands are localized by staining side strips with Coomassie blue as above (Section IV,A).

The proteins are eluted from the excised gel strips by electrophoresis into dialysis bags. This is achieved by first cutting the excised strip into small pieces and placing the pieces into a 10-ml disposable pipette closed at the bottom end with a 1-cm 5% acrylamide gel plug. The bottom of the pipette is inserted into a section of dialysis tubing which is secured by a rubber band. The tubing is knotted in the other end. The pipette with its attached dialysis tubing is placed in an electrophoresis apparatus with upper and lower buffers identical to the buffer of the last dimension. To minimize protein losses, the dialysis tubing should be arranged so that both ends are above the buffer level. In the case of acid–urea gels, the proteins are eluted at 120 V (constant voltage) for 24 hours. For the SDS dimensions, electrophoresis is continued at 10 mA/per tube (constant current) for 24–48 hours.

The two-dimensional system has been used in purification of several nucleolar proteins. The amino acid compositions of four nucleolar proteins are

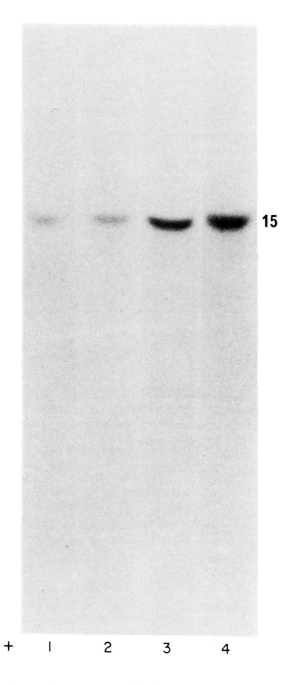

FIG. 15.   Electrophoretic patterns on analytical gels of four different concentrations of the protein of band 15. Slot 1, 20 μg; 2, 40 μg; 3, 60 μg; 4, 80 μg. From Knecht and Busch. (*14*). Reproduced with permission.

TABLE V

AMINO ACID ANALYSIS OF THE PURIFIED NUCLEOLAR PROTEINS

Amino acid	Mole %				
	Whole nucleolar acid-soluble proteins	Band 15[a]	B23[b]	C23[b]	C18[b]
Ala	7.0	9.8	6.5	9.3	10.3
Arg	6.0	4.3	4.2	2.7	5.0
Asx	10.0	11.2	10.0	11.5	9.2
Glu	13.8	15.9	13.2	17.7	13.6
Gly	7.6	10.6	15.1	12.9	13.1
His	2.4	1.0	2.0	0.3	1.7
Ileu	4.1	3.3	4.4	1.8	3.4
Leu	8.9	6.9	7.9	4.6	8.1
Lys	7.3	10.2	7.3	10.9	6.8
Met	2.1	0.8	0.2	0.8	1.0
Phe	3.6	3.1	2.7	2.5	3.0
Pro	5.3	4.7	2.9	6.2	4.4
Ser	7.6	6.2	8.5	7.6	8.5
Thr	5.1	5.3	3.8	5.3	5.3
Tyr	2.4	1.2	1.9	0.5	2.2
Val	6.0	5.6	5.1	4.4	4.5
Pser	—	—	0.5	1.2	1.3
Pthr	—	—	—	—	0.4
Trp	—	—	0	0	—

[a] Knecht and Busch (14).
[b] Compositions of proteins B23, C23, and C18 were kindly supplied by Mark Mamrack.

shown in Table V. Band 15, which was isolated in high purity (Fig. 15), contains a single amino-terminal serine determined by the $[^3H]$flurodinitrobenzene method (14). The protein has a molecular weight of about 65,000. Proteins B24, C23, and C18 are phosphorylated nucleolar proteins also isolated by preparative gel techniques.

2. COMBINED PURIFICATION METHODS—PROTEIN A24

Protein A24 is a nonhistone chromatin protein with histonelike characteristics (55a). Although it is not uniquely localized to the nucleolar fraction of chromatin, it decreases in amount in nucleoli undergoing hypertrophy induced by thioacetamide or liver regeneration (18,56). The purification procedure for protein A24 combines a number of methods (55a).

Chromatin is prepared from rat liver or other tissues, as in Section III, A, 1, by washing nuclei with saline–EDTA and dilute Tris buffers. The chromatin is then pre-extracted 3 times with 0.35 M NaCl–10 mM Tris–1 mM PMSF (pH 8.0) to remove the loosely bound nonhistone proteins. The product is then washed twice with 0.5 M perchloric acid to remove H1 histones. The residue is finally extracted with $0.4\,N\,H_2SO_4$ (see Section II,D), and the acid-soluble proteins are ethanol-precipitated.

This A24-enriched fraction is dissolved at a concentration of 100 mg/ml in 50 mM HCl–1 M urea–1% β-mercaptoethanol and clarified by centrifugation. The supernatant is applied to a Sephadex G-100 column (295 × 2.5 cm) and eluted with 10mM HCl (57). The fractions containing A24 are pooled, made 0.4 N with respect to H_2SO_4, and precipitated with 80% ethanol. The proteins are washed with absolute ethanol, dried, and dissolved in 0.9 N acetic acid–10 M urea–1% β-mercaptoethanol at a concentration of 20 mg/ml. The solution is then loaded onto 10% polyacrylamide slab gels (10 × 7.5 cm, 0.5 ml of protein/slab) and subjected to electrophoresis at 120 V for 6 hours. Vertical sections (0.5 cm wide) are cut from the sides and the center of the slabs and stained for 1 hour with 1% Amido black in 7% acetic acid. The band containing protein A24 is cut from the unstained slabs and eluted by electrophoresis as described above (Section IV,B,1).

When protein A24 is purified by the above method, it migrates as a single

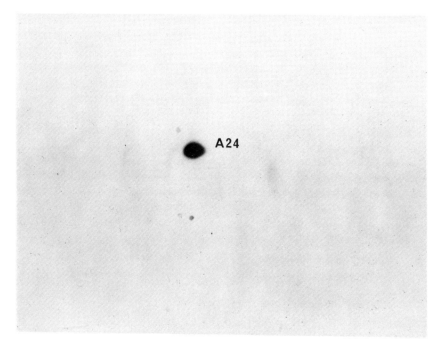

FIG. 16. Protein A24 purified as described in the text. From Goldknopf *et al.* (*55a*). Reproduced with permission.

TABLE VI

AMINO ACID COMPOSITION AND NH$_2$-TERMINAL
AMINO ACID OF A24[a]

Amino acid	Mole percent
Ala	9.6
Arg	7.4
Asx	7.3
Gly	9.2
Glx	12.3
His	2.4
Ile	5.8
Leu	10.9
Lys	11.3
Met	0.3
Phe	0.9
Pro	5.6
Ser	4.5
Thr	6.5
Trp	0.0
Tyr	1.3
Val	4.9
Lys + His + Arg	21.1
Glx + Asx	19.6
Glx + Asx/Lys + His + Arg	0.93
NH$_2$-terminal	Methionine

[a] From Goldknopf et al. (55a).

spot in two-dimensional electrophoresis (Fig. 16). The amino acid composition is shown in Table VI. This interesting protein contains the tryptic peptides of histone 2A (58), including the amino terminal Acetyl-Ser-Gly-Arg peptide and the His-His-Lys-Ala-Lys-Gly-Lys-carboxyl terminal sequence (59,60). However, the protein also contains an additional polypeptide chain with an amino terminal methionine. The amino terminal 37 residues have been determined by automated Edman degradation (60; Fig. 17). This sequence is not homologous to any known histone sequence. The site of attachment of the two chains is under investigation.

C. Nucleolar Phosphoproteins

Recently the nucleolus was found to contain numerous phosphorylated proteins (2, 15, 16). Approximately 40 nucleolar acid-soluble proteins were found to be ^{32}P-labeled in vivo (16). Most of these proteins are also labeled in vitro with $[\gamma\text{-}^{32}P]$ATP (17). The nucleolar preribosomal particles contain approximately 19 phosphorylated proteins (15). Four of these are ribo-

```
    1           5              10
Met-Gln-Ile-Phe-Val-Lys-Thr-Leu-Thr-Gly-
   11          15              20
Lys-Thr-Ile-Thr-Leu-Glu-Val-Glu-Pro-Ser-
   21          25              30
Asp-Thr-Ile-Glu-Asn-Val-Lys-Ala-Lys-Ile-
   31          35
Gln-Asp-Lys-Glu-Gly-Ile-Pro-
```

FIG. 17. The amino terminal sequence of protein A24 (60).

somal proteins, and the additional 15 are uniquely nucleolar. In addition, at least three phosphorylated nonhistone proteins are localized in the nucleolar fraction of chromatin (2) (Fig. 6). Therefore, the separate section on nucleolar phosphoproteins is presented in this chapter.

1. ^{32}P LABELING OF PROTEINS

Labeling is achieved either by *in vitro* or *in vivo* incubations. Novikoff hepatoma ascites cells are conveniently labeled in cell suspensions using the method described by Mauritzen et al. (61). In this system, up to 100 mCi of [^{32}P]orthophosphate/30 gm of cells is used. Our laboratory has found 2-hour incubations at 37° satisfactory for labeling most nucleolar proteins (2). Labeling of liver or Novikoff hepatoma in whole animals is achieved by intraperitioneal injection of up to 50 mCi per rat of carrier-free [^{32}P]orthophosphate in 0.9% NaCl (16).

Nucleolar proteins may also be labeled by incubation of nucleoli with [γ-^{32}P]ATP (17,62). Isolated nucleoli containing about 10 mg of protein are incubated in a medium containing 0.25 M sucrose–5 mM MgCl$_2$–12.5 mM NaCl–50 mM Tris-HCl (pH 7.3) and 0.125 mCi of [γ-^{32}P]ATP (64–120 Ci/mmol) in a total volume of 2.1 ml. The mixture is gently shaken at 37° for 10–20 minutes. The reaction is stopped by cooling on ice and addition of H$_2$SO$_4$ to a concentration of 0.4 N. The nucleolar acid-soluble proteins are then extracted and precipitated with 4 volumes of absolute ethanol (Section II,D). Other divalent metals such as Zn^{2+} may be added to provide altered profiles of phosphorylated proteins (17).

2. DETECTION OF ^{32}P-LABELED PROTEINS

^{32}P-Labeled phosphoproteins are subjected to the same analytical and preparative methods as described above. With *in vivo* labeled proteins it is important to remove ^{32}P-labeled nucleic acids or to provide some method of distinguishing between protein and nucleic acid. This problem was found to be minimal when ^{32}P-labeled proteins were analyzed by two-dimensional electrophoresis (16), i.e., the vast bulk of the radioactivity in spots cored from the gels was in the form of phosphoserine and phosphothreonine.

The radioactively labeled proteins may be detected by slicing one-dimension gels or by coring out spots followed by scintillation counting. Alternatively, gels may be dried and subjected to autoradiography on x-ray film (2).

3. CHARACTERISTICS OF NUCLEOLAR PHOSPHOPROTEINS

Several phosphorylated nucleolar proteins have been purified by the preparative two-dimensional technique. The amino acid compositions and approximate molecular weights are given in Table V. These proteins contain from 0.5–1.3 mol % phosphoserine. Interestingly, these nonhistone proteins are also relatively rich in glutamic and aspartic acid.

V.   Nucleolar Enzymes

This group of proteins constitutes a large number of proteins of which only a few have been even partially characterized. In addition, there are probably numerous enzymes involved in the assembly and processing of preribosomal particles for which no functional assay is available. This class of proteins therefore remains largely unexplored.

A.   DNA-Dependent RNA Polymerase

Of the nucleolar enzymes, RNA polymerase I is the most extensively studied. The structure and function of this and other RNA polymerases have been thoroughly reviewed (63–65). Therefore, only special aspects of this enzyme are covered in this chapter.

Previous studies on the RNA polymerase of normal liver and of tumors in our laboratory (66) demonstrated that this enzyme has a pH optimum of 8.2, requires the presence of DNA and all four types of ribonucleoside triphosphates, is dependent on the divalent ions Mg^{2+} and Mn^{2+}, and is inhibited by actinomycin D. The kinetics of labeling or biosynthetic reactions are linear for 10–12 minutes, and labeling is linear with respect to enzyme concentration within restricted limits. The nucleolar enzyme is stable at $-20°C$ in 3 mM 2-mercaptoethanol–1 mM EDTA–50% glycerol when buffered at pH 7–8 with Tris. The enzyme activity is enhanced by polyamines, including spermine, spermidine, and putrescine, and by ammonium sulfate in concentrations up to 30 mM. Similar to the nuclear enzyme studied by Cunningham and Steiner (67) and by Goldberg et al. (68), the nucleolar RNA polymerase is solubilized at pH 10. Simply by isolation of nucleoli, there was a purification of 32-fold over the nuclear activity. By treatment of the nucleoli with Tris buffer at pH 10, the enzyme is purified by a factor of 1.5. Saturation with ammonium sulfate to precipitate the enzyme (30–45% ammonium sulfate) results in a purification of approximately 4- to 4.5-fold over the isolated nucleoli and of 143-fold over the whole nuclear preparation.

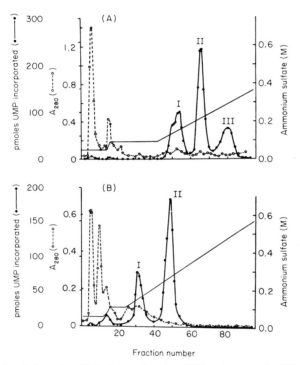

FIG. 18. Resolution of multiple forms of nuclear RNA polymerase by DEAE–Sephadex chromatography. (A) Sea urchin gastrula (52-hour development). A sample containing 11.4 mg of protein (0.07 units/mg at 50 mM ammonium sulfate) was applied to a 1 × 12 cm column equilibrated with 50 mM ammonium sulfate in standard buffer [50 mM Tris-HCl (pH 7.9)–25% (v/v) glycerol–5 mM MgCl$_2$–0.1 mM EDTA–0.5 mM dithiothreitol]. After a washing with 0.1 M ammonium sulfate, the enzyme activities were eluted with linear gradients of ammonium sulfate (in standard buffer) as indicated. One-milliliter fractions were collected, and 0.05-ml aliquots were assayed. The activity is expressed in picomoles of UMP incorporated into RNA/10 minutes/ml. The absolute activities for the peak tubes from I, II, and III were 5.5-, 8.7-, and 5.4-fold higher, respectively, when reassayed at the standard uridine 5′-triphosphate (UTP) concentration. (B) Adult rat liver. A sample containing 4.6 mg of protein (0.104 units/µg at 50 mM ammonium sulfate) was chromatographed as in (A). The activities of the peak tubes were 4.8- and 5.5-fold higher for I and II, respectively, when reassayed at the standard UTP concentration. From Roeder and Rutter. (69). Reproduced with permission.

DNA-dependent RNA polymerase has been purified from sea urchin embryos and rat liver (69) and separated into two or three peaks on O-(diethylaminoethyl) (DEAE)–Sephadex (Fig. 18). The enzymes have been designated polymerase I, II, and III in order of increasing concentration of ammonium sulfate at which they are eluted from the column. Polymerase I is the nucleolus-localized enzyme which is slightly more active in the presence of Mn^{2+} than in Mg^{2+} (70). Polymerase II is an extranucleolar nuclear enzyme which has a definite Mn^{2+} ion requirement and is inhibited by low concentrations of α-amanitin. Polymerase III is inhibited only by high concentrations of α-amanitin (71) and is implicated in the synthesis of 5 S and tRNA (71, 72).

Under nondenaturating conditions the polymerases electrophorese as single bands on polyacrylamide gels. Band AI is the nucleolus-localized polymerase and bands BI and BII are extranucleolar nuclear enzymes. The terminology for nuclear RNA polymerases is summarized in Table VII. The purification and structural analysis of the three classes of RNA polymerase activity from calf thymus have firmly established the multiplicity of RNA polymerases in animal tissues.

The activities of polymerases I and II are influenced by ionic strength. Form I exhibits optimal activity at less than 40 mM salt. Form II exhibits a sharp maximum at about 0.1 M salt. At saturating levels of template and at 40 mM ammonium sulfate, the activity ratios for denatured DNA/native DNA are 1:3 and 3:1 for rat enzymes I and II, respectively.

The molecular structures of the calf thymus RNA polymerases have been studied extensively (73, 74). The subunit composition of the three poly-

TABLE VII

NOMENCLATURE AND LOCALIZATION OF ANIMAL DNA-DEPENDENT RNA POLYMERASES[a]

Class of enzyme	Enzymes	Enzymes	Class of enzyme	Principal localization
A (insensitive to amanitin)	AI(a + b)[b]	I[b]		Nucleolar
	AII	I$_B$	I	Nucleolar
B (sensitive to low concentrations of amanitin, 10^{-9}–10^{-8} M)	BO[b]	II$_0$[b]		
	BI[b]	II$_A$[b]	II	Nucleoplasmic
	BII(a + b)[b]	II$_B$[b]		
C (sensitive to high concentrations of amanitin, 10^{-5}–10^{-4} M)	CI			
	CII			
	CIIIa	III$_A$[b]		Nucleoplasmic,
	CIIIb	III$_B$[b]	III	cytoplasmic

[a] From Chambon (64).
[b] Enzyme for which the subunit pattern is known.

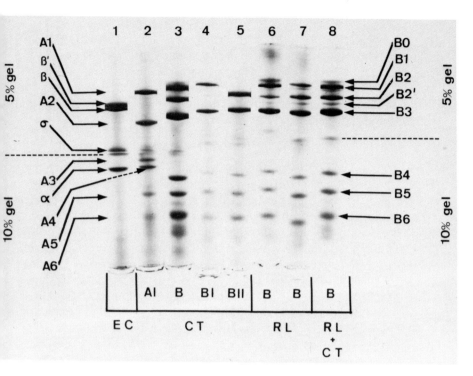

FIG. 19. Polyacrylamide gel electrophoresis showing the multiple subunits of RNA polymerase from various tissues. EC, *Escherichia coli*; CT, calf thymus; RL, rat liver. From Chambon *et al.* (*63*). Reproduced with permission.

TABLE VIII
SUBUNITS OF CALF THYMUS RNA POLYMERASES AI AND B[a]

Form AIb		Form AIa		Form BI		Form BIIa or BIIb	
Subunit	Molecular weight[b]	Subunit	Molecular weight[b]	Subunit	Molecular weight[b]	Subunit	Molecular weight[b]
SA1	197(1)[c]	SA1	197(1)	SB1	214(1)	—	—
SA2	126(1)	SA2	126(1)	—	—	SB2a or SB2b	180(1)
SA3	51(1)	—	—	SB3	140(1)	SB3	140(2)
SA4	44(1)	SA4	44(1)	SB4	34(1–2)	SB4	34(1–2)
SA5	25(2)	SA5	25(2)	SB5	25(2)	SB5	25(2)
				SB5′	20(1)	SB5′	20(1)
SA6	16.5(2)	SA6	16.5(2)	SB6(a + b)	16.5(3–4)	SB6(a + b)	16.5(3–4)

[a] From Chambon (*64*).
[b] Molecular weight: daltons × 10⁻³.
[c] Numbers in parentheses correspond to molar ratios.

merases was investigated by polyacrylamide gel electrophoresis on SDS gels. Figure 19 shows the resolution of the different subunits. This work has been summarized by Chambon (64) and is outlined in Table VIII. The nucleolar enzymes have 7–8 subunits ranging in molecular weight from 16,500 to 197,000. The basic structure of mammalian RNA polymerase is similar to that of prokaryotic enzymes in that it consists of two high-molecular-weight subunits accompanied by several small subunits.

1.   RNA POLYMERASE ASSAY

After separation of multiple forms of nuclear RNA polymerase activity by DEAE–Sephadex chromatography, the fractions are assayed according to Roeder and Rutter (69). The standard assay mixture for RNA polymerase contains in a volume of 0.125 ml: 20 μg native calf thymus DNA (Sigma, type 1); 2.5 μg pyruvate kinase; 7 μmoles Tris-HCl (pH 7.9); 0.2 μmoles MnCl$_2$; 1.0 μmole KCl; 0.75 μmole NaF; 0.5 μmole phosphoenolpyruvate; 0.2 μmole 2-mercaptoethanol; 0.075 μmole each of guanosine, cytidine, and adenosine 5'-triphosphates (GTP, CTP, ATP); 0.0125 μmole unlabeled UTP; 0.0005 μmole [^{3}H]UTP (2.0 Ci/mmol; Schwarz Bioresearch); 0.05 M ammonium sulfate; and 0.05 ml of the enzyme solution. After incubation at 30°C for 10 minutes, TCA-insoluble radioactivity (in RNA) is determined by standard procedures. One unit of activity is the amount of enzyme that incorporates 1 pmol of uridine 5'-monophosphate (UMP) into RNA per minute.

2.   TRANSCRIPTION OF NUCLEOLAR GENES

Recently it was shown that rat liver nucleoli are capable of synthesizing *in vitro* high-molecular-weight RNA which resembles ribosomal precursor RNA in base composition, sedimentation pattern, and in hybridization competition with 45 S RNA (75). In addition, RNA molecules larger than 45 S were found. In our laboratory, nucleoli were found to transcribe high-molecular-weight RNA which contained large oligonucleotide markers of 18 S and 28 S RNA and the spacer region of 45 S precursor RNA as identified by homochromatography (76). The availability of the system for transcription should greatly aid in the isolation of the regulatory factors for RNA polymerase I and the proteins involved in control of preribosomal RNA synthesis.

B.   RNA Methylase

The *in vivo* methylation of ribosomal precursor RNA has been studied in many animal cells (3). Isolated nucleoli have recently been shown to contain

RNA methylase activity. Both isolated rat liver nucleoli (77) and Novikoff hepatoma nucleoli (78) contain an RNA methylase which can transfer *in vitro* the methyl groups from 5-adenosylmethionine to the 2'-position of the ribose of preribosomal RNA. In addition, it was shown that Novikoff hepatoma nucleoli contain a methylase that can transfer methyl groups to the purine and pyrimidine bases of exogenous *E. coli* tRNA (79). This base-methylating activity was extracted from nucleoli with 0.5 M NH$_4$Cl–0.5 mM MgCl$_2$–10 mM dithiothreitol–10 mM Tris–HCl (pH 7.8). The tRNA-methylating activity of nucleoli appears to be different from the tRNA base-methylating activity of the cytoplasmic pH 5 fraction.

Recently the endogenous ribosomal RNA methylase activity was studied by Liau and Hurlbert (80) and Grummt et al. (81). A good substrate for this enzyme is nuclear RNA prepared from Novikoff hepatoma cells cultured in the absence of methionine (80). Methylation of ribose rather than bases was found to predominate by both groups. In addition, nucleolar RNA polymerase and RNA methylase appear to function independently.

C. Nucleolar Phosphoprotein Kinase

The presence of phosphorylated proteins in the nucleolus (15, 16) has led to the search for phosphokinase activity capable of phosphorylating nucleolar proteins. Kang et al. (17) and Grummt (82) showed that nucleolar proteins were phosphorylated when isolated nucleoli were incubated in the presence of [γ-^{32}P]ATP, thereby establishing that the nucleolus contained kinase activity. Recently, Wilson and Ahmed (83) showed that protein phosphokinase was not only present in prostate nucleoli but differed from extranucleolar kinase in pH optima and in metal ion requirements. The activity was not significantly affected by cyclic adenosine 5'-monophosphate (AMP).

Nucleolar phosphoprotein kinase is assayed using histones, casein, or phosvitin as substrates. Ideally, one would prefer to use the natural substrate, nucleolar proteins. Recent studies in this laboratory have used nucleolar proteins soluble in 0.4 N sulfuric acid as substrates. Numerous assay systems are available. A convenient method employed in this laboratory is the filter paper disc technique described by Corbin and Reimann (84). The methods described for chromatin kinases by Kish and Kleinsmith (85) should also be generally applicable to the nucleolar enzymes.

Although the function of the nucleolar protein phosphorylation is unknown, it has been suggested that specific kinases could control the assembly of nucleolar preribosomes (15) or the activity of RNA polymerase (86).

D. Phosphoprotein Phosphatases

Recently phosphatase activity that acts upon ^{32}P-labeled nucleolar protein substrates was found in nucleoli of Novikoff hepatoma ascites cells (87). The activity is optimal near pH 7.0 and is inhibited by increasing concentrations of NaCl. The activity is moderately inhibited by the divalent cations of $CaCl_2$, $MnCl_2$, and $CoCl_2$. $ZnCl_2$ completely inhibited the enzyme at 2 mM while EDTA and $MgCl_2$ had little effect. The activity also requires free sulfhydryl groups.

Phosphoprotein phosphatase assay.   Release of ^{32}P as inorganic phosphate is measured as the phosphomolybdate complex (88). A typical substrate mixture contains 0.1 M of 33 mM Bistris, pH 7.2; 0.1 ml of water or substances to be tested for stimulatory or inhibitory activity, and 50 μl of ^{32}P-labeled 0.4 N H_2SO_4-extracted nucleolar proteins (20,000–180,000 cpm/mg of protein) at a concentration of 1.0 mg/ml. The reaction is initiated by the addition of 50 μl of nucleoli homogenized in water at a protein concentration of 1.6 mg/ml (determined by A_{280} nm : A_{260} nm in 6 M guanidine hydrochloride). After various times of incubation at 37°, the reaction is stopped by the addition of 25 μl of 0.1 M silico-tungstic acid in 0.1 N H_2SO_4 and then cooled in an ice bath. To this is added 100 μl of 5% $(NH_4)_6Mo_7O_{24} \cdot 4 H_2O$ in 4 N H_2SO_4 followed by 100 μl of bovine serum albumin (10 mg/ml in water). The mixture is extracted with 0.5 ml of isobutanol : benzene, 1:1, and aliquots of the upper organic phase are counted in the scintillation counter.

The specificity and function of this phosphatase activity remain to be determined. Preliminary data indicate that multiple enzymes are present.

E. Endoribonuclease

A nucleolar endoribonuclease was extracted from HeLa cell nucleoli by Mirault and Scherrer (89). For this purpose isolated nucleoli are resuspended in 5 ml of 10 mM triethanolamine–10 mM KCl–0.1 mM $MgCl_2$–10 mM dithiothreitol–0.1% Brij 35 (pH 7.4) and homogenized 10 minutes at 25°C in a glass Dounce homogenizer. After addition of 1 ml of triethanolamine (1 M) at pH 8 and 6 ml of 2 M NH_4Cl–20 mM EDTA (pH 8.0), the mixture is homogenized at 25°C for 5 minutes and centrifuged for 20 minutes at 20,000 g at 0°C. To the 12 ml of crude nucleolar supernatant, 4.3 gm of solid ammonium sulfate are added. After 30 minutes of stirring at 0°C, the precipitate is removed by sedimentation for 20 minutes at 20,000 g and the supernatant is dialyzed extensively against a dialysis buffer [50 mM triethanolamine–50 mM KCl–1 mM β-mercaptoethanol (pH 8.0)]. The precipitate is removed by sedimentation and the supernatant containing the endoribonuclease activity (designated "crude enzyme") is kept frozen at −20°C.

A further purification was made by adding DEAE–cellulose equilibrated with dialysis buffer to the crude enzyme. This procedure, as described by Lazarus and Sporn (90), separates endoribonuclease activity from exoribonuclease activity. The nonabsorbed fraction contains a 3- to 5-fold increase in specific activity of the endoribonuclease. Figure 20 shows the ability of the nuclease to "process' high-molecular-weight RNA by incubation of the enzyme with isolated nucleolar preribosomal particles containing intact 45 S RNA. Endonucleolytic cleavage of 45 S nucleolar RNA has also been observed on incubating isolated nucleoli under various conditions (91,92).

Studies from our laboratory have shown that an RNP-associated nucleolar endoribonuclease is capable of degrading 45 S nucleolar RNA to fragments

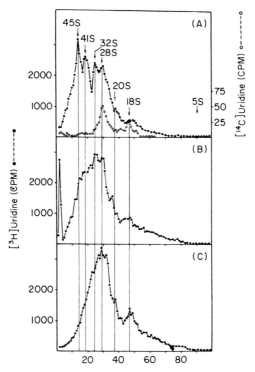

FIG. 20. Enzyme dependence of "processing" of high-molecular-weight RNA. Preribosomes (0.2 A_{260} units) were incubated at 37°C for: (A) 3 minutes with 60 μl DEAE–enzyme equivalent to 0.5 μg protein; (B) 3 minutes with 120 μl DEAE–enzyme fraction; (C) 6 minutes with 240 μl DEAE–enzyme fraction. The RNA was extracted (89) and analyzed by electrophoresis on exponential gels (37). [¹⁴C]uridine-labeled HeLa cell rRNA was used as internal marker. Closed circles, ³H radioactivity; open circles, ¹⁴C radioactivity. From Mirault and Scherrer. (89). Reproduced with permission.

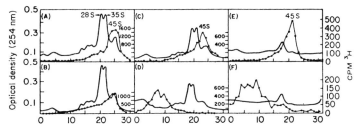

FIG. 21.   Temperature effects on endonuclease activity associated with 78 S RNP particles. Sucrose density gradient centrifugation of 45 S ³H-RNA (closed circles) after incubation with RNP particles and carrier nucleolar RNA (solid line) was carried out on 5–40% sucrose gradients containing 0.1 M NaCl–10 mM sodium acetate (pH 5.1)–1 mM EDTA. The samples were centrifuged in a Spinco SW-27 rotor at 113,000 g for 15 hours.   (A) Control incubation of ³H-labeled 45 S RNA in the absence of 78 S RNP particles at 0°C.   (B) Incubation of ³H-labeled 45 S RNA in the presence of 78 S RNP particles at 0°C.   (C) Control incubation of ³H-labeled 45 S RNA in the absence of 78 S RNP particles at 25°C.   (D) Incubation of ³H-labeled 45 S RNA in the presence of 78 S RNP particles at 25°C.   (E) Control incubation of ³H-labeled 45 S RNA in the absence of 78 S RNP particles at 37°C.   (F) Incubation of ³H-labeled 45 S RNA in the presence of 78 S RNP particles at 37°C. In (A–D) the carrier RNA is whole nucleolar RNA; in (E) and (F) the carrier is 28 and >28 S nuclear RNA. From Prestayko *et al.* (*35*). Reproduced with permission.

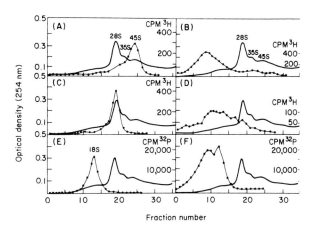

Fraction number

FIG. 22.   Effect of nucleolar RNP-associated RNase on 45, 28, and 18 S RNAs. Nucleolar particles were incubated with isotope-labeled RNA at 25°C, and the samples were layered onto sucrose gradients and centrifuged.   (A) Nucleolar ³H-labeled 45 S RNA incubated without RTP particles.   (B) Nucleolar ³H-labeled 45 S RNA incubated with RNP particles.   (C) Nucleolar ³H-labeled 28 S RNA incubated without RNP particles.   (D) Nucleolar ³H-labeled 28 S RNA incubated with RNP particles.   (E) Nuclear ³²P-labeled 18 S RNA incubated without RNP particles.   (F) Nuclear ³²P-labeled 18 S RNA incubated with RNP particles. Closed circles, radioactivity; solid line, absorbance of nuclear carrier RNA. From Prestayko *et al.* (*35*). Reproduced with permission.

which sediment in the 10–16 S regions of a sucrose gradient. These fragments are produced after hydrolysis of 45 S RNA at 25° or 37°C. The enzyme appears to be inactive at 0°C (Fig. 21).

The RNP-associated endoribonuclease is more reactive with 45 S ribosomal precursor RNA than with 28 and 18 S RNA, as shown in Fig. 22. Larger fragments are produced with the 28 and 18 S RNAs than with the 45 S RNA. These results imply that the 45 S RNA has specific binding sites for the enzyme which allow the enzyme to cleave the RNA more efficiently at specific sites on the molecules. Subsequent studies have shown that the low-molecular-weight 4 and 5 S RNAs are nonreactive with the enzyme under conditions that result in degradation of 45 S RNA.

Column chromatographic procedures have been utilized to purify the enzyme. The RNP particles are extracted with 2 M LiCl at 0°C for 15 hours. After centrifugation at 20,000 g for 15 minutes, the supernatant containing the enzyme is dialyzed against a buffer containing 6 M urea–10 mM Tris-acetate (pH 7.5)–5 mM EDTA–0.1 mM dithiothreitol. The sample is then applied to a DEAE–cellulose column which absorbs the acidic proteins; the enzyme activity is recovered in the void volume and no nuclease activity, either endonucleolytic or exonucleolytic, is retained on the DEAE–cellulose column. Presumably, the procedure for isolation of RNP particles removes the exonuclease activity reported in nucleoli by Kelley and Perry (*93*).

The sample containing the enzyme is applied to a carboxymethyl cellulose (CMC) column equilibrated with a buffer containing 30 mM methylamine–3 mM β-mercaptoethanol–6 M urea adjusted to pH 5.6 with acetic acid. A step by step gradient of sodium acetate in the above buffer is applied, and the enzyme is eluted with 0.2 M sodium acetate. Analysis of this fraction on polyacrylamide gels shows two to three bands present, and this fraction is called "purified enzyme." Studies with purified enzyme showed that it (1) does not cleave any of the following dinucleotides: UpU, ApU, GpU, UpC, GpC, GpA; (2) is active at pH 5–8 and inactive in 0.5 M sodium acetate (pH 5.6); (3) is inactive at 0°C; and (4) is inactivated by heating to 100°C.

Endoribonuclease assay. Several methods used to assay for endoribonuclease activity have been reviewed by Klee (*94*). The endoribonuclease activity associated with nucleolar RNP particles was assayed by sedimentation on sucrose gradients after the reaction. The extent of degradation was detected by a decrease in the sedimentation coefficient of the RNA.

RNA labeled with ^{3}H or ^{32}P is added to a 1-ml sample of RNP particles (50 μg RNA/100 μg RNP particles), ribosomal subunits, 10 μg of 2 M LiCl protein extract of the RNP particles, or 1–2 μg of purified enzyme. Incubation is carried out at 37°C for 15 minutes in a 5 mM Tris-acetate buffer (pH 7.4). Unlabeled nuclear or nucleolar RNA (150–200 μg) is added to each sample as a carrier. The samples are adjusted to 0.3% SDS for direct layering on gradients, or precipitated for 15 hours with 2 volumes of absolute ethanol at $-20°C$. After ethanol precipitation and centrifugation, the pellet is suspended in 0.5 ml of 50 mM sodium acetate (pH 5.1)–0.14 M NaCl–0.3% SDS. The samples are layered on 5–40% sucrose density gradients containing 0.1 M NaCl–20 mM sodium acetate (pH 5.0)–1 mM EDTA and centrifuged for 15 hours at 95,000 g (average) in a Spinco SW-27 rotor. The gradients are fractionated with an ISCO gradient fractionator, and the 1-ml fractions are acidified with 5 M HClO$_4$ and heated at 70°C for 20 minutes. After cooling, 10 ml of scintillation solution are added and the samples are counted in a Packard Tri-Carb liquid scintillation counter.

F.   Exoribonuclease

The requirement for a specific endonuclease is indicated in the scheme for processing ribosomal precursor RNA (3). The presence of an exoribonuclease in nucleoli was demonstrated by Kelley and Perry (93). They postulated that this enzyme is involved in exonucleolytic trimming in the processing of 45 S RNA, releasing 5'-mononucleotides as the nucleolytic product.

Exoribonuclease assay. Nucleolar exonuclease (93) is assayed by chromatography of the acid-soluble components after enzyme reaction. Chromatography is carried out with plastic-backed 5 cm × 20 cm strips of cellulose impregnated with polytheneimine and a two-step elution process of 1.0 M acetic acid followed by 0.3 M LiCl (95–97). The products of the exonuclease are identified by the mobility of 2'-, 3'-, or 5'-mononucleotides in the same system.

G.   Other Nucleolar Enzymes

Studies on highly purified rat liver nucleoli (98) showed that the enzymes diphosphopyridine nucleotide pyrophosphorylase and adenosine triphosphatase A are localized to nucleoli. The differences in substrate specificities of RNase, adenosine triphosphatase A, and adenosine triphosphatase B of nuclear and nucleolar origin point to qualitative differences between the nuclear and nucleolar enzymes.

VI.   Methods for Studies on Nucleolar Antigens

A.   Immunization

Nucleoli were isolated from normal liver and Novikoff hepatoma cells by the sucrose–Ca^{2+} method (3). New Zealand white rabbits were injected intramuscularly, intradermally, and subcutaneously in 3 weekly doses with 10–60 mg of whole nucleoli suspended in 1 ml of Freund's complete adjuvant diluted 1:2 with 0.15 M NaCl. The blood was collected from the ear vein 7–10 days following the third inoculation (*101–103*).

B.   Nucleolar Immunofluorescence

The procedure used for detecting immunofluorescence was modified from the indirect immunofluorescence technique of Hilgers *et al.* (*99*). The fixed specimens were overlaid with the antinucleolar antisera diluted 1:10 with 0.15 M NaCl. The slides were placed in a moist chamber and incubated for 40 minutes at 37°. The slides were washed 3 times in phosphate buffered saline (PBS) for 5, 50, and 5 minutes and were air-dried. Fluorescein-labeled caprine antirabbit 7 S globulin diluted 1:5 (Hyland Laboratories, Los Angeles, California) was added, and the slides were incubated in a moist chamber for 30 minutes at room temperature. The slides were rinsed, washed overnight in PBS at 4°, dried, and counterstained with Evans blue (0.06%). Coverslips were mounted using glycerol: PBS (1:1). After sealing, the slides were examined using a Zeiss fluorescence microscope (HB200, mercury bulb) with a BG 12 excitation filter and a 530 nm secondary filter (*101–103*).

C.   Immunodiffusion Techniques

The immunodiffusion plates were prepared from 0.8% agarose dissolved in 0.15 M NaCl–0.15 M phosphate buffer (pH 7.4). Each antibody well contained 30–35 μl of normal control serum or specific antiserum. Each antigen well (30–35 μl) contained approximately 450–500 μg (15 mg/ml) of protein.

The proteins used as the antigen in the immunodiffusion test were extracted from chromatin prepared from Novikoff hepatoma and normal liver nuclei by a modification of Yeoman *et al.* (*30*). The precipitin bands were allowed to develop at room temperature in a moist chamber for 48 hours (*100*).

D.   Immunofluorescence Analysis of Nucleolar Antigens

In studies on antinucleolar antibodies, antinucleolar antisera were produced in rabbits immunized with whole isolated nucleoli from normal rat liver and Novikoff hepatoma ascites cells. These antisera produced positive nucleolar fluorescence of varying degrees in nucleoli (Fig. 23) of Novikoff tumor, rat liver, kidney, and Walker tumor cells. Nucleolar specificity of the antibodies was demonstrated by inhibition of fluorescence following pretreatment of the immune sera with whole nucleoli or nucleolar 0.15 M NaCl-soluble protein fractions (*101*).

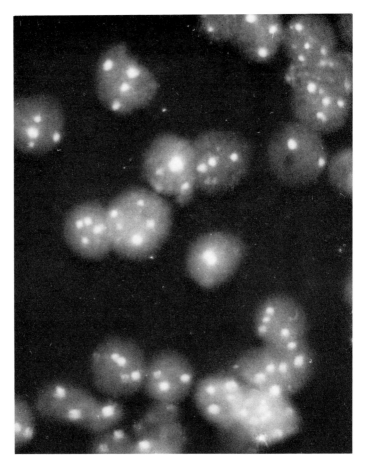

FIG. 23. Photomicrograph of nuclei tested with antisera by the indirect immuno-fluorescent technique. Liver antinucleolar antisera were incubated with liver nuclei. The immunofluorescent spots are seen as the white nucleoli against the gray nuclear background.

E.   Immunodiffusion of Nucleolar Antigens

Antibodies to tumor nucleoli demonstrate the presence of antigens in tumor chromatin that are absent in normal liver chromatin (Fig. 24). In addition, liver chromatin contained antigens that formed precipitin bands with antiserum to liver nucleoli. These bands differed from the precipitin bands formed with the tumor nucleolar (and nuclear) antisera. Interestingly, none of the corresponding antigens were present in tumor polysomes, indicating that in the Novikoff hepatoma the nuclear antigens do not emerge in either the ribosome precursors or the mRNA-associated proteins of tumors.

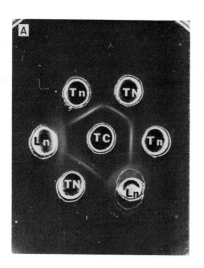 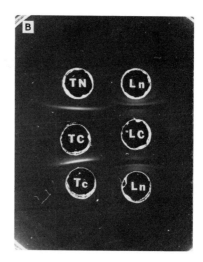

FIG. 24. (A) Ouchterlony immunodiffusion plate showing formation of a common band between tumor chromatin proteins (TC) and antibodies to tumor nuclei (TN) and tumor nucleoli (Tn). No identity was found with antibodies produced to liver nucleoli (Ln). (B) Ouchterlony immunodiffusion plate showing that antibodies to liver nucleoli (Ln) produce 4 immune complex bands with liver chromatin proteins (LC). None of these overlap the bands formed between tumor chromatin proteins (TC) and tumor chromatin antibodies (Tc) or tumor nuclear antibodies (TN). (Courtesy of R. K. Busch.)

VII. Discussion

Important advances in this field offer opportunities for elegant analysis of biochemical and morphological aspects of nucleolar function. Along with the very important new procedures for structural and sequence analysis of proteins by [³H]dansyl and [³⁵S]phenylthiohydantoin (PTH) methods, opportunities now exist for development of much more precise information on the nucleolar proteins.

The improvements being made in two-dimensional (or three-dimensional) electrophoretic techniques have demonstrated that the number of polypeptides in nucleoli is greater than 100, and accordingly a very large effort will be necessary to elucidate their structures and functions.

The recent use of immunologic methods detailed in the latter portion of this chapter offers a beginning of very important advances in studies on tissue and tumor specificity of nucleolar proteins (*103*).

ACKNOWLEDGMENTS

The original studies reported here were supported by Cancer Center Grant Ca-10893, awarded by the National Cancer Institute, DHEW, the Pauline Sterne Wolff Memorial Foundation, the Davidson Fund, and a generous gift from Mrs. Jack Hutchins.

The authors wish to express their sincere appreciation to Dr. I. Daskal for his help in providing the electron micrographic figures and to Mrs. Rose K. Busch for the figures relating to the antigenic activity of the nucleolar proteins.

REFERENCES

1. Orrick, L. R., Olson, M. O. J., and Busch, H., *Proc. Natl. Acad. Sci. U.S.A.* **70**, 1316 (1973).
2. Olson, M. O. J., Ezrailson, E. G., Guetzow, K., and Busch, H., *J. Mol. Biol.* **97**, 611 (1975).
3. Busch, H., and Smetana, K., "The Nucleolus." Academic Press, New York, 1970.
4. Muramatsu, M., and Busch, H., *Methods Cancer Res.* **2**, 303 (1967).
5. Busch, H., and Mauritzen, C. M., *Methods Cancer Res.* **3**, 392 (1967).
6. Busch, H., Ballal, N. R., Olson, N. O. J., and Yeoman, L. C., *Methods Cancer Res.* **11**, 43 (1975).
7. Birnstiel, M. L., Chipchase, M., and Flamm, W. G., *Biochim. Biophys. Acta* **87**, 111 (1964).
8. Chakravorty, A. K., and Busch, H., *Cancer Res.* **27**, 780 (1967).
9. Grogan, D. E., and Busch, H., *Biochemistry* **6**, 573 (1967).
10. Grogan, D. E., Desjardins, R., and Busch, H., *Cancer Res.* **26**, 775 (1966).
11. Fogel, S., and Sypherd, P. S., *Proc. Natl. Acad. Sci. U.S.A.* **59**, 1329 (1968).
12. Shapiro, A. L., Vinuela, E., and Maize, J. V., *Biochem. Biophys. Res. Commun.* **28**, 815 (1967).
13. Gronow, M., and Thackrah, T., *Arch. Biochem. Biophys.* **158**, 377 (1973).
14. Knecht, M. E., and Busch, H., *Life Sci.* **10**, 1297 (1971).
15. Olson, M. O. J., Prestayko, A. W., Jones, C. E., and Busch, H., *J. Mol. Biol.* **90**, 161 (1974).
16. Olson, M. O. J., Orrick, L. R., Jones, C., and Busch, H., *J. Biol. Chem.* **249**, 2823 (1974).
17. Kang, Y. J., Olson, M. O. J., and Busch, H., *J. Biol. Chem.* **249**, 5580 (1974).
18. Ballal, N. R., Kang, Y.-J., Olson, M. O. J., and Busch, H., *J. Biol. Chem.* **250**, 5921 (1975).
19. Warner, J. R., *J. Biol. Chem.* **246**, 447 (1971).
20. Kumar, A., and Warner, J. R., *J. Mol. Biol.* **63**, 233 (1972).
21. Maizel, J. U., *Science* **151**, 988 (1966).
22. Flemming, W., "Zellsubstanz, Kern & Zeiltheiling." Vogel, Leipzig (1882).
23. Marushige, K., and Bonner, J., *J. Mol. Biol.* **15**, 160 (1966).
24. Taylor, C. W., Yeoman, L. C., Daskal, I., and Busch, H., *Exp. Cell Res.* **82**, 215 (1973).
25. Wilhelm, J. A., Ansevin, A. T., Johnson, A. W., and Hnilica, L. S., *Biochim. Biophys. Acta* **272**, 220 (1972).
26. Chauveau, J., Moule, Y., and Rouiller, C., *Exp. Cell Res.* **11**, 317 (1956).
27. Spelsberg, T. C., and Hnilica, L. S., *Biochim. Biophys. Acta* **228**, 202 (1971).
28. Johns, E. W., and Forrester, S., *Eur. J. Biochem.* **8**, 547 (1969).
29. Wilson, E. M., and Spelsberg, T. C., *Biochim. Biophys. Acta* **322**, 145 (1973).
30. Yeoman, L. C., Taylor, C. W., Jordan, J. J., and Busch, H., *Biochem. Biophys. Res. Commun.* **53**, 1067 (1973).
31. Warner, J. R., and Soeiro, R., *Proc. Natl. Acad. Sci. U.S.A.* **58**, 1984 (1967).
32. Penman, S., *J. Mol. Biol.* **17**, 117 (1966).
33. Prestayko, A. W., Klomp, G. R., Schmoll, D. J., and Busch, H., *Biochemistry* **13**, 1945 (1974).
34. Liau, M. C., and Perry, R. P., *J. Cell Biol.* **42**, 272 (1969).
35. Prestayko, A. W., Lewis, B. C., and Busch, H., *Biochim. Biophys. Acta* **269**, 90 (1972).
36. Shinozawa, T., Yahara, I., and Imahori, K., *J. Mol. Biol.* **36**, 305 (1968).
37. Daskal, I., Prestayko, A. W., and Busch, H., *Exp. Cell Res.* **88**, 1 (1974).
38. Mirault, M. E., and Scherrer, K., *Eur. J. Biochem.* **23**, 372 (1971).
39. Koshiba, K., Chandra, T., Daskal, I., and Busch, H., *Exp. Cell Res.* **68**, 235 (1971).

40. Tsurugi, K., Morita, T., and Ogata, K., *Eur. J. Biochem.* **32**, 555 (1973).
41. Soeiro, R., and Basile, C., *J. Mol. Biol.* **79**, 507 (1973).
42. Higashinakagawa, T., and Muramatsu, M., *Eur. J. Biochem.* **42**, 245 (1974).
43. Warner, J. R., in "Ribosomes" (M. Nomura, A. Tissieres, and P. Lengyel, eds.), p. 461. Cold Spring Harbor Lab., Cold Spring Harbor, New York, 1974.
44. Das, N. K., Micou-Eastwood, J., Ramamurthy, G., and Alfert, M., *Proc. Natl. Acad. Sci. U.S.A.* **67**, 968 (1970).
45. Granboulan, N., and Granboulan, P., *Exp. Cell Res.* **38**, 604 (1965).
46. Unuma, T., Arendell, J. P., and Busch, H., *Exp. Cell Res.* **52**, 429 (1968).
47. Kaltschmidt, E., and Wittmann, H. G., *Anal. Biochem.* **36**, 401 (1970).
48. Biekla, H., Stahl, J., and Welfle, H., *Arch. Geschwulstforsch.* **38**, 109 (1971).
49. Martini, O. H. W., and Gould, H. J., *J. Mol. Biol.* **62**, 403 (1971).
50. Delaunay, J., and Shapira, G., *Biochim. Biophys. Acta* **259**, 243 (1972).
51. Sherton, C. C., and Wool, I. G., *J. Biol. Chem.* **247**, 4460 (1972).
52. Howard, G. A., Traugh, J. A., Crosen, E. A., and Traut, R. R., *J. Mol. Biol.* **93**, 391 (1975).
53. Busch, G. I., Yeoman, L. C., Taylor, C. W., and Busch, H., *Physiol. Chem. Phys.* **6**, 1 (1974).
54. Peterson, J. L., and McConkey, E. H., *J. Biol. Chem.* **251**, 548 (1976).
55. Peterson, J. L., and McConkey, E. H., (1976). *J. Biol. Chem.* **251**, 555 (1976).
55a. Goldknopf, I. L., Taylor, C. W., Baum, R. M., Yeoman, L. C., Olson, M. O. J., Prestayko, A. W., and Busch, H., *J. Biol. Chem.* **250**, 7182 (1975).
56. Ballal, N. R., Goldknopf, I. L., Goldberg, D. A., and Busch, H., *Life Sci.* **14**, 1835 (1974).
57. Goldknopf, I. L., and Busch, H., *Physiol. Chem. Phys.* **5**, 131 (1973).
58. Goldknopf, I. L., and Busch, H., *Biochem. Biophys. Res. Commun.* **65**, 951.
59. Goldknopf, I. L., Olson, M. O. J., James, G. T., Mays, C. J., and Guetzow, K., *Fed. Am. Soc. Exp. Biol.* **35**, 1722 (1976).
60. Olson, M. O. J., Goldknopf, I. L., Guetzow, K. A., James, G. T., Hawkins, T. C., Mays, C. J., and Busch, H., *J. Biol. Chem.* **251**, 5901 (1976).
61. Mauritzen, C. M., Choi, Y. C., and Busch, H., *Methods Cancer Res.* **6**, 253 (1970).
62. Kang, Y. J., Olson, M. O. J., Jones, C., and Busch, H., *Cancer Res.* **35**, 1470 (1975).
63. Chambon, P., Gissinger, F., Kedinger, C., Mandel, J. L., and Meilhac, M., in "The Cell Nucleus" (H. Busch, ed.), Vol. 3, p. 269. Academic Press, New York, 1974.
64. Chambon, P., *Annu. Rev. Biochem.* **44**, 613 (1975).
65. Jacob, S. T., *Prog. Nucleic Acid Res. Mol. Biol.* **13**, 93 (1973).
66. Ro, T. S., and Busch, H., *Cancer Res.* **24**, 1630 (1964).
67. Cunningham, D. D., and Steiner, D. F., *Biochim. Biophys. Acta* **145**, 834 (1967).
68. Goldberg, M. L., Moon, H. D., and Rosenau, W., *Biochim. Biophys. Acta* **171**, 192 (1969).
69. Roeder, R. G., and Rutter, W. J., *Nature (London)* **224**, 234 (1969).
70. Roeder, R. G., and Rutter, W. J., *Proc. Natl. Acad. Sci. U.S.A.* **65**, 675 (1970).
71. Weinman, R., and Roeder, R. G., *Proc. Natl. Acad. Sci. U.S.A.* **71**, 1790 (1974).
72. Weil, P. A., Benson, R. H., and Blatti, S. P., *Fed. Am. Soc. Exp. Biol.* **33**, 1349 (1974).
73. Gissinger, F., and Chambon, P., *Eur. J. Biochem.* **28**, 277 (1972).
74. Kedinger, C., Gissinger, F., and Chambon, P., *Eur. J. Biochem.* **44**, 421 (1974).
75. Grummt, I., *Eur. J. Biochem.* **57**, 159 (1975).
76. Ballal, N. R., and Rogachevsky, L., *Proc. Am. Assoc. Cancer Res.* **17**, 162 (1976).
77. Culp, L. A., and Brown, G. M., *Arch. Biochem. Biophys.* **137**, 222 (1970).
78. Liau, M. C., Flatt, N. C., and Hurlbert, R. B., *Biochim. Biophys. Acta* **224**, 282 (1970).
79. Liau, M. C., O'Rourke, C. M., and Hurlbert, R. B., *Biochemistry* **11**, 629 (1972).
80. Liau, M. C., and Hurlbert, R. B., *Biochemistry* **14**, 127 (1975).
81. Grummt, I., Loening, U. E., and Slack, J. M. W., *Eur. J. Biochem.* **59**, 313 (1975).
82. Grummt, I., *FEBS Lett.* **39**, 125 (1974).

83. Wilson, M. J., and Ahmed, K., *Exp. Cell Res.* **93**, 261 (1975).

84. Corbin, J. D., and Reimann, E. M., *in* "Methods in Enzymolology," Vol. 38: Hormone Action, Part C, Cyclic Nucleotides (J. G. Hardman and B. W. O'Malley, eds.), p. 287. Academic Press, New York, 1974.

85. Kish, V. M., and Kleinsmith, L. J., *in* "Methods in Enzymology," Vol. 40: Hormone Action, Part E, Nuclear Structure and Function (B. W. O'Malley and J. G. Hardman, eds.) p. 198. Academic Press, New York, 1975.

86. Dahmus, M. E., *Biochemistry* **15**, 1821 (1976).

87. Olson, M. O. J., and Guetzow, K., *Biochem. Biophys. Res. Commun.* **70**, 717 (1976).

88. Ullman, B., and Perlman, R. L., *Biochem. Biophys. Res. Commun.* **63**, 424 (1975).

89. Mirault, M. E., and Scherrer, K., *FEBS Lett.* **20**, 233 (1972).

90. Lazarus, H. M., and Sporn, M. B., *Proc. Natl. Acad. Sci. U.S.A.* **57**, 1386 (1967).

91. Choi, Y. C., Smetana, K., and Busch, H., *Exp. Cell Res.* **53**, 582 (1968).

92. Liau, M. C., Craig, N. C., and Perry, R. P., *Biochim. Biophys. Acta* **169**, 196 (1968).

93. Kelley, D. E., and Perry, R. P., *Biochim. Biophys. Acta* **238**, 357 (1971).

94. Klee, W., *in* "Proceedings in Nucleic Acid Research" (G. L. Cantoni and D. R. Daives, eds.), p. 20. Harper, New York, 1966.

95. Randerath, K., and Randerath, E., *J. Chromatogr.* **16**, 111 (1964).

96. Randerath, E., and Randerath, K., *Anal. Biochem.* **12**, 83 (1965).

97. Randerath, K., and Randerath, E., *in* "Methods in Enzymology," Vol. 12: Nucleic Acids (L. Grossman and K. Moldave, eds.) Part A, p. 323. Academic Press, New York, 1967.

98. Seibert, G., Villalobos, J., Jr., Ro., T. S., Steele, W. J., Lindenmayer, G., Adams, H., and Busch, H., *J. Biol. Chem.* **241**, 71 (1966).

99. Hilgers, J., Nowinski, R. C., Geering, G., and Hardy, W., *Cancer Res.* **32**, 98 (1972).

100. Crowle, A. J., "Immunodiffusion." Academic Press, New York, (1972).

101. Yeoman, L. C., Taylor, C. W., Jordan, J. J., and Busch, H., *Exp. Cell Res.* **91**, 207 (1975).

102. Busch, R. K., Daskal, I., Spohn, W. H., Kellermayer, M., and Busch, H., *Cancer Res.* **34**, 2362 (1974).

103. Busch, R. K., and Busch, H., *Tumori*, in press.

Part C.   Fractionation and Characterization of Histones. II

Chapter 15

Electrophoretic Fractionation of Histones Utilizing Starch Gels and Sodium Dodecyl Sulfate–Urea Gels

LUBOMIR S. HNILICA, SIDNEY R. GRIMES, AND JEN-FU CHIU

Department of Biochemistry,
Vanderbilt University School of Medicine,
Nashville, Tennessee

I.   Introduction

Among the techniques for protein separation and fractionation, electrophoresis in polyacrylamide gels and, to a lesser extent, in starch gels have contributed enormously to the rapid advancement of histone biochemistry and biology. Histone separation in polyacrylamide gels is clearly superior to any other method for their qualitative and quantitative analysis, and without doubt it should be standard procedure in every laboratory studying these proteins. While polyacrylamide gels containing urea in addition to the employed buffers produce the best separation of the individual histone fractions, the somewhat inferior systems containing either polyacrylamide gels with sodium dodecyl sulfate (SDS) or starch gels with urea can be useful under special circumstances.

II.   Starch Gel Electrophoresis

The starch gel electrophoresis was developed by Smithies (*1*) who showed the exceptional resolution power of this new method in his studies on proteins of human plasma and serum. The starch gel electrophoresis was first applied to the histones by Neelin and Neelin (*2*) who resolved calf thymus histone preparations into 16–22 protein bands. Since the number and intensity of several of the calf thymus histone bands varied with changing conditions of the electrophoresis (pH of the buffer, starch concentration, presence or absence of urea, etc.), it was concluded that some of the electrophoretic bands must be aggregates of some of the histone fractions and that there are probably only a few histone species. This conclusion was supported by the experiments of Johns *et al.* (*3*) who used a simplified horizontal version of starch gel electrophoresis. With 10 m*M* HCl as the only electrolyte, these authors separated calf thymus histones into 5–6 reproducible bands, all of which corresponded to the major histone fractions obtained by chemical fractionation procedures.

Although starch gel electrophoresis of histones is now almost completely abandoned to a much more elegant, rapid, and sensitive analysis in acrylamide polymers, it is still the method of choice for metabolic studies on histones in higher animals, where low specific activities of the fractions exclude their accurate radioactivity analysis by polyacrylamide gels. It is possible to accommodate and resolve 1–2 mg of whole histone mixture in one starch gel slot as compared with 25–50 μg tolerated by each standard polyacrylamide gel.

The method described here (*4*) was adapted from the horizontal system of Johns *et al.* (*3*). Hydrolyzed starch (Connaught Medical Research Laboratories, Toronto, Canada), 12% suspension in 10 m*M* HCl–0.2 m*M* AlCl$_3$, is heated to 85–90° and the hot viscous mixture is poured into a plastic mold permitting the formation of at least eight strips, each measuring 15 × 6 × 250 mm. The strips are formed by septa separating each slot longitudinally. These individual troughs communicate at both ends through a horizontal cut across all the septa dividing the individual slots. Two rectangular pieces of Whatman 3 MM filter paper (double thickness) measuring approximately 120 × 180 mm are inserted into the cross cuts on both ends of the mold to serve as connectors (bridges) to electrode vessels (beakers). They are held in place by solidified starch gel. After cooling to room temperature, the elevated gel surface is removed by pulling thin molybdenum wire (3 mil) across the gel using the surface of the mold for guidance. The upper starch layer is discarded and histone samples, dissolved in 0.1 *M* HCl, are applied soaked into rectangular pieces (5 × 14 mm) of Whatman No. 17 filter paper by inserting the rectangles into narrow cuts in the individual gels, about 3–4 cm from the anodic end. To prevent evaporation, the gel surface and filter paper bridges (wicks) connecting the beakers with electrodes are covered with a sheet of plastic (Saran Wrap). The electrode vessels are two 1000-ml beakers containing about 900 ml of 10 m*M* HCl and 0.2 or 0.6 m*M*

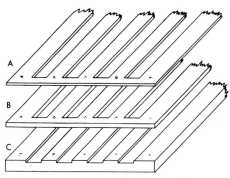

FIG. 1. Frame for slicing central portions out of starch gels for quantitative evaluation. The frame is made from Plexiglas, and the individual layers serve as leading edges for the cutting wire which is drawn between them.

AlCl$_3$. The higher concentration of AlCl$_3$ increases the separation of H2A and H4 histones. The separation is obtained at 40 mA of constant current for 22 hours (5 mA per slot) at room temperature. After electrophoresis, the filter-paper bridges are removed and the mold is placed in a refrigerator for about 30 minutes to facilitate transfer of the individual gel slabs. Approximately 16-cm-long pieces of gels are lifted from their troughs with a spatula and transferred into individual test tubes containing 0.1% Amido black in a mixture of methanol–water–acetic acid (5:5:2, v/v/v). After staining for about 18 hours (gentle rocking accelerates the staining), the gels are partially destained in the methanol–water–acetic acid solution without the dye approximately 2 hours. The partially destained gels are transferred into a cutting frame (Fig. 1), where approximately 1-mm-thick portions are sliced off the top and bottom of each gel using a thin molybdenum wire (3 mil). The central gel portions measuring 3 × 15 × ~150 mm are placed into test tubes containing methanol–water–acetic acid mixture and destained under gentle rocking motion. The destaining solution is changed 2–3 times until the gels show dark blue-stained histone bands against an almost white or pale blue background (Fig. 2).

For quantitative evaluation, the individual protein bands as well as a piece of gel without any protein (ahead of the fastest H4 fraction) are cut from the gels with a razor blade. The cut pieces of gels are placed into individual test tubes and dried with acetone (three changes, each 4–6 hours), in a stream of dry air, and finally in an oven at 105° for 3–4 hours. The dry pieces are weighed and each is dissolved in 7 ml of concentrated formic acid by heating in a boiling water bath for 20 minutes. After rapid cooling in ice water, 7 ml of 1 M NaOH are added and the mixture is mixed vigorously.

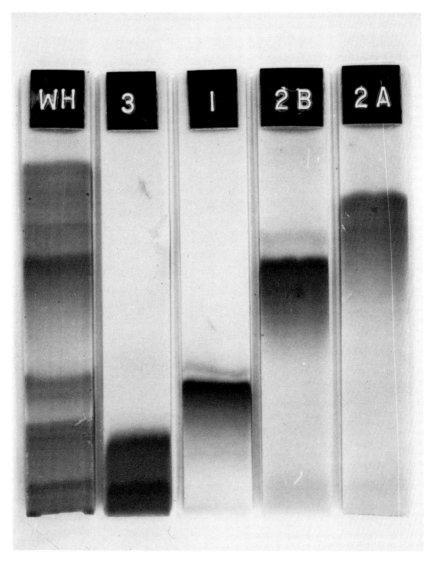

FIG. 2. Starch gel electrophoregram of calf thymus histone fractions. The electrolyte was 0.2 m*M* AlCl₃ in 10 m*M* HCl. WH = unfractionated histone; 3 = H3 (oxidized dimer); 1 = H1; 2B = H2B; and 2A = mixture of H2A and H4.

The absorbancy determined at 630 nm is corrected for the background:

$$A_{corr} = A_s \left[(A_b \times W_s)/W_b \right]$$

where A_{corr} = corrected absorbance, A_s and A_b are absorbances of the

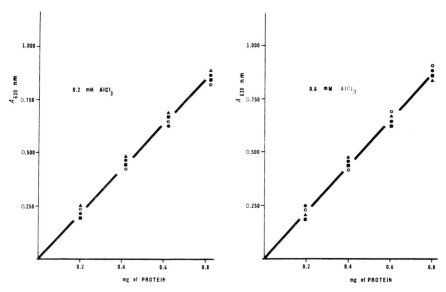

FIG. 3.   The absorbancy plot of four major calf thymus histone fractions after staining with Amido black 10 B. (●) H1; (○) H2A and H4; (▲) H2B; and (■) H3.

sample and blank gel, respectively, and W_s and W_b are dry weights of the sample and blank, respectively. The absorbance remains linear for the amount of histone between 0.05 and 1.0 mg of protein in each band. The amount of dye bound to the protein in starch gel is independent from the composition of histone fractions and from the concentration of $AlCl_3$ (Fig. 3). However, since occasional batches of Amido black show differences in their binding to the Hl histones, it is recommended that each new batch of this dye be calibrated with isolated histone fractions.

The radioactivity of histones labeled with $[^{14}C]$- or $[^3H]$amino acids or acetate is determined by combusting the dried gel pieces in oxygen (5–7). The method of Kalberer and Rutschmann (5) can be followed with excellent results. Dried gels are wrapped into small strips of Whatman No. 1 filter paper, placed into platinum baskets, and combusted in special flasks filled with oxygen under atmospheric pressure. The radioactive products of combustion are trapped in a mixture of monoethanolamine in methyl cellosolve (1:8, v/v), aliquots are mixed with a toluene-based scintillation fluid, and their radioactivity is determined in a liquid scintillation spectrometer. A combustion apparatus made by Packard Co. can be used with great convenience to process large numbers of samples in a single day. Because of the yellowish coloring of the samples, construction of quench curves to measure the extent of quenching in the individual samples is recommended.

Usually one sample of histone is electrophoresed in at least eight slots (troughs). One-half (four gels) is used for quantitative determination of protein, and the other half (four gels) is combusted to determine the radioactivity. All values are averaged, yielding quite accurate figures of specific activities for each histone band. Histone samples with specific activities of about 100 cpm/mg or more can be analyzed with confidence. It is preferable to use chemically prefractionated histone samples for electropheresis instead of whole extracts. In a typical determination, chromatin or nuclei are first washed with 80% aqueous ethanol. The arginine-rich histones are extracted with absolute ethanol: 1.25 M HCl (4:1, v/v) followed by extraction of the lysine-rich histones with 0.25 M HCl (8). Histones recovered from extracts either by dialysis and lyophilization or by precipitation with a 10- to 15-fold excess of cold acetone are electrophoresed separately to obtain a better resolution. A modification of histone electrophoresis in starch gels was described by Sung and Smithies (9). This modification employs aluminum lactate buffer and 4 M urea for separation of histones similar to the method described here in detail.

III.   Histone Electrophoresis in Polyacrylamide Gels Containing SDS

As was already mentioned, the best method for rapid quantitative and qualitative separation of all histone fractions as well as their phosphorylated or acetylated derivatives is electrophoresis in polyacrylamide gels, which was first developed by Ornstein (10) and Davis (11). While the application of SDS to this technique results in some loss of resolution as compared to the now classical method of Panyim and Chalkley (12), the SDS gels are valuable when the mixtures of histones and nonhistone proteins are to be analyzed.

Extensive aggregation and insolubility of chromosomal nonhistone proteins in most conventional electrophoresis buffers led to the introduction of polyacylamide gel electrophoresis performed in the presence of detergents, most frequently SDS (13). Because of their interactions with this detergent, most proteins form negatively charged complexes which separate electrophoretically in polyacrylamide gels in order of their molecular weights (13, 14). Indeed, the electrophoretic mobility of proteins larger than 10,000 daltons can be used for fairly accurate molecular weight determination (13, 15).

SDS dissociates most proteins into their subunits (unless covalently linked) and, consequently, the number of proteins bands observed in poly-acrylamide gel electrophoresis performed in the presence of this detergent corresponds to the number of subunits and may be much larger than the true number of the *in vivo* protein species. On the other hand, polypeptide chains of similar or identical molecular weights will form a single protein band. Since histones are simple proteins, their electrophoretic homogeneity is not affected by exposure to SDS. However, their molecular weights just about approach the lowest range of confidence for molecular weight deter-minations of the SDS–protein complexes (*15*). The excessive electropositive charge of the carboxyterminal half of the H1 histones which results from a large accumulation of lysyl residues disturbs the distribution of the negatively charged SDS micelles along the H1 molecule. This disturbance is reflected in an anomalous electrophoretic mobility of H1 histones in SDS-containing gels. Consequently, the apparent molecular weight of about 34,000 calcul-ated for H1 histones from their mobilities in SDS gels differs considerably from their true molecular weights of 19,500–22,000 (average) as determined physicochemically (*16*). These disturbances are even greater when SDS interacts with more basic proteins of male gametes. Such complexes (e.g., of protamines or arginine-rich proteins from mammalian spermatids and spermatozoa) may become partially insoluble in electrophoretic buffers, and their electrophoretic separation in the presence of anionic detergents cannot be accomplished.

The electrophoresis in SDS-containing gels can be performed both with or without urea. Since SDS eliminates virutally all the protein–protein interactions, the presence of urea is not as beneficial as in the detergent-free polyacylamide gel systems. Elimination of the positive charges on histones by SDS makes impossible the electrophoretic separation of differentially acetylated or phosphorylated histone fractions.

A reasonably good separation of histones, even in the presence of chromo-somal nonhistone proteins, can be obtained by a modified procedure of Maizel and his associates (*13*).

Polyacrylamide gels (10% polyacrylamide containing 4 M urea) can be prepared by mixing the following stock solutions:

A Gel buffer: 0.2 M phosphate buffer–0.2% SDS–8 M urea (pH 7.0).

B. Acrylamide solution: 24% acylamide (recrystallized commercial or electrophoretic grade) and 0.6% Bis (electrophoretic grade N,N'-methylene-bisacylamide) in deionized water.

C. Ammonium persulfate, 90 mg in 10 ml of deionized water (must be made fresh before use).

D. TEMED (N,N,N',N',-tetramethylethylenediamine), undiluted re-agent.

The solutions are mixed in the ratio A:B:C:D = 6:5:1:0.06. For 12–14 gels

(10 × 0.6 cm each), 30 ml of A are mixed with 25 ml of B and 5 ml of C, and the mixture is cooled on ice under reduced pressure for 15 minutes. Finally, 0.03 ml of TEMED (solution D) are added and the gels are poured immediately into stoppered glass columns to the desired height. Chilling the monomer mixture delays the polymerization and allows the time necessary for the manipulation of the individual columns. Disposable plastic syringe or Pasteur pipettes can be used conveniently to transfer the monomer mixture. The filled columns are freed of any entrapped air bubbles and carefullly overlaid each with 0.1 ml of solution A diluted 1:1 with deionized water. The gels are polymerized for at least 2 hours at room temperature. In the presence of a large excess of nonhistone proteins or DNA, a 3% polyacrylamide stacking gel should be layered over the 10% gel. To prepare the stacking gel, a modified solution B containing 7.5% acrylamide and 1.2% Bis in deionized water is used instead of the regular solution.

The columns are made of fire-polished 0.6-cm-bore glass tubing, cut to the desired height, washed in chromic solution, rinsed well with water, and coated with silicone (the silicone coating is not important in SDS-containing gels because these can be removed from the columns with relative ease). To prepare for electrophoresis, the tubes are closed on one end with rubber caps and placed into a polymerization rack which can be a small test tube or a special stand made of Plexiglas or other material.

Samples containing 10–30 μg of unfractionated histone are dissolved in solution A, which has been made more dense by the addition of concentrated sucrose or urea, and a measured amount of the solution is carefully layered on top of each gel. The sample volume should not exceed 50 μl to assure good resolution of the histone fractions. After filling the columns with electrolyte buffer, the rubber caps are gently removed from the columns. The columns are inserted into the upper tray of an electrophoretic apparatus (some commercial instruments do not have a separable upper and lower tray), which is then filled with electrolyte buffer (solution A), and the samples are electrophoresed at 8–10 mA per column for 8–10 hours. Application of a tracking dye (Bromophenol blue) to at least one of the samples helps to monitor the migration. After electrophoresis, the gels are stained in the solution of Coomassie blue or Amido black (0.2% or 0.5%, respectively) in methanol–acetic acid–water mixture 5:1:5, or they can be first fixed in 20% aqueous sulfosalicylic acid or 15% trichloroacetic acid. This treatment also removes most of the detergent which may interfere with subsequent staining. Some batches of Coomassie blue are incompatible with trichloroacetic acid, and it may be necessary first to remove the acid by placing the fixed gels into 20% sulfosalicylic acid solution for several hours. A mixture of methanol–acetic acid–water (5:1:5) can be also used to remove the 15% trichloroacetic acid from the gels before their staining with Coomassie blue. The gels can be

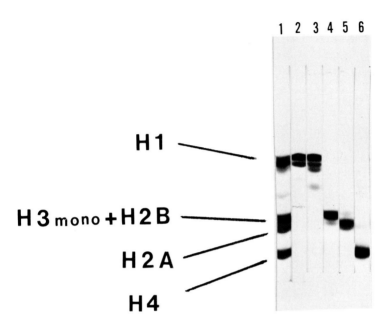

FIG. 4. Polyacrylamide gel electrophoresis in the presence of SDS of total rat thymus histones and calf thymus histone fractions. 1 = total rat thymus histone; 2 and 3 = H1 histones from rat thymus and liver, respectively; 4 = calf thymus H2B; 5 = calf thymus H2A; and 6 = calf thymus H4. The origin of migration is at the top of the gels.

destained by diffusion in an excess of methanol–acetic acid–water mixture or, much faster, by electrophoresis in 7% acetic acid.

In the presence of SDS, the following histone fractions, can be resolved in order of their increasing mobilities: oxidized H3 dimer, H1, a mixture of H3 monomer and the H2B fraction, the H2A, and finally the H4 bands (Fig. 4).

In addition to electrophoresis in polyacrylamide gels contained in glass tubing, electrophoresis in relatively thin gel slabs (1–5 mm) is rapidly gaining popularity. Besides the better resolution of the slab-gel electrophoresis, the uniformly thin sheets of polyacrylamide can be easily dehydrated and used for autoradiography or scanning. The stained gel can also be impregnated with 2,5-diphenyloxazole solution and dried for fluorography (17).

A good electrophoretic separation of all common histone fractions can be obtained by adaptation of the discontinuous method of Laemmli (18)

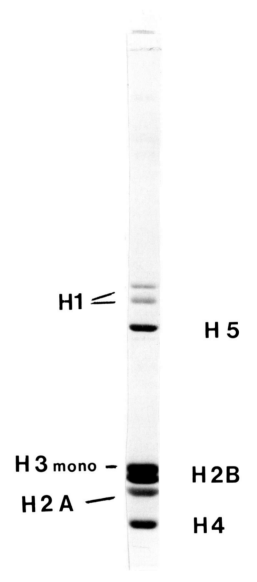

Fig. 5. Polyacrylamide gel electrophoresis in the presence of SDS in slab gels. This photograph illustrates the good resolution of all histones from chicken reticulocyte chromatin. The histone fractions are identified by their symbols. The origin of migration is at the top of the gel. (Courtesy of Dr. A. Beyer and Dr. K. Hardy.)

using a gel of 12% or 15% polyacrylamide (Fig. 5). The gel can be formed from the following solutions:

A. Acrylamide solution: 30% acrylamide and 0.8% Bis (*NN*-methylene-bisacrylamide) in deionized water.

B. 1.5 *M* Tris-HCl, pH 8.8 (ultrapure).

C. 10% SDS.

D. 0.5 *M* Tris-HCl, pH 6.8.

E. 1.5% ammonium persulfate (made fresh).

F. TEMED (*N,N,N',N'*-tetramethylethylenediamine), undiluted reagent. In order to prepare 50 ml of solution for a 12% polyacrylamide gel, 20 ml of solution A, 12.5 ml of B, and 16 ml of water are mixed and kept on ice under reduced pressure for 15 minutes. Finally 0.5 ml of C, 1.0 ml of E, and 12 μl of F are added.

A 10-ml pipette can be used to introduce 30 ml of the mixed solution into a 14 $\times$ 15.5 cm vertical gel chamber which is formed of two glass plates separated by plastic spacers. Two spacers for the sides are 1 $\times$ 19 cm and the bottom spacer is 1 $\times$ 17 cm. The gel is overlayered with 0.1% SDS until polymerization is complete. A spacer gel of 2 or 3% polyacrylamide is prepared by mixing 1.0 ml of solution A, 2.5 ml of D, 0.1 ml of C, 0.7 ml of E, 5 μl of F, and 6.3 ml of deionized water. The overlayer solution is removed and the gel surface is rinsed with the spacer gel solution. The upper gel is poured with a 10-ml pipette, and a plastic template is inserted to form sample wells 0.7 $\times$ 2.0 cm. The slab gel can be used for a preparative separation of proteins by omitting the template so that a single sample can be applied to the entire gel width.

After polymerization, the template and the bottom spacer are removed, and buffer solution [0.2 *M* glycine–0.025 *M* Tris-HCl (pH 8.3)–0.1% SDS] is added to the upper and lower reservoirs. Proteins (usually dissolved in water) are mixed with an equal volume of sample mix (2.5 ml of solution D, 4.0 ml of C, 2.0 ml of glycerol, 1.0 ml of 2-mercaptoethanol, 0.2 ml of 0.05% Bromophenol blue, and 0.3 ml of water) and heated in boiling water for 1 to 2 minutes or kept at 37° for 1 hour. Samples are centrifuged at 1000 *g* to remove any insoluble materials.

Samples containing 10–50 μg of unfractionated histone in a volume less than 100 μl are placed into each sample well. The samples are electrophoresed at 20 mA until the tracking dye reaches the bottom of the gel (about 6 hours). The gel is removed and stained overnight in 0.1% Coomassie blue in 25% isopropanol and 10% acetic acid as described by Fairbanks *et al.* (*19*). The gel is destained for 6 to 9 hours in 0.03% Coomassie blue in 10% isopropanol and 10% acetic acid followed by several changes in 10% acetic acid. The stained gel can be impregnated with 2,5-diphenyloxazole and dried for fluorography (*17*) or dehydrated onto cellophane sheets and cut for scanning or liquid scintillation counting.

The slab gels can also be adapted for preparative elution electrophoresis by forming a channel near the lower edge of the gel for a continuous flow of buffer to carry off proteins separated by electrophoresis. A lower gel is poured up to the elution part in the 3-mm plastic side spacers. A narrow band of glycerol is added to the surface of the polymerized lower gel, and the separating gel is layered carefully over the glycerol. A template forming one large well is inserted into the upper gel during polymerization. When electrophoresis is started the glycerol is forced out of the elution port as the sweeping buffer flow is started. This method allows isolation of relatively pure protein fractions for amino acid analysis or other chemical determinations. Various commercial apparatuses permitting the analytical as well as preparative slab-gel electrophoresis can be obtained from several suppliers. Exceptionally good results were obtained with an apparatus acquired from Hoefer Scientific Instruments, San Francisco, California. The slab-gel method described here was performed with this apparatus.

ACKNOWLEDGMENTS

This work was supported by U.S. Public Health Service Grant CA 18389 and by NCI contract NO1-CP-65730.

REFERENCES

1. Smithies, O. *Biochem. J.* **61**, 629 (1955).
2. Neelin, J. M., and Neelin, E. M., *Can. J. Biochem. Physiol.* **38**, 355 (1960).
3. Johns, E. W., Phillips, D. M. P., Simson, P., and Butler, J. A. V., *Biochem. J.* **80**, 189 (1961).
4. Hnilica, L. S., Edwards, L. J., and Hey, A. H., *Biochim. Biophys. Acta* **124**, 109 (1966).
5. Kalberer, F., and Rutschmann, J., *Helv. Chim. Acta* **44**, 1956 (1961).
6. Dobbs, H. E., *Anal. Chem.* **35**, 783 (1963).
7. Baggiolini, M., *Experientia* **21**, 731 (1965).
8. Johns, E. W., Phillips, D. M. P., Simson, P., and Butler, J. A. V., *Biochem. J.* **77**, 631 (1960).
9. Sung, M., and Smithies, O., *Biopolymers* **7**, 39 (1969).
10. Ornstein, L., *Ann. N. Y. Acad. Sci.* **121**, 321 (1964).
11. Davis, B. J., *Ann. N. Y. Acad. Sci.* **121**, 404 (1964).
12. Panyim. S., and Chalkley, R., *Arch. Biochem. Biophys.* **130**, 337 (1969).
13. Shapiro, A. L., Viñuela, E., and Maizel, J. V., *Biochem. Biophys. Res. Commun.* **28**, 815 (1967).
14. Dunker, A. K., and Rueckert, R. R., *J. Biol. Chem.* **244**, 5074 (1969).
15. Williams, J. G., and Gratzer, W. B., *J. Chromatogr.* **57**, 321 (1971).
16. Hnilica, L. S., "The Structure and Biological Functions of Histones" Chem. Rubber Publ. Co., Cleveland, Ohio, 1972.
17. Bonner, W. M., and Laskey, R. A., *Eur. J. Biochem.* **46**, 83 (1974).
18. Laemmli, U.K., *Nature* (*London*) **227**, 680 (1970).
19. Fairbanks, G., Steck, T. L., and Wallach, D. F. M., *Biochemistry* **10**, 2606 (1971).

Chapter 16

Resolution of Histones by Polyacrylamide Gel Electrophoresis in Presence of Nonionic Detergents

ALFRED ZWEIDLER

The Institute for Cancer Research,
The Fox Chase Cancer Center,
Philadelphia, Pennsylvania

I. Introduction

Histones are difficult to resolve by classical biochemical fractionation techniques because of their similarity in size and charge, their tendency to aggregate, and the high frequency of post-transcriptional charge modification (acetylation and phosphorylation). For recent reviews of histone chemistry see Elgin and Weintraub (1) and Zweidler (2). The most widely used analytical system for histones, polyacrylamide gel electrophoresis at low pH (3), resolves five major histone species and some of their modified forms. However, in most cases, it is not possible to resolve histones 2A, 2B, and 3 completely because of overlapping modified forms. The resolution of the histones can be improved significantly by the addition of nonionic detergents to the gels which results in a differential reduction in the electrophoretic mobility of different histones [Fig. 1 (4)]. This effect is due to the formation of mixed micelles between the detergent and the hydrophobic regions of protein molecules and is very sensitive to small differences in the hydrophobic properties of the proteins (5,6). This method has been valuable for:

1. The isolation and characterization of the histones of *Drosophila* (7) and histone variants as well as minor histone species in mammals (5,6).

2. The detection of switches in the synthesis of different histone variants during embryogenesis of the sea urchin (8).

3. The quantitation of the different histone components (2).

4. The identification and quantitation of histone modification (9–11).

5. Characterizing histone messenger RNAs (12,13).

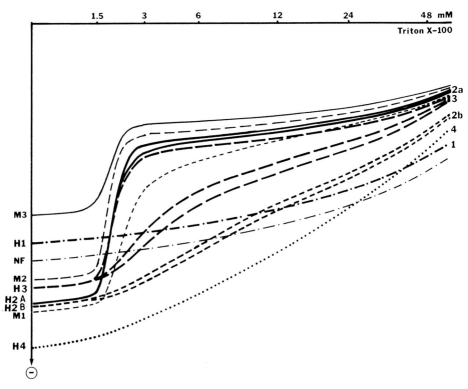

II. Method

A. General Considerations

Since the detergent affinity of different proteins varies widely, the electrophoretic conditions have to be varied for optimal resolution of different protein mixtures. For technical reasons it is in most cases advantageous to use the same detergent concentration (e.g., 6 mM = 0.375% Triton X-100) and to vary the concentration of urea (an inhibitor of detergent binding) in the gel (Fig. 2). The high sensitivity of the method for small changes in the hydrophobic properties of the protein requires some extra precautions in the preparation of the sample to prevent chemical modifications such as the oxidation of methionine, tryptophan, and cysteine residues. The proteins have to be isolated and stored in the presence of reducing agents (dithioerythritol, mercaptoethanol). All glassware (which is usually washed with detergents containing strong oxidizing agents) should be rinsed with reducing agents before use. The polyacrylamide gels (which are polymerized using oxidizing agents as catalysts) have to be scavenged with reducing agents after pre-electrophoresis. The alternative of oxidizing all methionine residues with hydrogen peroxide in acid pH (*14*) is not recommended as a general procedure because of the drastic conditions necessary for the complete oxidation of certain methionine residues.

B. Materials

In order to obtain reproducible results, purified chemicals should be used. Cation electrophoresis is especially sensitive to contamination of acrylamide with acrylic acid.

◀ FIG. 1. Differential change in the electrophoretic mobility of histones in the presence of increasing concentrations of nonionic detergents. Rat liver histones were loaded across the top of a 12% polyacrylamide slab gel containing 5% acetic acid–6 M urea and a transverse, concave gradient of 0 to 48 mM Triton X-100, made by filling the gel chamber left side with a density gradient formed by pumping 25 ml of gel solution containing 48 mM Triton X-100 and 0.5 M sucrose into the mixing chamber containing 75 ml of gel solution without detergent or sucrose at a pumping ratio of 1:4 (in to out of mixing chamber). The scale of Triton concentration was calibrated on the basis of the electrophoretic patterns of the same sample electrophoresed on individual disc gels containing the indicated detergent concentrations. Histone 1 does not bind any detergent under these conditions and serves as a control for the change in electrophoretic mobilities due to differences in viscosity. Note the continuous differential reduction in electrophoretic mobility of different histones relative to H1 as well as the resolution of H2A, H2B, and H3 subcomponents and several minor histones (M1, M2, M3 and NF).

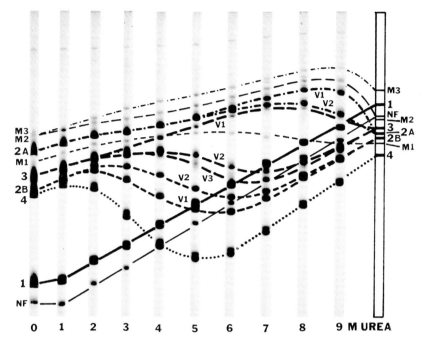

FIG. 2.   Dependence of detergent effect on the concentration of urea. The same sample of mouse liver histones was electrophoresed (from top to bottom) in 12% polyacrylamide gels containing 5% acetic acid–6 mM Triton X-100 and different concentration of urea, as indicated. On the extreme right indicated schematically is the position of the different components in the absence of detergent. Histone 1 does not bind detergent under these conditions and serves as a marker for the effect of viscosity on the electrophoretic mobility. Note the inhibition of the detergent effect with increasing concentrations of urea, and the different conditions necessary for optimal resolution of variants (V1, V2, V3) of histones 2A, 2B, and 3, as well as the minor histones M1, M2, M3, and NF.

STOCK SOLUTIONS

1. *Acrylamide* (60% acrylamide–0.4% bisacrylamide). Dissolve 60 gm of acrylamide (Sigma) in 44.4 ml of water. Add 3 gm of Amberlite MB-1 mixed bed resin (A. H. Thomas) and 1 gm of Darco G-60 activated charcoal (A. H. Thomas), stir for 30 minutes, and filter through a glass-fiber filter. Add 0.4 gm of N,N'-methylene bisacrylamide (electrophoresis grade).

2. *Urea–acetic acid* (10 M urea–6.667% acetic acid). Dissolve 60.06 gm of urea (absolute grade, R plus) in 48.32 ml of water and 6.67 ml of glacial acetic acid (reagent grade). Filter twice through a glass-fiber filter.

3. *Triton X-100* (0.3 *M*). Dissolve 18.75 gm of Triton X-100 (Sigma) in

83 ml of water. Check concentration by measuring the optical density of 1/1000 dilution (0.3 mM) at 275.5 nm (should be 0.400), and adjust the volume if necessary. Store in presence of chelating resin (Dowex Chelating resin 50–100 mesh, Sigma) to prevent growth of microorganisms.

C.   Gel Preparation

1. Mix the ingredients (except the ammonium persulfate) as described in Table I, and deaerate for a few minutes under vacuum.

2. Add the freshly prepared ammonium persulfate solution, and mix quickly and thoroughly without forming air bubbles.

3. Pour the gel solution into clean glass containers to the desired height (avoid bubbles), and overlayer quickly with 2–4 mm of water without mixing the interphase.

4. After 2–4 hours assemble the electrophoresis apparatus, and fill the electrode chambers with 5% acetic acid. Disc gels have to be supported by a nylon screen held in place with a small piece of silicon rubber tubing; this prevents them from slipping out of the tubes during electrophoresis.

5. Overlayer the gels with a 1–2 cm high layer of 5% acetic acid containing the same urea and detergent concentration as the gel.

6. Pre-electrophorese to constant wattage (3).

7. Overlayer the gels with a 5-mm high layer of 5% acetic acid containing

TABLE I

SOLUTIONS FOR MAKING DIFFERENT TYPES OF POLYACRYLAMIDE GELS

Stock solutions	Type of gel (M urea/mM Triton X-100)[a]			
	3/0	3/6	6/6	7.5/6
60% acrylamide, 0.4% bisacrylamide	0.200	0.200	0.200	0.200
10 M urea, 6.67% acetic acid	0.300	0.300	0.600	0.750
0.3 M Triton X-100	0	0.020	0.020	0.020
TMEDA[b]	0.005	0.005	0.005	0.005
Acetic acid, glacial	0.028	0.028	0.005	0
Water	0.442	0.442	0.150	0.008
Deaerate				
40 mg/ml ammonium persulfate	0.025	0.025	0.020	0.017

[a]Amounts are given as fractions of total volume.
[b]TMEDA = N,N,N',N'-tetramethylethylenediamine (Sigma).

urea and detergent as in the gel plus 0.5 M cysteamine–HCl. Continue pre-electrophoresis for another 30 minutes.

  8. Change the electrode buffer.

D. Sample Preparation, Electrophoresis, Staining

  9. Dissolve the protein sample at a concentration of ca. 1 mg/ml in 8 M urea–5% mercaptoethanol–0.01% Pyronin Y.

  10. After rinsing the top of the gel with 5% acetic acid, layer ca. 2 (1 to 5) μl of sample/mm^2 of gel surface directly onto the gel with a microsyringe.

  11. Electrophorese at ca. 4 mA/cm^2 gel surface area until the fastest component reaches a point ca. 2 cm from the bottom of the gel (determine in a trial run for your electrophoresis set-up).

  12. Remove the gel and stain in filtered 0.2% Amido black 10B (Merck) dissolved in 10% acetic acid–45% methanol (or 25% isopropanol) for 1–2 hours at 60°C (4–16 hours at room temperature). Destain electrophoretically or by diffusion in 5% acetic acid–25% methanol (or isopropanol). For better visualization of minor components, poststain with 0.1% Coomassie blue in 10% acetic acid–45% methanol. Destain by diffusion in 5% acetic acid–25% methanol.

III. Interpretation and Evaluation

A. Components of Mammalian Histone Fractions

  With very few exceptions, all tissues of all mammalian species which have been examined so far contain the same set of histones, although in different relative amounts (2). There are five major histone classes, four of which consist of a small number of closely related primary structure variants (2,6, 15), and at least five minor histone species, which have all the biochemical properties typical for histones but their amino acid compositions are not closely related to any of the major histone classes (2). The very lysine-rich histones (H1) do not bind nonionic detergents and are not well resolved by electrophoretic methods. They are best resolved by chromatography on Amberlite CG-50 resin (15). All the other histones are best resolved by polyacrylamide gel electrophoresis in the presence of nonionic detergents. Figure 2 illustrates the resolution of mouse liver histones by electrophoresis in the presence of 6 mM Triton X-100 and different concentrations of urea. The electrophoretic mobilities relative to H1 of the known mammalian histones in the four electrophoretic conditions described in the method section

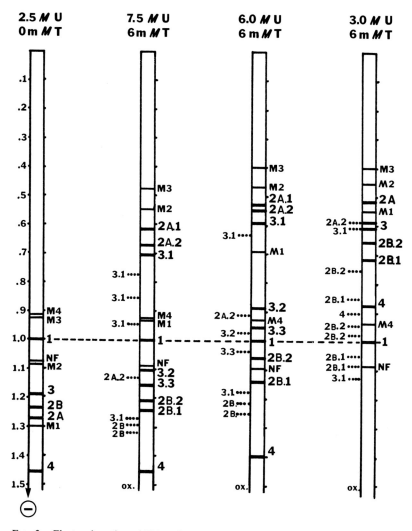

FIG. 3. Electrophoretic mobilities of mammalian histones in the different electrophoretic conditions described in the text. For simplicity, only the unmodified form of each histone species is indicated. The electrophoretic mobility is expressed relative to H1. The positions of oxidation products are indicated on the left in each case (...). Nomenclature as in Figs. 1 and 2.

are given in Fig. 3 together with components produced by the oxidation of methionine residues. Identification of individual components on the basis of electrophoretic mobilities should involve analyses in at least two different electrophoretic systems. Additional components found in mammalian histone preparation [except for testes-specific histones (16)] are likely to be either degradation products of histones or nonhistone proteins. Such components can usually be eliminated by isolating the histones in the presence of proteolytic inhibitors (phenylmethyl sulfonyfluoride, tosyl-lysine chloromethyl ketone) from purified chromatin extracted with 0.35 M NaCl.

B.   Histones from Other Sources

The histones of *Drosophila* embryos have also been analyzed and isolated by nonionic detergent electrophoresis (7). For a number of other species the major histones have been identified in other electrophoretic systems (3). In these cases the position of the major histones in the nonionic detergent electrophoresis system can be determined by two-dimensional electro-phoresis [Fig. 5 (2)]. A tentative identification of the major histones can be made on the bases of:

1. Staining properties in Amido black 10B after extensive destaining (H2A and H3 stain green, H1, H2B, and other lysine-rich histones stain purple, and H4 is intermediate).

2. The pattern of histone modification (H3 and H4 are usually highly acetylated and form triple bands, H2A is phosphorylated to form a double band, and H2B is only slightly modified and usually appears as a single band).

3. The detergent affinity (H2A highest, H1 lowest, and H2B, H3, and H4 intermediate).

4. Chemical fractionation (17).

5. Ion exchange chromatography (18).

For positive identification and classification, the electrophoretic components have to be isolated by preparative electrophoresis and characterized by amino acid analysis and, if necessary, even peptide analysis (6). Oxidation products can be identified by following the shift of certain histones after incubating aliquots of the sample in 8 M urea–50 mM perchloric acid containing different concentrations of hydrogen peroxide (e.g., 0.1 to 2%) for 10 minutes at room temperature (14). The reaction is stopped by adding 10 volumes of cold acetone containing 0.2% HCl and washing the precipitated histones twice with acetone. Analysis of oxidation products can be a valuable tool for distinguishing histones on the basis of methionine content [e.g., H2A variants (8)] and for comparing the location of methionine residues within the polypeptide chain [e.g., H2B of different species (7)]. Alternatively, some (e.g., those in H2A and H2B) but not all methionine residues can be reduced by treating the sample with 8 M urea containing 10% mercaptoacetic acid (under exclusion of oxygen) for 1 to 4 days at room temperature or several hours at 60°C.

IV. Applications

Polyacrylamide gel electrophoresis in the presence of nonionic detergents is capable of resolving variants or mutants differing only in the substitution of a single neutral amino acid (5, 6). It supplements existing electrophoretic methods which separate on the basis of charge and/or size only and constitutes a powerful new method for analyzing the purity of and detecting microheterogeneity in protein fractions isolated by ion exchange chromatography and gel filtration, or other methods separating on the basis of size and charge.

An additional application of the method is as a means for monitoring the oxidation of methionine residues, which could be a major cause of loss of biological activity. In the case of histones, the oxidation of methionine residues would probably interfere with the reconstitution of histone and nucleohistone complexes.

Polyacrylamide gel electrophoresis in the presence of urea and acetic acid (3) has a number of advantages (less aggregation, elimination of microheterogeneity due to deamidation) for resolving complex protein mixtures. The addition of nonionic detergents improves the method even further by increasing the solubility of many proteins, by producing more uniform polyacrylamide gel supports, and by offering the possibility of selectively changing the electrophoretic mobility of different components and therefore optimizing resolution. Parallel electrophoresis of the same sample in gels containing different urea and detergent concentrations (Fig. 4) combines the simplicity of one-dimensional electrophoresis (easier quantitation of stain

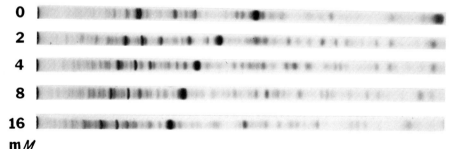

FIG. 4.   Effect of nonionic detergents on the resolution of complex mixtures of nonhistone proteins. Chromosomal proteins were extracted from isolated chromatin of mouse Ehrlich ascites tumor cells with 0.27 M NaCl–10 mM trisodium ethylenediamine tetracetic acid–10 mM mercaptoethanol and electrophoresed on 7.5% polyacrylamide gels containing 2.5 M urea and different concentrations of Triton X-100 as indicated. Note the shift in electrophoretic mobilities of the major bands and the increased resolution in the presence of detergent.

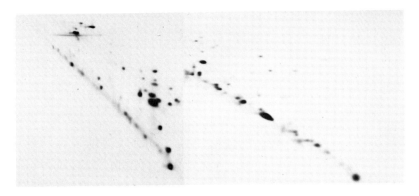

FIG. 5. Two-dimensional electrophoresis of the same nonhistone chromosomal protein fraction as in Fig. 4. First dimension (left to right): 7.5% polyacrylamide disc gel containing 5% acetic acid and 6 M urea. After electrophoresis in the first dimension, the gel was cut in half and the top part electrophoresed 2.7 times longer than the bottom part in the second dimension: 12% polyacrylamide slab gel containing 5% acetic acid, 4 M urea, and 8 mM Triton X-100. Note the diagonal line formed by the components which did not bind the detergent under these conditions. Many of these components can be resolved better in a gel containing either more detergent or less urea.

and radioactivity, better reproducibility) with the increased resolution of two-dimensional analysis. For qualitative comparison of complex protein mixtures, the electrophoresis system of Panyim and Chalkley (3) can be combined with our method in a simple and effective two-dimensional electrophoresis technique (Fig. 5), which uses the same buffer system in both directions and therefore avoids the precipitation problems and the possibility of differential loss of proteins during the transition from one buffer system to another.

ACKNOWLEDGMENTS

I wish to express my sincere thanks to Leonard Cohen for his continued interest and support, and to Dilip Londhe, Martha Honeycutt, and Judy Mamas for their assistance. This work was supported by USPHS grants CA-15135, CA-06927, and RR-05539 from the National Institutes of Health and by an appropriation from the Commonwealth of Pennsylvania.

REFERENCES

1. Elgin, S. C. R., and Weintraub, H., *Annu. Rev. Biochem.* **44**, 725 (1975).
2. Zweidler, A., in "Organization and Expression of Chromosomes" (V. G. Allfrey, E. K. F. Bautz, B. J. McCarthy, R. T. Schimke, and A. Tissieres, eds.), p. 187. Dahlem Konferenzen, Berlin, 1976.
3. Panyim, S., and Chalkley, R., *Arch. Biochem. Biophys.* **130**, 337 (1969).
4. Zweidler, A., and Cohen, L. H., *Fed. Proc, Fed. Am. Soc. Exp. Biol.* **31**, 926A (1972).

5. Franklin, S. G., and Zweidler, A., *J. Cell Biol.* **67**, 122a (1975).
6. Franklin, S. G., and Zweidler, A., *Nature (London)* **266**, 273 (1977).
7. Alfageme, C. R., Zweidler, A., Mahowald, A., and Cohen, L. H., *J. Biol. Chem.* **249**, 3729 (1974).
8. Cohen, L. H., Newrock, K. M., Zweidler, A., *Science* **190**, 994 (1975).
9. Balhorn, R., Oliver, D., Hohmann, P., Chalkley, R., and Granner, D., *Biochemistry* **11**, 3915 (1972).
10. Gurley, L. R., and Walters, R. A., *Biochem. Biophys. Res. Commun.* **55**, 697 (1973).
11. Zweidler, A., and Borun, T. W., *J. Cell Biol.* **63**, 386a (1974).
12. Borun, T. W., Gabrielli, F., Ajiro, K., Zweidler, A., and Baglioni, C., *Cell* **4**, 59 (1975).
13. Borun, T. W., Ajiro, K., Zweidler, A., Dolby, T. W., and Stephens, R. E., *J. Biol. Chem.* **252**, 173 (1977).
14. Neumann, N. P., *in* "Methods in Enzymology" Vol. 25: Enzyme Structure, Part B (C. H. W. Hirs and S. Timasheff, eds.), p. 393. Academic Press, New York, 1972.
15. Kinkade, J. M., and Cole, R. D., *J. Biol. Chem.* **241**, 5790, 5798 (1966).
16. Shires, A., Carpenter, M. P., and Chalkley, R., *Proc. Natl. Acad. Sci. U.S.A.* **72**, 2714 (1975).
17. Johns, E. W., *in* "Histones and Nucleohistones" (D. M. P. Phillips, ed.), p. 1. Plenum, New York, 1971.
18. Luck, J. M., Rasmussen, P. S., Satake, K., and Tsvetikov, A. N., *J. Biol. Chem.* **233**, 1407 (1958).

Chapter 17

Polyacrylamide Gel Electrophoretic Fractionation of Histones

ROSS HARDISON AND ROGER CHALKLEY

Department of Biochemistry,
School of Medicine,
University of Iowa, Iowa City, Iowa

I. Introduction

Histones are small, basic proteins associated with the nuclear DNA of eukaryotes. They possess no known enzymic activity, but are present in large amounts in the nucleus (1:1 mass ratio with the DNA). The most convenient means of assaying for histones is polyacrylamide gel electrophoresis. The basicity of histones means, of course, that they are positively charged at low pH; and routine electrophoretic analysis is often done at this pH, which affords a separation on the basis of charge, molecular weight, and shape. Electrophoresis in the presence of sodium dodecyl sulfate (SDS) at high pH provides a separation predominately on the basis of size. Electrophoresis at neutral pH permits an analysis of acid-labile histone modifications and some histone–histone complexes. Two-dimensional gels may be required to separate histones from some species, as well as to analyze the products from the reaction of histones with reversible cross-linking reagents. In this review we will discuss advantages and limitations of these four categories of gel systems, as well as explain how to make and run the gels.

II. Low pH Gel Systems

A. Acid–Urea Gels

1. CHARACTERISTICS

A convenient system for routine histone analysis is 0.9 M acetic acid–2.5 M urea in 15% polyacrylamide (*1*). In this system, the five major histones separate nicely on a 9-cm gel (Fig. 1A), and secondary bands due to covalent modifications and/or sequence heterogeneity are evident upon longer

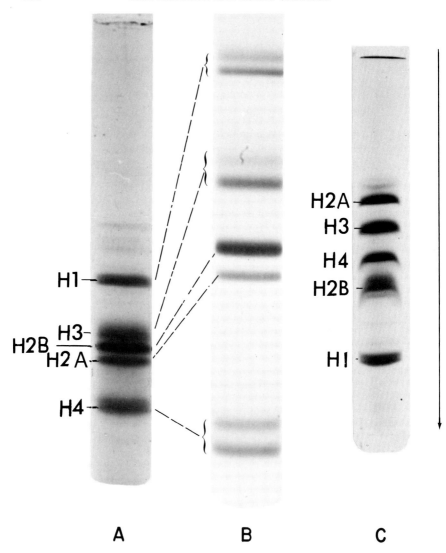

FIG. 1. Separation of histones on low pH gels: (A) 2.5 M urea–0.9 M acetic acid 9-cm gel; (b) 1.0 M urea–0.9 M acetic acid, 25-cm gel; (C) 2.5 M urea–0.9 M acetic acid–0.4% Triton X-100, 9-cm gel. The Triton-containing gel was 15% polyacrylamide and was electrophoresed for 4.5 hours at 130 V. All other methods are described in the text. Arrow indicated direction of mobility.

electrophoresis on 25-cm gels (Fig. 1B). Factors affecting mobility are charge density, size, and shape (frictional coefficient). Variation of the urea concentration alters the relative mobility of H2B (presumably due to an effect on the frictional coefficient) such that H2B and H3 comigrate in 1.0 M urea, whereas H2B and H2A comigrate in 6.0 M urea; optimal separa-

TABLE I
MOLAR STAINING COEFFICIENTS[a] FOR HISTONES

Histone	Amido black[b]		Fast Green[b]		Coomassie blue[b,c]	
	Molar staining coefficient	Upper limit of linearity (nmoles)	Molar staining coefficient	Upper limit of linearity (nmoles)	Molar staining coefficient	Upper limit of linearity (nmoles)
H1	1.00	2.5	1.00	0.4	1.00	3.6
H2A	0.49 ± 0.02[d]	4.4	0.672 ± 0.005	2.0	0.44 ± 0.05	13.9
H2B	0.59 ± 0.02	3.6	0.77 ± 0.02	1.1	0.22 ± 0.03	6.6
H3	0.78 ± 0.06	4.0	1.07 ± 0.04	0.8	0.65 ± 0.06	6.5
H4	0.47 ± 0.02	5.3	0.677 ± 0.007	1.4	1.67 ± 0.08	8.9

[a] The molar staining coefficient (msc) is defined as:

$$\text{msc} = \frac{\text{(weight of chart paper in an absorbance peak per nanomole of the designated histone)}}{\text{(weight of chart paper in an absorbance peak per nanomole of H1)}}$$

Thus, msc is a measure of total absorbance of a protein band per nmole (peak area) and not solely a measure of maximal absorbance (peak height). The extinction coefficient is related to H1 to avoid variability due to scanning speed, chart speed, length of time since destaining, or other ancillary considerations.

To calculate the molar quantity of each histone, run a sample of purified H1 of known molar concentration [we recommend an amino acid analysis of the H1 solution and calculation of molarity using the H1 amino acid composition data of Elgin and Weintraub (2)] on a gel simultaneously with the histone samples to be determined. By scanning the stained gels one obtains the weight of chart paper in the absorbance peak/per nmole H1 (from the H1 standard) and the weight of chart paper in each histone band (a Dupont curve resolver may facilitate this analysis). The molar quantity of each histone is then determined by:

$$\text{nmoles histone} = \frac{\text{weight of chart paper in histone absorbance peak}}{\text{(msc) (weight of chart paper per nmole H1)}}$$

[b] Acid–urea gels (Section II,A) were stained with Amido black or Fast green; pH = 8.8 SDS gels (Section III,A) were stained with Coomassie blue R. Amido black and Coomassie blue-stained gels were scanned at 600 nm; Fast-green stained gels were scanned at 625 nm.

[c] The data for Coomassie stained histones has been subject to the greatest degree of variability (2a), and we would recommend using either Amido Schwartz or Fast Green-stained acid-urea gels for analysis of molar amounts of histone.

[d] The error is calculated from the error in the slope of a least-squares fitted line to standard curves (amount of staining versus nmoles histones) for each histone in the three stains.

tion of calf thymus H2B is found in 2.5 M urea. H2B migration is also selectively increased at lower temperatures.

Besides assaying for the presence of the various histones, one can determine the amount of histone present by microdensitometric measurement of the total A_{600nm} of each band; this is related to the molar amount of stained histone by the extinction coefficients given in Table I. Radioactive label in the minor bands representing modified histone can be determined by slicing the long gels, digesting the slices in 30% H_2O_2, and analyzing in a scintillation counter.

2. RECIPE

a. Final concentrations in the gel.

0.9 M acetic acid
Variable urea
15% acrylamide
0.09% N,N'-methylenebisacrylamide (BIS)
0.12% w/v ammonium persulfate
0.4% N,N,N',N'-tetramethylethylenediamine (TEMED)
pH 3

b. Stock solutions.

A = 63.5% w/v acrylamide, 0.4% 3/v BIS in H_2O

B = 4% v/v TEMED, 43.2% v/v glacial acetic acid in H_2O

C = 40 mg ammonium persulfate and 2.04 gm urea for 1 M urea final concentration in gel, 5.10 gm urea for 2.5 M urea final concentration in gel, or 12.24 gm urea for 6.0 M urea final concentration in gel, dissolved in 22 mm (final volume) H_2O. Make fresh just prior to use.

Electrode buffers, both top and bottom, are 0.9 M acetic acid in H_2O.
Amido black (Napthol blue black) staining solution =

0.1% w/v Amido black
0.7% v/v acetic acid
30% v/v ethanol

c. Methods.

To make 12 gels, 9 cm long and 5 mm in diameter, place acid-washed glass tubes at least 10 cm long and 5 mm inner diameter in a convenient holder, e.g., rubber serum stoppers inverted and glued in place, or a commercially available gel rack. Glass tubes longer than this should be siliconized by boiling in a 1% solution of dichlorodimethylsilane in benzene. Combine the 22 ml of solution C with 4 ml of solution B in a side-arm suction flask, and place 8 ml of solution A in another side-arm flask. Deaerate the solutions by applying a vacuum to the suction flasks, combine solutions, and immediately pour into the glass tubes (a 50-ml disposable syringe with a length of Tygon tubing attached to a 16-gauge needle is useful for this). Slowly layer about $\frac{1}{2}$ cm of cold (4°) 0.9 M acetic acid onto the top of the polymerizing solutions; this produces a flat top and excludes atmospheric oxygen which interferes with polymerization.

After the gels have polymerized, they should be pre-electrophoresed to remove excess persulfate or anionic contaminants that would form an ion front moving in opposition to the histones. To do this, electrophorese the gels for ~ 4 hours at 130 V (or until constant current is reached) prior to addition of the sample.

Electrophoresis is from the anode (+) to cathode (−). Nine-centimeter gels are run at 130 V for 3.5 hours at room temperature; 25-cm gels are run at 4°, 200 V, for about 48 hours. It is wise to apply a control histone which can be removed after 24 hours to precisely predict the length of the electrophoresis. Alternately, one can use methyl green as a marker dye. The blue component, present as ∼30% of the dye, moves slightly faster than H4 in this gel system.

After electrophoresis the gels are stained in the Amido black staining solution for 3 hours at 80° or overnight at room temperature, and are then destained electrophoretically in very dilute aqueous acetic acid (∼0.006% acetic acid). The Amido black staining solution may be reused about 3 or 4 times.

B. Triton–Acid–Urea Gel System

1. CHARACTERISTICS

Addition of Triton X-100 (or Triton DF-16), a nonionic detergent, to the acid–urea gel system causes an alteration in the order of mobilities for the histones (Fig. 1C). This results from the fact that Triton binds histones proportionately to their degree of hydrophobicity, thus differentially altering the size and shape of the histones (3). Since the order of histone migration is different from that in acid–urea gels, this system has also been used for one dimension of a two-dimensional gel system (see Section V,A). It has also proved valuable in resolving questions as to whether H2B or H2A is phosphorylated *in vivo* (4) as well as for routine histone analysis (3). Electrophoresis of histones for a long distance (20 cm) in Triton gels containing 8 M urea reveals several minor species of H2A, H2B, and H3. The fact that these minor bands arise at different times of development of the sea urchin means they are not preparative artifacts (5). Preliminary data is available on sequence heterogeneity of H2B (6). Thus Triton gels would appear to be the system of choice for the study of these minor histone species not readily resolvable in other gel systems.

The major disadvantage to the use of Triton gels is the artifactual histone heterogeneity due to oxidation of methionine residues. (This heterogeneity is distinct from the histone variants mentioned above.) Oxidized histones have altered Triton-binding characteristics, and once a methionine is oxidized it can be very difficult to reduce (A. Zweidler, personal communication). As a precaution against being misled by this potential artifact one should keep the histones in a reducing environment (e.g., dithiothreitol) prior to analysis on Triton gels. Running the same sample on acid–urea (Section II,A) or SDS gels (Section III) should allow identification on nonhistone contaminants and degradation products.

2. Recipe

All solutions and procedures are the same as for acid–urea gels (Section II,A,2) except:

a. 0.4% Triton is included in the gel solution (add 1.36 ml of 10% Triton to the deaerated and combined solutions A, B, and C).

b. 0.1% Triton is present in the electrode buffers.

c. 0.1 ml of 1 M cysteamine (β-mercaptoethylamine), a free radical scavenger, in 5% acetic acid is electrophoresed through each pre-electrophoresed gel to reduce oxidation of methionine during electrophoresis.

d. A 1-cm plug of polyacrylamide without Triton in the bottom of the glass tube, or a nylon netting support, will prevent the gel from slipping out of the tube during electrophoresis.

e. An acrylamide concentration of 12% or less is used to avoid curved bands.

III. High pH SDS Gel Systems

The high pH gel systems described in this section both utilize SDS, an anionic detergent, which binds to proteins at a ratio of approximately one SDS molecule per two amino acids (7). This produces an overall negative charge on the protein, and the proteins separate primarily on the basis of size. However, because of the high initial positive charge residing on the histone, the final negative charge density after association with SDS is less than that found with other proteins, and there does appear to be a small contribution to separation based on shape (8); also see Mattice et al. (8a).

A. pH = 8.8 Gels

1. Characteristics

Thomas and Kornberg (9) have modified the Laemmli SDS gel (10) to produce SDS gel capable of separating the five histones, especially on long (25-cm) gels (Fig. 2A). One can visualize the oligomers produced by chemically cross-linking histones (Fig. 2B), although not as well as in the pH = 10 gel system (Section III,B). The molar amount of histone can be determined by scanning the gels in a microdensitometer and using the molar extinction coefficients given in Table I.

2. Recipe

a. *Final concentrations in gel.*

0.75 M tris(hydroxymethyl)aminomethane (Tris)

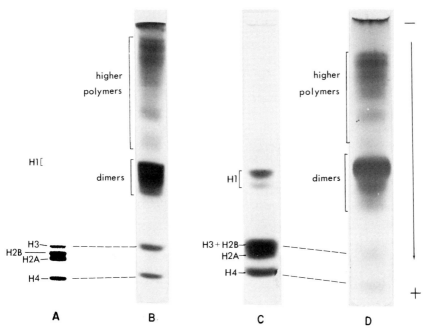

FIG. 2. Separation of histone on high pH gels containing SDS: (A) pH 8.8, 14-cm gel of monomer histones;   (B) pH 8.8, 10-cm gel of histones extracted after reacting chromatin with 1 mg/ml dimethylsuberimidate (pH 9)–0.1 M borate–1.9 M NaCl for 30 minutes at 22°; (C) pH 10, 10-cm gel of monomer histones; (D) pH 10, 10-cm gel of histones cross-linked under conditions similar to those described in (B).

0.1% SDS
18% acrylamide
0.12% BIS
0.025% TEMED
0.025% ammonium persulfate
pH 8.8

b. Stock solutions.

A = 63.5% w/v acrylamide, 0.4% w/v BIS in H_2O

B = 1.20 M Tris, 0.16% w/v SDS, 0.04% v/v TEMED, pH 8.8 (dissolve 14.52 gm Tris, 0.16 gm SDS, and 0.04 ml TEMED in 100 ml; adjust pH to 8.8)

C = 0.3% ammonium persulfate (make fresh just prior to use)

Electrode buffer, for both top and bottom reservoir, is 0.05 M Tris–0.38 M glycine–0.1% SDS (pH = 8.3).

Coomassie blue staining solution =
0.1% w/v Coomassie brilliant blue R

40% v/v methanol

10% v/v glacial acetic acid

c. *Methods.* To make 12 gels, 10 cm × 5 mm, mix 11.6 ml of solution A, 25 ml of solution B, and 3.4 ml of solution C, pour into 11 cm × 5 mm glass tubes, and carefully layer about ½ cm H_2O onto the top of the polymerizing gles. It is not necessary to deaerate the gel solution, nor is it necessary to pre-electrophorese the gels, probably due to the smaller amount of ammonium persulfate added to the gels.

Electrophoresis is from the cathode ($-$) to the anode ($+$) at 40 V for 16 hours. Bromophenol blue is a convenient marker dye. After electophoresis the gels are stained in Coomassie blue staining solution at room temperature overnight and are destained by diffusion in 8% acetic acid, 12% methanol. We find the staining solution loses its staining efficiency with repeated use, so we do not recycle the Coomassie blue stain.

d. *Stacking gel.* It sometimes enhances resolution to put a short (~ 1 cm) 5% acrylamide stacking gel atop the polymerized 18% acrylamide separating gel. To make a stacking gel solution, one should supplement the amount of BIS and TEMED to insure polymerization as follows:

3.13 ml solution B

0.015 ml TEMED

0.40 ml solution A

0.011 gm BIS

1.04 ml H_2O

0.43 ml solution C

5.0 ml total

B. pH 10 Gels

1. CHARACTERISTICS

Although this pH 10 system (*8*) does not afford quite the separation of histone monomers as that achieved with the pH 8.8 system (Fig. 2C, note the comigation of H2B and H3), the resolution of chemically cross-linked histone oligomers is somewhat improved (Fig. 2D).

2. RECIPE

a. *Final concentrations in the gel.*

0.025 *M* glycine	0.25% TEMED
0.025 *M* NaCl	0.016% ammonium persulfate
0.05% SDS	10% glycerol
16% acrylamide	pH 10
0.1% BIS	

b. Stock solutions.

0.5 M glycine, 0.5 M NaCl stock buffer (37.5 gm glycine, 29.25 gm NaCl in 1 liter of solution; adjust pH to 10)

Gel solution A = 63.5% w/v acrylamide, 0.4% w/v BIS in H_2O

Gel solution B = 0.2 M glycine–0.2 M NaCl–0.4% SDS–2% TEMED (40 ml glycine–NaCl stock buffer, 0.4 gm SDS, 2 ml neat TEMED; dilute to 100 ml with H_2O)

Gel solution C = 0.025% ammonium persulfate in 16% glycerol (dissolve 25 mg ammonium persulfate in 10 ml 16% glycerol; add 2.5 ml of this solution to 22.5 ml of 16% glycerol.)

16% v/v glycerol

Electrode buffer, both top and bottom reservoirs, is 0.025 M glycine–0.025 M NaCl–0.1% SDS–15% glycerol. (Dilute 150 ml glycine–NaCl stock buffer, 3 gm SDS, and 450 ml glycerol to 3 liters; adjust pH to 10 if necessary.)

c. Methods.   To make 12 gels, 10 cm × 5 mm, mix 10 ml of gel solution A, 5 ml of gel solution B, and 25 ml of gel solution C; pour into glass tubes and carefully layer ∼½ cm H_2O into the tops of the gels. The remaining methods are identical to those for the pH = 8.8 gels (Section III,A,2,c).

IV.   Neutral pH Systems

The previously described systems have used extremes of pH, urea, or an anionic detergent to prevent aggregation of the histones and to allow the histones to separate. A disadvantage of this treatment is that modifications labile to pH extremes, such as adenosine diphosphoribosylation (*11*), cannot be examined electrophoretically. Also, specific histone–histone interactions may be broken down by these denaturants, thus precluding their study by electrophoresis. Unfortunately, biologically meaningful interactions may exist only under conditions mimicking the high negative charge density of DNA; electrophoresis in such a high ionic strength will require extremely long running times. Nevertheless, it is clear that a neutral pH gel system would be useful for analyzing acid-labile histone modifications and histone–histone binding independent of DNA (*12*).

1.   CHARACTERISTICS

Gels buffered with glycylglycine operate at pH = 7.1 and separate H1, H2B, and H3 from H2A and H4 without the use of denaturants (Fig. 3A) (*13*). This gel system may then be useful for examining histone–histone com-

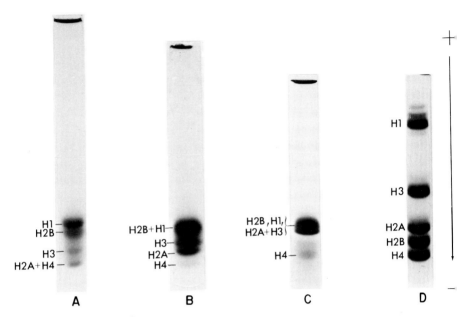

FIG. 3.   Separation of histones on neutral pH gels containing: (A) no added urea;
(B) 2.0 M urea;   (C) 6.0 M urea;   (D) 5.0 M urea and 0.5% CTAB.

plexes stable at low salt concentrations. The addition of urea to the system
alters the migratory behavior of H2A, H3, and H2B (Fig. 3A–C), presumably due to an unfolding of the histone molecules, so that at an appropriate
urea concentration any one histone can be separated from the others. Thus
labile histone modifications could be studied using gels with the proper urea
concentration, i.e., 6 M urea for H4 modifications and 0 M urea for H1
modifications.

A further enhancement of the resolution of glyclyglycine–urea gels is
obtained by electrophoresis in the presence of cetyl trimethylammonium
bromide (CTAB) (Fig. 3D). This cationic detergent binds histones reversibly,
giving them a uniform positive charge density, thus accentuating the separation of the baisis of the size of the histone–CTAB complex. Since the detergent binding is reversible, histones are kept in a CTAB environment by
including 0.5% CTAB in the top electrode buffer. The gels are stained with
Fast green since Amido black stains CTAB, thus masking the histone bands.
As seen in Fig. 3D, all five histone bands are clearly resolved, and there is no
complication from differential detergent binding of a single histone fraction,
as can be the case with Triton gels (see Section II,B). The glycylglycine–urea
CTAB gel system may then be the method of choice for the study of histone

modifications labile to pH extremes. In fact, although this latter method is a new discovery and is still being improved, it seems highly likely that this may turn out to be the most versatile method for histone analysis as well as the most highly resolving.

2. RECIPE

a. Final concentrations in gel.
0.18 M glycylglycine
variable urea
15.6% acrylamide
0.1% BIS
0.37% TEMED
0.008% ammonium persulfate
pH 7.8 (decreases to pH = 7.1 during electrophoresis)
b. Stock solutions.
Gel solution A = 63.5% w/v acrylamide, 0.4% w/v BIS in H_2O
Gel solution B = 0.586 M glycylgycine, pH 7.8 (15.48 gm glycylglycine per 200 ml; adjust pH to 7.8 with NaOH; store at 4°)
Gel solution C = 3% v/v TEMED. Store at 4°.
Gel solution D = 5 mg/ml ammonium persulfate in H_2O. Make fresh just prior to use.
Electrode buffers:
Top buffer (anodic) = 0.03 M imidazole-HCl, pH 7.2 (2.05 gm imidazole in 1 liter of solution; adjust pH to 7.2 with concentrated HCl)
Bottom buffer (cathodic) = 0.03 M glycylglycine, 0.01 M imidazole, pH 7.2–7.4 (4.0 gm glycylglycine and 0.7 gm imidazole in 1 liter of solution; no adjustment of pH is necessary)
Fast-green staining solution = 1% w/v Fast green, 30% v/v ethanol, 7% v/v acetic acid.
c. Methods. To prepare 12 gels, 9 cm × 5 mm, mix in one side-arm suction flask:
1. 8 ml solution A
2. 10 ml solution B
3. 4 ml solution C
4. 10 ml H_2O, 0 gm urea for 0 M final concentration,
   7.14 ml H_2O, 3.84 gm urea for 2 M urea final concentration,
   2.86 ml H_2O, 9.60 gm urea for 5 M urea final concentration, or
   1.43 ml H_2O, 11.52 gm urea for 6 M urea final concentration.
When the urea has dissoved, degas by applying a vacuum to the suction flask. Then add 0.5 ml of the persulfate solution (gel solution D) and immediately pour into glass tubes. Carefully layer 3 M urea into the top of the gel (use H_2O for 0 M urea gels), and allow to polymerize for ~2 hours.

Electrophoresis is from anode (+) to cathode (−) at a constant current of 2.5 mA/gel (30 mA/12 gels) for 6 hours at room temperature. Pre-electrophoresis is not necessary. Cytochrome C (30 μg/tube) may be used as a marker; it migrates almost identically to H4. Stain with Amido black and destain electrophoretically (see Section II,A, 2).

To run neutral gels in the presence of CTAB, simply include 0.5% w/v CTAB in the top buffer (5 gm/liter) and 2% CTAB in the histone sample applied to a 5 M urea–0.18 M glycylglycine gel. The CTAB migrates faster then the histone and thus permeates the gel during electrophoresis. Inclusion of CTAB in the polymerizing gel leads to precipitation. One achieves good separation using gels 16 cm in length; to make these simply double the volumes described previously. Electrophoresis is for 9 hours at 1.67 mA/gel (20 mA/12 gels), still from (+) to (−). One can monitor the course of electrophoresis by watching the progress of a line resulting from a sharp refractive index gradient at the CTAB front. Stain with Fast-green staining solution and destain by diffusion in 0.7% acetic acid or by electrophoresis in very dilute acetic acid. Several changes of buffer are necessary to obtain complete destaining.

V.   Two-Dimensional Gel Systems

A.   Triton–Acid–Urea Two-Dimensional Gels

1.   CHARACTERISTICS

Histones from some species do not completely resolve on the gel systems developed for analysis of vertebrate histone, e.g., H2A from peas comigrates with H2B in acid–urea gels. This problem can sometimes be solved by running the histones in two dimensions, the first in an acid–urea and the second in a Triton–acid–urea gel (13a). Since the order of mobility of the histones is different in the two systems, the result is a good separation of the five histones (Fig. 4). Some microheterogeneity in H2A is apparent; since the H2As separated in the acid–urea dimension, the distinction is not due to differential Triton binding and may be due to a slight sequence variation. The separation of the five histones seems to be greater than that achieved by other two-dimensional systems (14) for analysis of whole nuclear protein.

2.   RECIPE

a. Final concentrations in gels.
0.9 M acetic acid

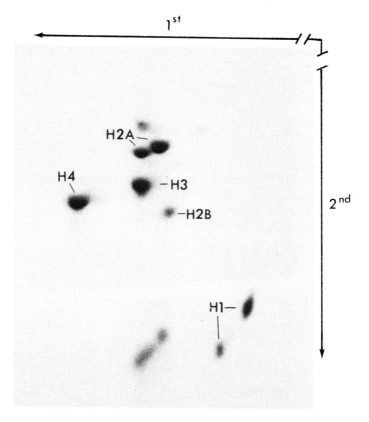

FIG. 4. Two-dimensional gel separation of histones extracted from *Triticale hexaploide*, an intergeneric hybrid of rye and wheat. The first dimension (horizontal) is the acid-urea system described in Section II,A; the second dimension (vertical) is the acid–urea–Triton system described in Section II,B. Only the part of the gel containing histones is shown; the rapidly moving, unlabeled spots are presumably histones that have not bound Triton. (Photograph furnished courtesy of Dr. Steven Spiker.)

   2.5 M urea
   15% acrylamide
   0.1% BIS
   1% Triton (second dimension only)
   b.   A = 40% w/v acrylamide–0.267% w/v BIS
       B = 43.2% v/v glacial acetic acid–4% v/v TEMED
       C = 5 M urea–0.2% ammonium persulfate; add 2% Triton X-100
for Triton dimension (make fresh just prior to use)
   Electrode buffer is 0.9 M acetic acid. 1% Triton X-100 is present in the top buffer for the second dimension.

c. Methods.   Gels for both dimensions are made by mixing solutions A, B, and C in proportions of 3:1:4. First-dimension cylindrical gels are cast in tubes 3 × 70 mm, as described in Section II,A,2, and pre-electrophoresed; histone samples are run at 1.5 mA/gel for 2 hours at room temperature. The second-dimension slab gel is formed in a BioRad Model 220 Dual Vertical Gel Electrophoresis Cell, dimensions 3 × 140 × 100 mm, and pre-electrophoresed at 20 mA for 12 hours. A 1-cm plug of acrylamide without the Triton X-100 can be used to prevent slippage of the gel during electrophoresis. The first-dimension gel is placed on the slab gel and polymerized in position by adding 2 ml of the Triton gel solution and layering 0.9 M acetic acid–1% Triton X-100 onto that. This added gel solution is allowed to polymerize for 1 hour, and then electrophoresis is carried out, (+) to (−), for 15 hours at 20 mA. The gels are stained in Amido black and destained by diffusion in 7% acetic acid–20% ethanol.

B.   Acid–Urea Two-Dimensional Gels

1.   CHARACTERISTICS

The recent interest in histone–histone interactions as potential determinants of chromatin structure has generated a proliferation of studies involving covalent cross-linking of histones. If the cross-link is reversible, then one can analyze the monomeric components of the oligomers by separating the oligomers in the first-dimension gel, reversing the cross-links, and running this gel in a second dimension. This has been done in our laboratory using acid–urea gels in both dimensions (Fig. 5) and by Thomas and Kornberg (*16*) using SDS gels in both dimensions. In both cases the reversible cross-link was through disulfide bonds. The acid–urea system emphasizes the heterogeneity of the dimeric and trimeric products of cross-linking, whereas the SDS system is more useful for analyzing the composition of higher oligomers.

2.   METHODS

The acid–urea gels are made according to the recipes given in Section II,A,2. Gels for the first dimension are cast in glass tubes 14 cm long and 3 mm in diameter, precoated with $(CH_3)_2Cl_2Si$, to a height of 12.5 cm. These gels are pre-electrophoresed and the histone sample run at 100 V for 13 hours. The gels are removed from the tubes and stored at − 10° until ready to be run in the second dimension.

The disulfide cross-linked oligomers are reduced by placing the gel in a solution of 2 M β-mercaptoethanol–4 M urea–0.09 N acetic acid (3.5 ml β-mercaptoethanol, 6.0 gm urea, and 2.5 ml 0.9 N acetic acid in 25 ml final

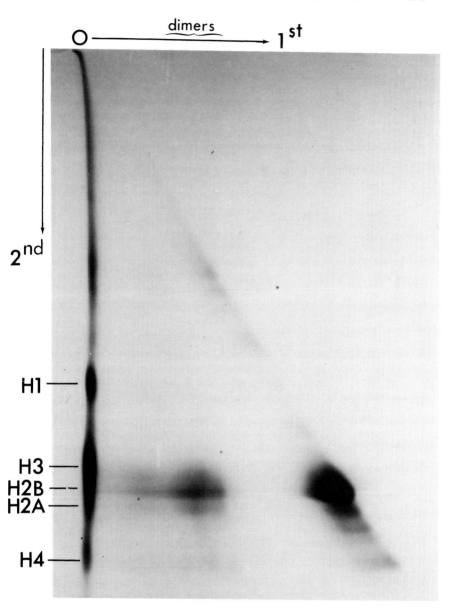

FIG. 5. Two-dimensional gel showing the composition of histone oligomers produced by reacting nuclei with methyl-4-mercaptobutyrimidate as described by Hardison *et al.* (*15*); the histone products accumulating in the H2B fraction are shown. O refers to the origin, which is the top of the first-dimension gel. The bands in the area marked "dimers" in the first dimension are the monomeric components of the histone dimers. Gels for both dimensions are 3.25 *M* urea–0.9 *M* acetic acid.

volume) for 2 hours at room temperature. The reducing solution is then replaced by 95% ethanol for ~30 minutes to shrink the gel, thus facilitating addition of this gel to the top of a slab gel 14 cm × 16.5 cm × 3 mm (made using twice the volumes given in the recipe in Section II,A,2). The second dimension is then run at 100 V for 18 hours. The gel is then stained in Amido black and destained electrophoretically.

C. Other Two-Dimensional Systems

It should be obvious that using the methodologies described in this section, one could run two-dimensional gels which are combinations of any of the systems described above. The potential then exists for great refinement of the resolution of histones from various species and our ability to analyze histone modifications and histone–histone interactions.

VI. Concluding Remarks

We now have a veritable plethora of electrophoretic systems with which to separate histones in seemingly whatever guise they may appear. However, the reader is cautioned to be highly discriminating in selection of an appropriate method. Thus SDS electrophoresis is useless for analyzing modifications to histones such as phosphorylation or acetylation, whereas it becomes the method of choice for looking at high-molecular-weight cross-linked polymers. Acid-labile modifications should of course be studied on neutral pH gels. If one wishes to analyze histones on neutral gels it is necessary to extract at neutral pH. Elevated ionic strength effectively extracts histones, but the sample cannot be applied directly to gels and proteases are almost invariably activated; thus one should extract with protamine or CTAB and apply the sample directly. It is also possible to extract histones with SDS, as addition of a 15 to 20-fold excess of CTAB will reverse the SDS–histone interaction and permit analysis either on neutral pH or even on acid–urea gels. Likewise, histones bound to CTAB can be treated with a 10-fold excess of SDS and analyzed on SDS gels. On the other hand, it is not usually a good idea to extract with SDS and analyze on SDS gels, as contamination with nonhistone proteins can make interpretation difficult. Further, if ^{32}P is present, it becomes impossible to disentangle phosphorylated histones from other phosphorylated proteins and from comigrating nucleic acids.

We note that one can quantitatively analyze histones without recourse to first isolating the proteins. Thus small quantities of chromosomal material can be treated with SDS and run on SDS gels, treated with CTAB and run on neutral gels or on acid–urea gels, or even extracted with 0.4 $N\,H_2SO_4$, titrated with a small volume of $Ba(OH)_2$, and analyzed on acid–urea gels. If the original chromosomal material is of low concentration (as low as 10 $\mu g/ml$), one simply modifies the strategy. Extract with acid, remove DNA, and dialyze against 0.9 N acetic acid containing 40% sucrose to concentrate the sample before applying directly to acid–urea gels. In extreme cases we have surrounded the dialysis bag with solid sucrose (remove H_2SO_4 first) and achieved a 10 to 20-fold concentration without any handling losses.

Thus the basic technologies for histone isolation and analysis are at hand; all that remains is to exploit them properly.

ACKNOWLEDGMENTS

We thank Drs. S. Spiker, P. Hoffmann, and A. Zweidler for information, helpful discussions, and photographs of some of the gels seen in the figures. We are grateful to various members of this laboratory for some other gels seen in the figures. Work done in this laboratory was supported by NIH grants 5T32 GM 07228–01, CA 10871, and GM 46410.

REFERENCES

1. Panyim, S. and Chalkley, R., *Arch. Biochem. Biophys.* **130**, 337 (1969).
2. Elgin, S. and Weintraub, H., *Annu. Rev. Biochem.* **44**, 725 (1975).
2a. Hoffmann, P. Eichner, M., and Chalkley, R., *Nuclei Acids Res.* in press.
3. Alfageme, C., Zweidler, A.,Mahowald. A. and Cohen, L., *J. Biol. Chem.* **249**, 3729 (1974).
4. Balhorn, R., Oliver, D., Hohmann, P., Chalkley, R., and Granner, D., *Biochemistry* **11**, 3915 (1972).
5. Cohen, L., Newrock, K., and Zweidler, A., *Science* **190**, 994 (1975).
6. Franklin, S., and Zweidler, A., *J. Cell Biol.* **67**, 112a (1975).
7. Reynolds, J., and Tanford, C., *Proc. Natl. Acad. Sci. U.S.A.* **66**, 1002 (1970).
8. Panyim, S., and Chalkley, R., *J. Biol. Chem.* **246**, 7557 (1971).
8a. Mattice, W. L., Riser, J. M., and Clark, D. S., *Biochemistry* **15**, 4264 (1976).
9. Thomas, J., and Kornberg, R., *Proc. Natl. Acad. Sci. U.S.A.* **72**, 2626 (1975).
10. Laemmli, U.K., *Nature (London)* **227**, 680 (1970).
11. Ueda, K., Omachi, A., Kawaichi, M., and Hayaishi, O., *Proc. Natl. Acad. Sci. U.S.A.* **72**, 205 (1975).
12. D'Anna, J., and Isenberg, I., *Biochemistry* **13**, 4992 (1974).
13. Hoffmann, P., and Chalkley, R., *Anal. Biochem.* **76**, 539 (1976).
13a. Spiker, S., *Nature (London)* **259**, 418 (1976).
14. Orrick, L., Olson, M., and Busch, H., *Proc. Natl. Acad. Sci. U.S.A.* **70**, 1316 (1973).
15. Hardison, R., Eichner, M., and Chalkley, R., *Nucleic Acids Res.* **2**, 1751 (1975).
16. Thomas, J., and Kornberg, R., *FEBS Lett.* **58**, 353 (1975).

Part D. Fractionation and Characterization of Nonhistone Chromosomal Proteins

Chapter 18

Affinity Chromatography of DNA-Binding Proteins on DNA Covalently Attached to Solid Supports

VINCENT G. ALLFREY AND AKIRA INOUE

The Rockefeller University,
New York, New York

I. Introduction

The use of columns containing DNA and RNA for the selective binding of proteins or complementary polynucleotide sequences has opened an approach of major significance to the analysis of gene function and its regulation. The affinity concept has been employed for the separation of DNA polymerases (*1–3*), RNA polymerases (*4*), ribonucleases (*5*), reverse transcriptase (*6*), and DNA-binding proteins with sequence (*7–10*) or strand specificity (*11*). This chapter will deal with the preparation and use of DNA covalently linked to solid supports for studies of eukaryotic DNA-binding proteins.

II. Preparation of DNA Columns

A variety of methods exists for the preparation of affinity columns bearing double-stranded and single-stranded DNA chains. The adsorptive binding of DNA to cellulose (*12*) has been widely employed, as has ultraviolet (UV) fixation of the adsorbed polynucleotide to improve the capacity of the columns (*1*). DNA has been immobilized in agar (*13*) and in polyacrylamide (*14*), or covalently linked to solid supports such as agarose (Sepharose) (*3*), aminoethyl-Sepharose (*7*), Sephadexes (*15,16*) cellulose (*17–21*), and diazo-cellulose (*22*), as well as to soluble polysaccharides such as Ficoll (*23*).

Methods for the covalent linkage of DNA to solid supports offer several options; some favor attachment through the DNA bases (e.g., 22), while others provide a phosphodiester bond between the 5′-phosphoryl group of the DNA and hydroxyl groups on the polysaccharide matrix (*7,15,17–21*). Poly (*dT*) cellulose containing a 3′-phosphodiester bond has also been prepared (*24*). Some methods favor attachment of single-stranded DNA molecules to the solid support (*3,22*), while others (*7,15,23*) permit the coupling of the 5′-ends of double-stranded polynucleotides to polysaccharide supports without degradation or loss of template function. Most of the methods involving formation of a phosphodiester bond through the 5′-phosphoryl group of DNA employ a carbodiimide coupling reaction originally introduced by Gilham (*17*). Some recent modifications of this technique which permid the attachment of native or single-stranded DNAs to Sepharose, aminoethyl-Sepharose, or Sephadexes will now be presented in detail.

III.   Covalent Coupling of DNA to Polysaccharide Supports

Solid supports suitable for DNA coupling through phosphodiester linkages include the polymerized dextrans (Sephadex G-25 and Sephadex G-200; Pharmacia, Ltd., Uppsala, Sweden), agarose derivatives such as Sepharose 4B (Pharmacia, Ltd.), and Sepharose 4B activated with cyanogen bromide and aminoethylated according to Cuatrecasas (*25*). Each of these materials is swollen in H_2O, poured into columns, and washed with 10 volumes of 40 mM Na-2-[N-morpholino]-ethane sulfonate buffer (pH 6.0) (MES buffer). The reagent used for the coupling reaction is the water-soluble carbodiimide, 1-cyclohexyl-3-[2-morpholinoethyl]-carbodiimide metho-p-toluene sulfonate (CMC) (Pierce Chemical Co., Rockford, Illinois) (*7,16, 19,21,23*). At pH 6.0 it specifically activates the monoesterified phosphate groups of nucleotides or polynucleotides (*17*), permitting their interaction with the hydroxyl groups of the solid supports.

IV. Deproteinization of DNA Samples

It is advisable, in most studies of DNA–protein interactions, to remove all contaminating polypeptides from the DNA preparations before the coupling reaction with the solid support. This can be accomplised by chemical or enzymic methods, as follows. DNA is dissolved at a concentration of 2 mg/ml in 40 mM MES buffer (pH 6.0), and solid $NaClO_4$ is added to obtain a final concentration of 1 M. The solution is deproteinized by vigorous shaking for 10 minutes with an equal volume of a 24:1 (v/v) mixture of chloroform and isopentanol. After centrifugation at 2000 g for 5 minutes, the aqueous phase is collected and shaken with $CHCl_3$–isopentanol as before. The deproteinization step is repeated 4 times. The final DNA solution is dialyzed against 40 mM MES buffer (pH 6.0) to remove $NaClO_4$.

Alternatively, contaminating proteins can be removed by digestion with Pronase or a similar protease preparation with multiple specificities, such as proteinase K (26). Pronase is dissolved as a stock solution containing 10 mg/ml in 0.15 M NaCl–0.015 M Na citrate–2 mM ethylendiaminetetraacetic acid (EDTA) (pH 7.0). This is autodigested at 35°C for 2 hours to remove possible nuclease contamination. The enzyme preparation is then added to the DNA solution to a final concentration of 0.5 mg/ml and incubated with shaking for 2 hours at 35°C. For proteinase K digestions, the enzyme is dissolved in 10 mM Tris-(HCl) (pH 8.0) containing 10 mM EDTA– 10 mM NaCl–0.5% sodium dodecyl sulfate (SDS), and used at a concentration of 50 μg/ml for 12-hour incubations at 37°C. The DNA can then be purified by phenol treatments as described by Gross-Bellard et al. (26). The final DNA solution is dialyzed against 40 mM MES buffer (pH 6.0) prior to use in the coupling reaction.

V. DNA Shearing, Sizing and C_0t Fractionation

Depending upon the aims of the affinity studies, the DNA may need to be processed in additional ways. For example, total DNA may be fractionated to yield satellites of different buoyant density, or it may be sheared to shorter chain-lengths to facilitate C_0t fractionation. For the latter purpose, deproteinized DNA is dissolved in 0.12 M sodium phosphate buffer (pH 6.8) at a concentration of 2 mg/ml. In a large-scale preparation (15), 45-ml aliquots are placed in ice-jacketed 100-ml beakers and subjected to sonication (Sonifier Model W185; Heat Systems–Ultrasonic, Inc., Plainview, New

York) with a $\frac{1}{2}$ inch disruptor horn placed 1 cm from the bottom of the beaker. Sonication is carried out for 1 munite at 75 W output, for 1.5 minutes at 88 W, and for two 1-minute periods at 88 W, allowing 1-minute cooling intervals between bursts. The solution is centrifuged at 10,000 g for 30 minutes.

The average lengths of the DNA fragments can be determined by electron microscopy (27) and expressed in terms of nucleotide pairs (28). Shearing of calf thymus total DNA by the above procedure gives fragments ranging in size from 200–400 nucleotide paris, with a double-stranded content of 89%, as judged by hyperchromicity analysis (7, 15).

Separation of these fragments according to C_0t value is achieved by denaturation at 100° for 5 minutes, followed by a program of renaturation steps in which the reannealed DNA molecules are successively removed by chromatography on hydroxyapatite (29). Details for the C_0t fractionation of calf thymus DNA are also presented in (7) and (15).

Because reassociation of DNA strands broken at random by sonication is not complete, single-stranded sequences may need to be removed by single-strand-specific nucleases. Nuclease S1 from *Aspergillus oryzae* is prepared from Sanzyme R (Calbiochem, La Jolla, California) by chromatography on O-(diethylaminoethyl) cellulose (DEAE-cellulose) (30). The reannealed DNA fractions are dialyzed against 0.1 M NaCl–3 $\times$ 10^{-5} M ZnSO$_4$–0.03 M Na acetate (pH 4.5) and treated with 800 units of S1 nuclease/mg of DNA for 30 minutes at 50 °C. The solution is chilled, $\frac{1}{10}$ volume of 1.5 M NaCl–0.15 M Na citrate (pH 7.0) is added, and the DNA is deproteinized as described above. The DNA is dialyzed against 0.12 M Na phosphate buffer (pH 6.8) and applied to hydroxyapatite columns [which have been pretreated with boiling 0.14 M Na phosphate buffer (pH 6.8) for 20 minutes and equilibrated with 0.12 M Na phosphate buffer at 60 °C to reduce binding of single-stranded DNA molecules]. The columns are washed with 4 volumes of 0.12 M Na phosphate buffer before removing the double-stranded DNA in 0.4 M Na phosphate buffer at 60 °C. The eluted double-stranded DNA is finally dialyzed against 40 mM MES buffer (pH 6.0) prior to the coupling reaction. Hyperchromicity analysis should indicate a double-stranded content in excess of 96%.

VI.   Coupling Reaction for Double-Stranded DNA

In preparing solid supports bearing double-stranded DNA, the coupling reaction is carried out at temperatures which avoid heat denaturation, as follows: 120 A_{260nm} units of DNA, dissolved in 3.6 ml of 40 mM MES buffer (pH 6.0), are mixed with 1.5 gm of gravity-packed Sephadex, Sepharose, or aminoethyl-Sepharose, and excess water is removed by evaporation under a warm air (or nitrogen) stream, keeping the temperature of the paste below 40° and mixing frequently with a spatula. [The drying can be scaled up to accomodate larger amounts of DNA (e.g., 400 mg) by using a rotary evaporator.] It is essential that most of the water be removed at this time to avoid side reactions which diminish DNA-coupling yields. Eighty milligrams of CMC in 1 ml of H_2O are added, and the paste is maintained at 45–50°C for 8 hours. The coupling reaction is repeated twice adding 80-mg portions of CMC dissolved in 1-ml aliquots of 40 mM MES buffer. The reaction is allowed to proceed with intermittent drying for 8 hours at 45–50 °C each time. Finally, 5 ml of cold 0.1 M NaCl–1 mM EDTA–10 mM Na phosphate buffer (pH 7.0) are added, and the matrix is allowed to swell for 12 hours at 4 °C. After decantation of excess fluid, the slurry is resuspended in 20 volumes of 0.2 M NaHCO$_3$ (pH 10.2) for 10 hours at 4°C in order to hydrolyze any adducts formed between DNA bases and the coupling reagent. The polymer-linked, double-stranded DNA is washed 6 times with 10 volumes of 0.1 M NaCl–1 mM EDTA–10 mM Na phosphate buffer (pH 7.0) and centrifuged at 1000 g for 3 minutes to remove unbound DNA. The DNA matrix is stored in

TABLE I

COVALENT LINKAGE OF DNA TO SOLID POLYSACCHARIDE SUPPORTS

	Percent Binding[a] of		
	Low-molecular-weight DNA[b]		High-molecular-weight DNA[c]
Matrix	Double-stranded	Single-stranded	Double-stranded
Sephadex G-25	61 ± 1 (2 ± 0)[d]	53 ± 11	64 ± 2 (18 ± 5)[d]
Sephadex G-200	27 ± 7	33 ± 6	64 ± 6
Sepharose 4B	45 ± 6	47 ± 7	84 ± 2
Aminoethyl-Sepharose 4B	58 ± 5 (24 ± 2)[e]	56 ± 3 (13 ± 1)[d,f]	87 ± 5

[a]Yields are expressed as % of original DNA (120 A_{260nm} units) bound to 1.5 gm of gravity-packed solid support. All yields represent coupling observed after treatment of DNAs with CMC for 4 hours at 45–50°, followed by two successive treatments for 7 hours at 45–50°.

[b]DNA sheared to average lengths of 200–400 nucleotide pairs.

[c]Nonsheared calf thymus DNA 0.5–1.5 × 10⁴ nucleotide pairs in length.

[d]Values in parentheses indicate % binding to Sephadex G-25 and Aminoethyl-Sepharose 4B in the absence of the coupling reagent (CMC).

[e]Values in parenthesis indicate DNA content after washing with 3 M NaCl–1 mM EDTA–10 mM Na phosphate buffer (pH 7.0) to remove DNA adsorbed to aminoethyl-Sepharose by ionic interactions.

[f]DNA coupling yield after reaction at 100° was 84%.

sealed bottles in 5 ml of the washing buffer at $4°C$; 2–3 drops of $CHCl_3$ are used as a preservative.

Yields of this reaction, using high- and low-molecular-weight DNA preparations and four different polysaccharide supports are shown in Table I. These yields are calculated from the total amount of double-stranded DNA used for the reaction and the amount of DNA released from measured small aliquots of the DNA matrix by treatment with deoxyribonuclease I, as described below. (The percent binding of DNA can be increased by 10–25% by doubling the amount of solid support used in the coupling reaction, but this produces columns with fewer DNA molecules per unit volume.)

Release of matrix-bound DNA (for estimates of coupling yields, or for demonstrations that adsorbed proteins are bound to DNA and not to the polysaccharide support) is achieved by digestion with DNAse I. Columns containing 1 mg of DNA in a bed volume of 1.1 ml are washed at $25°C$ with 2 ml of a 0.5 mg/ml solution of proteinase-free pancreatic DNAse I in 10 mM NaCl–10 mM Tris-HCl (pH 7.5) containing 10 mM $NaHSO_3$–1 mM 2-mercaptoethanol–2 mM $MgCl_2$–0.4 mM $CaCl_2$–10% (v/v) glycerol. The enzyme solution is passed through the column at a flow rate of 1 ml/90 minutes. The amount of DNA released is determined by measuring the absorbancy at 260 nm.

VII. Coupling Reaction for Single-Stranded DNA

For linking single-stranded DNA molecules to solid supports, the rate of the reaction is accelerated by raising the temperature. Equivalent amounts of DNA (120 $A_{260 \text{ nm}}$ units) dissolved in 3.6 ml of 40 mM MES buffer (pH 6.0) are heated to $100°$ for 5 minutes and rapidly chilled in an ice bath prior to adding 1.5 gm of gravity-packed solid support. The excess liquid is removed by evaporation in an air stream, and 80 mg of CMC in 1 ml of H_2O are added. The paste is heated at $100°C$ for 7 minutes and then rapidly chilled to $O°C$. An additional treatment with the coupling reagent is carried out, adding 80 mg of CMC in 1 ml of MES buffer. After thorough mixing the paste is heated to $100°C$ for 7 minutes and then chilled. Subsequent operations in the preparation of polymer-linked, single-stranded DNAs are the same as those described for double-stranded DNA columns. Coupling yields for single-stranded DNA binding to Sephadex G-200, Sepharose 4B, and aminoethyl-Sepharose 4B at 45–50° are presented in Table I. (In our experience the coupling of single-stranded DNAs to Sephadex G-25 has generally resulted in low or highly variable yields.)

TABLE II

DNA CONTENT OF COVALENTLY LINKED SOLID SUPPORTS

Matrix	Low-molecular-weight DNA[a]				High-molecular-weight DNA[b] double-stranded	
	Double-stranded		Single-stranded[c]			
	mg/ml[d]	mg/g[e]	mg/ml	mg/g	mg/ml	mg/g
Sephadex G-25	1.19	6.1	–	–	1.24	6.4
Sephadex G-200	–	18.8	–	23.0	–	44.7
Sepharose 4B	–	22.5	–	23.5	–	42.0
Aminoethyl-Sepharose 4B	1.58	29.0	1.53	28.0	2.37	43.5

[a]DNA sheared to average lengths of 200–400 nucleotide pairs.
[b]Nonsheared calf thymus DNA 0.5–1.5 $\times$ 10^4 nucleotide pairs in length.
[c]Single-stranded calf thymus DNA of C_0t greater than 200.
[d]Concentration expressed as mg DNA/ml of gravity-packed solid support.
[e]Concentration expressed as mg DNA/gm (dry weight) of solid support.

VIII.   Properties of DNA-Affinity Columns

The covalent linkage of double-stranded and single-stranded DNA molecules to Sephadexes, Sepharose 4B, and aminoethyl-Sepharose 4B provides an assortment of solid supports of different physical properties and DNA contents. The DNA concentration is expressed in terms of the volume or dry mass of the different supports in Table II.

The stability of the linkage between DNA and Sephadex G-25 has been examined after prolonged storage in 0.1 M NaCl–1 mM EDTA–10 mM Na phosphate buffer (pH 7.4) using CHCl$_3$ as a preservative. Less than 1% of the DNA originally present is lost after 6 months' storage at 4°C. The corresponding figure for aminoethyl-Sepharose–DNA is 2.6%. Tests for DNA "leakage" from affinity columns show that losses are minimal (usually less than 0.3% of the DNA originally present) over a wide range of salt concentrations (50 mM to 2.0 M NaCl) and despite the presence of protein denaturants such as 6 M urea and 0.4 M guanidine–HCl (15).

The DNA columns can be regenerated after use in protein-binding studies by washing in 4 M guanidine-HCl–1 mM EDTA–0.1 M Tris-HCl (pH 8.0) followed by 0.1 M NaCl–1 mM EDTA–10 mM Na phosphate (pH 7.0). It is advisable to measure the DNA content of the regenerated column before use.

IX.   DNA-Affinity Chromatography of Nuclear Proteins

Affinity columns prepared by the above procedures have been used to fractionate both histones and nonhistone nuclear proteins. Histone H1 fractions which differ in their degree of phosphorylation have been separated on Sephadex G-25–DNA (31), and acetylated forms of histone H4 have been found to elute before the nonacetylated H4 molecules when chromatographed on DNA–Sephadex (32).

Fractionations of nuclear nonhistone proteins have been employed to test the reproducibility and the resolving power of DNA-affinity columns, as well as to compare the DNA-binding properties of different classes of chromosomal proteins (7, 15). A brief account of procedures for the fractionation of calf thymus nuclear proteins will serve to illustrate the general applicability of the method, while more specific examples, such as the separation of the regulatory (non-DNA-binding) and catalytic subunits of protein kinase (which selectively bind to DNA of the species of origin) are described elsewhere (33, 34).

Although there have been a number of studies of proteins extracted under denaturing conditions and then recombined with DNA under conditions that permit chromatin reassociation and restoration of specific template functions (35–38), the question of incomplete reversibility or protein conformational changes induced by high concentrations of urea of guanidine-HCl complicates interpretations of differential DNA binding. Complex mixtures of fully and partially renatured protein classes might be expected to differ in their DNA-binding properties. For this reason, emphasis will be placed on the affinity chromatography of nuclear proteins which are extractable in neutral salt solutions, avoiding exposure to urea, guanidine-HCl, SDS, ionic detergents, or other denaturing agents commonly employed to extract chromosomal proteins.

Calf thymus nuclei, purified by differential centrifugation through sucrose solutions (39) and washed extensively with 0.25 M sucrose–3 mM MgCl$_2$–0.5 mM CaCl$_2$, are extracted 3 times with 10 volumes of 0.14 M NaCl containing 10 mM NaHSO$_3$ (added as a protease inhibitor). The nuclear residue is then suspended in 10 volumes of 0.35 M NaCl–1 mM 2-mercaptoethanol–50 mM Tris-HCl (pH 8.0) and stirred for 30 minutes at 4°C. After centrifugation at 200,000 g for 2 hours, the clear extract is applied to a BioRex 70 column which had been equilibrated earlier with the extracting buffer. The run-off peak containing the less basic nonhistone proteins of the 0.35 M NaCl extract is collected and concentrated to 1 mg protein/ml by ultrafiltration through a Diaflo PM 30 membrane filter (Amicon Corp., Lexington, Massachusetts).

To facilitate monitoring of the elution of proteins from DNA columns it is convenient to radioactively label a fraction of the total protein to be analyzed. Aliquots of the 0.35 M NaCl extract containing about $\frac{1}{10}$ of the total sample have been labeled to specific activities in excess of 10^5 cpm/mg by reaction with $[^3H]$formaldehyde and subsequent reduction with sodium borohydride (40). The radioactive proteins are extensively dialyzed against 0.35 M NaCl–1 mM 2-mercaptoethanol–50 mM Na phosphate buffer (pH 7.5), and measured aliquots are added to the corresponding unlabeled fraction before chromatography on DNA columns.

In studies of the DNA-binding properties of the nuclear proteins, the salt concentration of the 0.35 M NaCl extract is first reduced by dialysis against 50 mM NaCl in a low-ionic-strength buffer containing 10 mM Tris-HCl (pH 7.5)–10 mM NaHSO$_3$–0.1 mM EDTA–1 mM 2-mercaptoethanol–10% (v/v) glycerol (buffer A). After centrifugation at 15,000 g for 20 minutes, a 2-ml aliquot containing 1 mg of protein is applied to a 4.5-ml Sephadex (or Sepharose) column containing 4 mg of DNA. The proteins are subsequently eluted using a salt gradient or by step-by-step increments in the NaCl concentration of the eluting buffer A. The column is finally washed with 6 M urea–0.4 M guanidine-HCl–0.1 M Na phosphate buffer (pH 7.5) to remove a small amount (less than 1%) of tightly adsorbed protein. Recovery of the 0.35 M NaCl-soluble proteins from DNA–Sephadex columns under these conditions is 99–100% complete.

Control experiments have included tests for protein binding to Sephadex G-25 columns which were treated with CMC but not reacted with DNA. Less than 0.8% of the protein is adsorbed to the DNA-free solid support. Less than 2% of the applied proteins adsorb to control (DNA-free) columns of aminoethyl Sepharose 4B. Both these figures are considerably lower than the 6.3% nonspecific adsorbtion of proteins to a blank cellulose column under the same conditions.

The extent to which nuclear proteins bind to DNA columns is a complex function of ionic strength, DNA/protein ratio, DNA sequence complexity, DNA strandedness, and the differential DNA affinities of different proteins in the extract. Some of these variables will now be considered.

X. Concentration Dependence and Saturability of Protein–DNA Binding

The proportion of nuclear nonhistone proteins that adsorb to DNA columns at low ionic strengths varies with the DNA/protein ratio. This dependency is illustrated for proteins in the 0.35 M NaCl extract in Fig. 1. In this experiment the proteins were applied to parallel DNA–Sephadex G-25 columns containing increasing amounts of double-stranded calf thymus DNA. After application in 50 mM NaCl in buffer A, and washing to remove

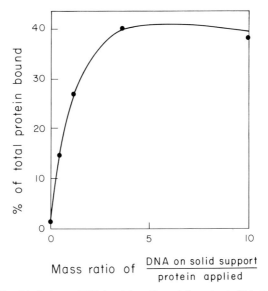

FIG. 1. Relationship between DNA/protein ratio and the extent of binding of calf thymus nuclear proteins to DNA–Sephadex G-25. The proteins were extracted in buffered 0.35 M NaCl, separated from basic components by chromatography on Bio-Rex 70, and applied to the DNA matrix in 50 mM NaCl. After washing the columns with 50 mM NaCl to remove unadsorbed proteins, the DNA-bound proteins were eluted and measured. The percentage of protein bound increases until the DNA/protein ratio reaches 4:1.

the unadsorbed proteins, the DNA-bound proteins were quantitatively removed by a single-step elution in 6 M Urea–0.4 M guanidine-HCl–0.1 M Na phosphate buffer (pH 7.5). The percent of the total protein bound to DNA is plotted as a function of the DNA/protein ratio. It is significant that protein binding to DNA under these conditions is saturable; i.e., adding more DNA to the columns does not lead to proportionate and ever-increasing associations between the nonhistone proteins and the DNA matrix, a fact which argues for selectivity rather than simple electrostatic attraction as the basis for the binding interaction.

XI. Ionic-Strength Dependence of Protein–DNA Binding

Even at DNA/protein ratios of 4:1, more than 60–65% of the protein in the 0.35 M NaCl extract passes directly through the DNA column. The proportion of the total protein which binds to DNA is strongly dependent upon the ionic strength of the medium, as shown in Fig. 2. In these experiments, the protein extract was dialyzed against buffers of increasing salt

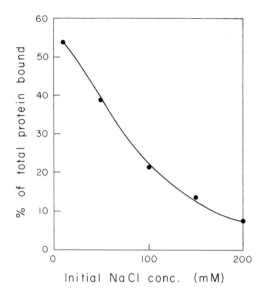

FIG. 2. Ionic-strength dependency of protein binding to DNA columns. The nuclear proteins extracted in buffered 0.35 M NaCl were applied to parallel columns of double-stranded DNA linked to Sephadex G-25, and the salt concentration was varied from 10 mM to 200 mM NaCl. After being washed at the same salt concentration to remove unadsorbed proteins, the DNA-bound proteins were eluted and measured. The percent binding is expressed as a function of the NaCl concentration during the initial contact with the DNA column.

concentration ranging from 50 mM to 200 mM NaCl, and equal aliquots were applied to parallel DNA–Sephadex G-25 columns at a DNA/protein ratio of 4:1. After washing each column with 7 volumes of buffer containing NaCl at the appropriate molarity, the adsorbed proteins were eluted quantitatively in 6 M urea–0.4 M guanidine-HCl–0.1 M Na phosphate buffer (pH 7.5). The results (Fig. 2) indicate a strong dependency of DNA binding on ionic strength. They also show that a substantial portion (about 8%) of the proteins in the 0.35 M NaCl extract are strongly adsorbed to DNA at the highest concentration tested (200 mM). Some of this binding appears to be species- and sequence-specific, as indicated by criteria to be described below.

XII. DNA Binding by Nuclear Proteins Extractable in 0.35 M NaCl—A Basis for Fractionation

As noted above, the 0.35 M NaCl-soluble protein fraction of calf thymus lymphocyte nuclei includes components which adsorb to DNA columns

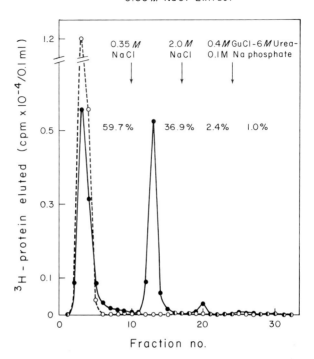

0.35 *M* NaCl Extract

FIG. 3A.   Affinity chromatography of calf thymus nuclear nonhistone proteins on double-stranded calf thymus DNA covalently linked to Sephadex G-25. The proteins in the 0.35 *M* NaCl extract were separated from basic components by chromatography on Bio-Rex 70 and applied to the DNA column in 50 m*M* NaCl. After washing the column with 50 m*M* NaCl to remove the unadsorbed proteins, the salt concentration was raised to 0.35 *M* NaCl to elute most of the DNA-bound proteins. A very small fraction was subsequently eluted in 2.0 *M* NaCl. The solid line shows the distribution of the ³H-labeled nuclear nonhistone proteins during the elution sequence. The dashed line shows the corresponding elution diagram from Sephadex G-25 which had been treated with the coupling reagent (CMC) but not reacted with DNA.

and proteins which do not. A simple fractionation of these components is illustrated in Fig. 3. In these experiments, the proteins were applied to DNA–Sephadex in 50 mM NaCl in buffer A, and the column was washed with buffered 50 m*M* NaCl to remove unadsorbed material. Under these conditions, about 60% of the applied protein passes through the column, but 37% remains bound to the DNA and can be eluted by raising the NaCl concentration of the eluting buffer to 0.35 *M*, the concentration origi-nally used to extract the proteins from the isolated lymphocyte nuclei (Fig. 3A). The control experiment shows that there is no significant

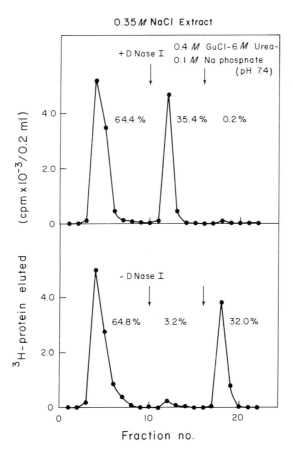

FIG. 3B.   Release of DNA-binding proteins by treatment of the calf thymus DNA–Sepha-dex G-25 column with DNAse I. The 0.35 M NaCl-soluble nuclear proteins were applied to parallel columns of DNA–Sephadex. After washing with 50 mM NaCl to remove the un-adsorbed proteins, one of the columns was treated at 15° with 1 mg of bovine pancreatic DNAse I in 2 ml of 10 mM NaCl in buffer A containing 2 mM MgCl$_2$ and 0.4 mM CaCl$_2$. Chromatography was continued in 50 mM NaCl. The upper panel shows the release of the ³H-labeled DNA-binding proteins after DNAse I treatment, although 50 mM NaCl does not remove the DNA-bound proteins from the control column (lower panel). The proteins attached to DNA–Sephadex in the control column were subsequently eluted in 6 M urea–0.4 M guanidine-HCl–0.1 M Na phosphate (pH 7.4) for analysis.

adsorption of the nuclear proteins to DNA-free Sephadex G-25, even though the polysaccharide matrix had been reacted with the carbodiimide reagent (Fig. 3A). Further evidence for DNA binding is provided in Fig. 3B. The nuclear proteins which adsorb to DNA–Sephadex at 50 mM NaCl are completely displaced when the column is treated with bovine pancreatic deoxyribonuclease I, even at salt concentrations (10 mM NaCl) which would not suffice to displace the adsorbed proteins from intact DNA–Sephadex columns.

XIII. Evidence for Sequence Specificity in DNA Binding by Nuclear Nonhistone Proteins

The recognition of specific nucleotide sequences in DNA by regulatory proteins of prokaryotes and by a wide variety of specific endonucleases ("restriction" enzymes) is now widely accepted as a model for genetic control in eukaryotic cells (*41*). The implication that certain chromosomal proteins of higher organisms interact differentially with different DNA sequences has been tested directly by DNA-affinity chromatography. Using techniques to fractionate calf thymus DNA into "highly reiterated" (low $C_0 t$), "moderately repetitive" (intermediate $C_0 t$), and "unique" (high $C_0 t$) DNA sequences, columns could be prepared which differed in their DNA sequence complexity. The binding affinities of calf thymus nuclear proteins for DNAs of different $C_0 t$ value, all derived from the same species and tissue, could then be compared.

Affinity chromatography was carried out by applying equal aliquots of the nuclear nonhistone protein fraction to parallel columns containing equal amounts of DNA–aminoethyl Sepharose 4B of differing $C_0 t$ values. The proteins were then eluted from each of the columns by a carefully matched series of step-by-step increments in the salt concentration of the eluting buffer, as shown in Fig. 4. The electrophoretic patterns of proteins eluted at each salt concentration are presented on the right of the appropriate peak in the elution diagram. The major conclusions of these experiments can be summarized briefly:

1. Different nuclear nonhistone proteins bind to DNA with different affinities. Most nuclear proteins have little or no DNA affinity and pass directly through the column. The DNA-binding proteins are heterodisperse. They elute characteristically at the indicated salt concentrations; some bind so strongly that they can be removed only at very high ionic strengths (or with the aid of protein denaturants).

2. The sets of proteins released at different salt concentrations differ in their complexity and molecular-weight distribution. An important point is that this type of nuclear protein fractionation is highly reproducible; the same sets of proteins emerge from the DNA column at a particular ionic strength when the experiment is repeated many times (*15*). Moreover, sets

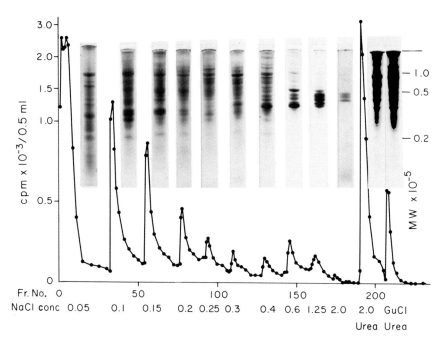

FIG. 4.   Fractionation of nuclear nonhistone proteins on double-stranded low C_0t (less than $C_0t = 6$) calf thymus DNA covalently linked to aminoethyl-Sepharose 4B. The ³H-labeled proteins were eluted with a discontinuous salt gradient and finally with denaturing buffers [2.0 M NaCl–5 M urea in buffer A, followed by 6 M urea–0.4 M guanidine-HCl in 0.1 M Na phosphate (pH 7.4)]. Each fraction was monitored for radioactivity, and the proteins eluted at each salt concentration were analyzed by electrophoresis on 10% polyacrylamide–0.1% SDS gels. The protein banding patterns are shown for each protein set at the appropriate peak of the elution diagram. The corresponding molecular-weight scale is shown on the right.

of proteins which emerge from the DNA column at a given salt concentration are eluted at the same salt concentration when they are rechromatographed on a new DNA column.

3. Comparisons of the electrophoretic patterns of nuclear proteins emerging at a given salt concentration from parallel columns containing identical amounts of high C_0t DNA, intermediate C_0t DNA, and low C_0t DNA show characteristic and reproducible differences. Several examples of differential binding of nuclear proteins to DNA fractions of different C_0t value are presented in Fig. 5. Analyses of the complete sets of electrophoretic patterns from parallel DNA columns differing in C_0t value show whether a given protein band interacts preferentially with "reiterated," "moderately repetitive" or "unique" DNA sequences. For example,

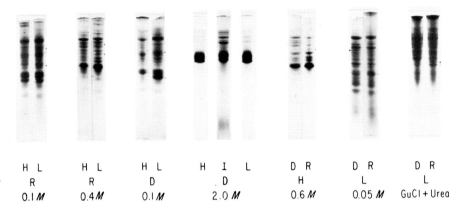

H L	H L	H L	H I L	D R	D R	D R
R	R	D	D	H	L	L
0.1 M	0.4 M	0.1 M	2.0 M	0.6 M	0.05 M	GuCl + Urea

FIG. 5. Side-by-side comparisons of the electrophoretic banding patterns of protein sets eluted at corresponding salt concentrations from parallel columns of aminoethyl-Sepharose 4B linked to DNA subfractions differing in C_0t value or strandedness. The first two panels compare the proteins eluted at 0.1 M NaCl and 0.4 M NaCl from double-stranded high C_0t (H: $C_0t = 2.25 \times 10^2$ to 4×10^4) thymus DNA and double-stranded low C_0t (L: C_0t less than 6) DNA. Examples of differential binding to high C_0t, intermediate C_0t (I: $C_0t = 6$ to 225), and low C_0t single-stranded DNA are shown in the third and fourth panels. The other panels illustrate differences in protein elution patterns from single-stranded and double-stranded DNA subfractions of the same C_0t value (R: reassociated DNA; D: denatured DNA). Major differences are indicated by the arrows. Many smaller differences in relative concentrations, indicative of differential DNA binding, are also present.

nuclear protein bands of mol wt 34,000 and 61,000 have greater affinity for double-stranded low C_0t DNA sequences and thus require higher salt concentrations to effect their displacement from the reiterated DNA column than from the unique sequences of the high C_0t DNA column. Conversely, a protein of molecular weight 50,000 binds more strongly to the high C_0t sequences (7,15). Figure 5 also includes an example of a protein (78,000 MW) which binds preferentially to intermediate C_0t DNA. Similar techniques have been employed to compare nuclear protein interactions with single-stranded and double-stranded DNA molecules of the same C_0t value. Some examples of preferential binding related to DNA strandedness are also shown in Fig. 5.

The differential affinities of nuclear proteins for particular DNA sequences have just begun to be analyzed. The resolution of DNA-affinity chromatography is likely to be improved by the use of cloned DNAs, ribosomal DNA satellites, and other well-defined DNA matrices. In addition, the use of new two-dimensional gel electrophoretic techniques coupled with autoradiographic procedures (42) should greatly improve the detection and measurement of individual DNA-binding proteins.

REFERENCES

1. Litman, R., *J. Biol. Chem.* **243**, 6222 (1968).
2. Jovin, T., and Kornberg, A., *J. Biol. Chem.* **243**, (1968).
3. Poonian, M. S., Schlabach, A. J., and Weissbach, A., *Biochemistry* **10**, 424 (1971).
4. Schaller, H., Nusslein, C., Bonhoeffer, F. J., Kurz, C., and Nietzschmann, I., *Eur. J. Biochem.* **26**, 474 (1972).
5. Weatherford, S. C., Weisberg, L. S., Achord, D. T., and Apirion, D., *Biochem. Biophys. Res. Commun.* **49**, 1307 (1972).
6. Gerwin, B. L., and Milstein, J. B., *Proc. Natl. Acad. Sci. U.S.A.* **69**, 2599 (1972).
7. Allfrey, V. G., Inoue, A., Karn, J., Johnson, E. M., and Vidali, G., *Cold Spring Harbor Symp. Quant. Biol.* **38**, 785 (1973).
8. Kleinsmith, L. J., Heidema, J., and Carroll, A., *Nature (London)* **226**, 1025 (1970).
9. van den Broek, H. W. J., Nooden, L. D., Sevall, J. S., and Bonner, J., *Biochemistry* **12**, 229 (1973).
10. Sevall, J. S., Cockburn, A., Savage, M., and Bonner, J., *Biochemistry* **14**, 782 (1975).
11. Alberts, B. M., and Frey, L., *Nature (London)* **227**, 1313 (1970).
12. Alberts, B. M., and Herrick, G., *in* "Methods in Enzymology," Vol. 21: Nucleic Acids, Part D (L. Grossman and K. Moldave, eds.), p. 198. Academic Press, New York, 1971.
13. Bendich, A., and Bolton, E. T., *in* "Methods in Enzymology," Vol. 12: Nucleic Acids (L. Grossman and K. Moldave, eds.), Part B, p. 635. Academic Press, New York, 1968.
14. Cavalieri, L., and Carroll, E., *Proc. Natl. Acad. Sci. U.S.A.* **67**, 802 (1970).
15. Allfrey, V. G., Inoue, A., and Johnson, E. M., *in* "Chromosomal Proteins and Their Role in the Regulation of Gene Expression" (G. S. Stein and L. J. Kleinsmith, eds.), pp. 265–300. Academic Press, New York, 1975.
16. Weissbach, A., and Poonian, M. S., *in* "Methods in Enzymology," Vol. 34: Affinity Techniques: Enzyme Purification, Part B (W. B. Jakoby and M. Wilchek, eds.), p. 463. Academic Press, New York, 1974.
17. Gilham, P. T., *Biochemistry* **7**, 2809 (1968).
18. Astell, C. R., and Smith, M., *Biochemistry* **11**, 4114 (1972).
19. Astell, C. R., Doel, M. T., Jahnke, P. A., and Smith, M., *Biochemistry* **12**, 5068 (1973).
20. Shih, T. Y., and Martin, M. A. *Biochemistry* **13**, 3411 (1974).
21. Rickwood, D., *Biochim. Biophys. Acta* **269**, 47 (1972).
22. Noyes, B. E., and Stark, G. R., *Cell* **5**, 301 (1975).
23. Scheffler, I. E., and Richardson, C. C., *J. Biol. Chem.* **247**, 5736 (1972).
24. Panet, A., and Khorana, H. G., *J. Biol. Chem.* **249**, 5213 (1974).
25. Cuatrecasas, P., *J. Biol. Chem.* **245**, 3059 (1970).
26. Gross-Bellard, M., Oudet, P., and Chambon, P., *Eur. J. Biochem.* **36**, 32 (1973).
27. Doerfler, W., and Kleinschmidt, A. K., *J. Mol. Biol.* **50**, 574 (1970).
28. Eigner, J., *in* "Methods in Enzymology," Vol. 12: Nucleic Acids (L. Grossman and K. Moldave, eds.), p. 386. Academic Press, New York, 1968.
29. Kohne, D. E., and Britten, R. J., *in:* "Procedures in Nucleic Acid Research" (G. L. Cantoni and D. R. Davies, eds.), p. 500. Harper, New York, 1971.
30. Sutton, W. D., *Biochim. Biophys. Acta* **240**, 522 (1971).
31. Fasy, T. M., Inoue, A., Johnson, E. M., and Allfrey, V. G., *J. Supramol. Struct.* **6**, Suppl. 1 (1977).
32. Fasy, T. M., Ph.D. Thesis, Columbia University, New York, 1976.
33. Johnson, E. M., Hadden, J. W., Inoue, A., and Allfrey, V. G., *Biochemistry* **14**, 3873 (1975).
34. Johnson, E. M., Inoue, A., Crouse, L. J., Allfrey, V. G., and Hadden, J. W., *Biochem. Biophys. Res. Commun.* **65**, 714 (1975).

35. Bekhor, I., Kung, G. M., and Bonner, J., *J. Mol. Biol.* **39**, 351 (1969).
36. Gilmour, R. S., and Paul, J., *FEBS Lett.* **9**, 242 (1970).
37. Barrett, T., Maryanka, D., Hamlyn, P. H., and Gould, H. J. *Proc. Natl. Acad. Sci. U.S.A.* **71**, 5075 (1974).
38. Stein, G. S., Mans, R. J., Gabbay, E. J., Stein, J. L., Davis, J., and Adawadkar, P. D., *Biochemistry* **14**, 1859 (1975).
39. Allfrey, V. G., *in* "Methods in Enzymology," Vol. 31: Biomembranes, Part A (S. Fleischer and L. Packer, eds.), p. 246. Academic Press, New York, 1974.
40. Rice, R. H., and Means, G. E., *J. Biol. Chem.* **246**, 831 (1971).
41. Allfrey, V. G., *in:* "Acidic Proteins of the Nucleus" (I. L. Cameron and J. L. Jeter, Jr., eds.), p. 2. Academic Press, New York, 1974.
42. O'Farrell, P. H., *J. Biol. Chem.* **250**, 4007 (1975).

Chapter 19

DNA-Binding Proteins

GRETCHEN H. STEIN

Department of Molecular, Cellular and Developmental Biology,
University of Colorado,
Boulder, Colorado

I. Introduction

Proteins involved in the regulation or process of DNA replication or RNA transcription are likely to possess an ability to bind reversibly to DNA. If these proteins are only transiently bound to DNA *in vivo*, then they may be present primarily in the soluble portion of a cell. One way to look for such proteins is to examine the DNA-binding proteins, which are isolated on the basis of their affinity for purified DNA *in vitro*. Previous studies have shown that a number of proteins that interact with DNA *in vivo* also recognize purified DNA *in vitro*; these include proteins involved in DNA replication, such as DNA polymerases and DNA-unwinding proteins, proteins involved in RNA transcription, such as RNA polymerases and factors which enhance RNA polymerase activity, and proteins involved in gene regulation, such as bacterial *lac* repressor, λ repressor, and bacterial cyclic adenosine 5'-monophosphate-receptor protein (*1–9*). It seems quite probable that eukaryotic DNA-binding proteins, such as the steroid hormone-receptor proteins, are also involved in gene regulation, but the available evidence is not as definitive as it is for the prokaryotic regulatory proteins (*10–13*).

DNA-binding proteins are isolated by affinity chromatography using purified DNA immobilized on a neutral substrate. DNA can be bound to cellulose by drying the two together (*1, 14*), by a combination of drying and ultraviolet irradiation (*2*), or by chemical coupling using activating agents, such as water-soluble carbodiimide (*15,16*). DNA can also be immobilized by trapping it in an agarose gel (*5*) or an acrylamide gel (*17*). Alternatively, it can be covalently attached to agarose beads by activating the DNA with carbodiimide (*18*) or by activating the agarose beads with cyanogen bromide (19). These various methods all yield a DNA–substrate complex with which known DNA-binding proteins can interact. However, they vary in the fraction of input DNA bound, in the stability of the complex formed, and in whether large or small, single- or double-stranded DNA molecules can be immobilized.

DNA–cellulose prepared by drying is well suited for most studies of DNA-binding proteins. Its advantages are that it is relatively simple to prepare, both unmodified single- or double-stranded DNA can be bound, there is no steric hindrance to the movement of proteins through the DNA–cellulose matrix, and there is no significant binding of protein to plain cellulose. However, its disadvantages are that only about half the input DNA is bound, small DNA fragments do not bind well, and the complex is slowly dissociated in aqueous solution. Therefore, for specific purposes, other DNA–substrate complexes may be preferable. DNA molecules too small to bind efficiently to cellulose by drying can be chemically linked to cellulose or agarose. A much larger fraction of the input DNA can be bound to cellulose if it is irradiated with ultraviolet light after drying. For the isolation of preparative amounts of DNA-binding proteins, large quantities of single-stranded DNA–agarose are probably the easiest to prepare. In some cases, a specific DNA-binding protein may bind more efficiently to a particular kind of DNA substrate. For example, the binding of HeLa DNA polymerase to DNA–Sepharose, prepared by cyanogen bromide activation, was approximately 50 times more efficient than to a corresponding ultraviolet light irradiated DNA–cellulose (19).

II.   DNA–Cellulose Chromatography

A.   Preparation of DNA–Cellulose

The preparation of DNA–cellulose by drying was developed by Alberts et al. (1) and described in greater detail by Alberts and Herrick (14). A highly purified grade of cellulose such as Munktell 410 (also sold as Cellex 410 by Bio-Rad), Whatman CF-11, or Machery-Nagel 2200ff cellulose is used. It is washed several times with boiling ethanol to remove pyridine. Then it is washed in succession with 0.1 N NaOH, 1 mM EDTA, and 10 mM HCl solutions at room temperature. Finally, the clean cellulose is rinsed to neutrality with distilled water and lyophilized to dryness.

DNA at a concentration of 1–3 mg/ml is dissoved in 10 mM Tris-HCl (pH 7.4)–1 mM ethylenediaminetetraacetic acid (Tris–EDTA). The solution may be stirred vigorously since Alberts et al. (1) suggest that DNA of about

10^7 daltons is probably optimal for preparing DNA–cellulose. Single-stranded DNA is prepared by heating the DNA solution to 100°C for 15 minutes and cooling it rapidly. Washed cellulose powder is added to the DNA solution in a glass beaker until the mixture is a very thick paste (this requires approximately 1 gm of cellulose per 3 ml of solution). The beaker is covered with gauze and kept at room temperature until the DNA–cellulose mixture is completely dry and powdery when pulverized with a glass rod. This drying step is very important for obtaining the maximum amount of DNA bound to the cellulose and may require 3–7 days. The drying process can be hastened by spreading the mixture in an evaporating dish or by blowing cool air over it.

The air-dried DNA–cellulose mixture is lyophilized to complete dryness, suspended in 20 volumes of Tris–EDTA at 4°C overnight, and then washed several times with Tris–EDTA to remove the free DNA. The DNA–cellulose may be collected from the washes by either suction filtration (e.g., using Whatman No. 2 paper) or centrifugation at approximately 500 g. The washed DNA–cellulose is suspended in an equal volume of Tris–EDTA containing 0.15 M NaCl and stored as a frozen slurry, where it keeps for at least 1 year.

The amount of DNA bound to cellulose can be determined in several ways. A measured amount of DNA–cellulose can be digested with DNase I (e.g., 100 μg/ml in buffer containing 10 mM MgCl$_2$ for 20 minutes at room temperature) or hydrolyzed with 0.5 M perchloric acid (30 minutes at 100°C). Then the cellulose is collected in a pellet by centrifugation and the amount of DNA determined by measuring the absorbance, at 260 nm, of the supernatant. Alternatively, an aliquot of DNA–cellulose can be suspended in 10 volumes of Tris–EDTA and heated to 100°C for 20 minutes to release the DNA. This procedure is useful for freshly prepared DNA–cellulose, but Fox and Pardee (20) found that some DNA apparently becomes irreversibly bound to the cellulose during storage at -20°C. An indirect estimate of the amount of bound DNA can be obtained by measuring the A_{260} due to free DNA in the washes. Approximately half the input DNA is expected to bind to the cellulose.

Lypholized DNA–cellulose prepared as above, but not washed of free DNA, can be irradiated with ultraviolet light according to the procedure of Litman (2). This step increases the amount of DNA bound to cellulose to approximately 90%. The dry DNA–cellulose is suspended in about 5 volumes of absolute ethanol and stirred to break up any lumps. The alcohol suspension, added to a beaker to a depth of about 1 cm, is slowly stirred and exposed to approximately 100,000 ergs/mm² of ultraviolet light (254 nm). Litman (2) suggests that this dose can be administered by a 15-minute period of irradiation with a Mineralight (Ultraviolet Products) positioned 10 cm above the surface of the alcohol. Following irradiation, the alcohol is removed by suction filtration and the DNA–cellulose is washed and stored either as a frozen slurry or as a lyophilized powder. The amount of DNA bound to cellulose can be determined by DNase digestion or perchloric acid hydrolysis, as described above.

B. Preparation of Cell Extract

DNA-binding proteins are isolated from whole cell extracts or subcellular fractions that contain undenatured proteins and are free of endogenous DNA. Removal of endogenous DNA is necessary to eliminate competition between it and the DNA–cellulose.

We use the following procedure, based on the method of Alberts and Herrick (14), to prepare cell extracts from human cells grown in monolayer culture. The cell sheets are washed twice with Dulbecco's phosphate buffered saline and once with TMEMC buffer [20 mM Tris-HCl (pH 8.1)–1 mM β-mercaptoethanol–1 mM EDTA–10 mM MgCl$_2$–2 mM CaCl$_2$]. The cells are harvested using a rubber policemen to scrape them into a small volume of TMEMC (about 2 ml/10^7 cells), thus avoiding exposing the cells to trypsin. Once harvested, the cells are kept at 0–4°C unless otherwise indicated. Since the scraped cells are too clumped to be counted, the cell number can be estimated by counting a trypsinized sister culture. Alternatively, the size of the cell pellet obtained by low-speed centrifugation of the cell suspension in a graduated tube can be calibrated to give a rough estimate of cell number or amount of protein. Alberts and Herrick (14) recommend the addition of 10% glycerol and 500 μg/ml bovine serum albumin (BSA) to mammalian cell extracts. However, we omit the BSA when our cell concentration is > 5 × 10^6/ml.

The cell suspension is passed through a 20-gauge needle several times to disperse any clumps. Then the cells are sonicated at 4°C by pulsing 3 times, for 15 seconds each, at power settling 1 of a Branson Sonifier equipped with a microtip. These conditions were established by monitoring the breakage by phase-contrast microscopy. Almost all the cells and nuclei are broken during the first 15-second pulse. The cell suspension is cooled and mixed between the pulses.

The cell sonicate is treated with DNase I (Worthington) at 100 μg/ml for 20 minutes at room temperature. Most of the DNA is digested into acid-soluble fragments in the first 5 minutes, but 3–5% of the DNA remains insoluble even after a 20 minute incubation. Therefore, it is possible that some species of DNA-binding proteins are not present in this soluble protein extract, but rather remain complexed to the residual endogenous DNA. After DNase digestion, the cell extract is centrifuged at 20,000g for 45 minutes at 4°C to pellet particulate matter. The supernatant is dialyzed 3 times against 200 volumes of NMET buffer [50 mM NaCl–1 mM β-mercaptoethanol–1 mM EDTA–20 mM Tris-HCl (pH 8.1)] in preparation for DNA–cellulose chromatography.

Several other methods have been used to prepare DNA-free protein extracts. Each one may permit a slightly different subset of DNA-binding proteins to be present in the final soluble protein preparation. One variation of the procedure described above is to treat the cell sonicate with DNase I in the presence of higher salt concentrations in order to dissociate any residual DNA–protein complexes. Alberts and Herrick (14) proposed this procedure and found that the residual DNA was reduced to < 2% when the DNase digestion was carried out in the presence of 0.6 M NaCl. Another alternative is to prepare the cell extract in NMET buffer with high salt (e.g., 1.7 M NaCl), centrifuge it at low speed to remove cell debris, and add to it polyethylene glycol 6000 (30% w/w stock solution) to a final concentration of 10%. After 30 minutes at 0°C, the precipitated DNA is pelleted by a 15-minute centrifugation at 10,000g. This procedure has the advantage that 98% of the DNA is precipitated, but has the disadvantage that approximately 12% of the soluble protein is also precipitated (14).

Additional insoluble proteins may be included in the cell extract by treating the cell sonicate with solubilizing agents such as nonionic detergents or urea. Control experiments using bacterial cell extracts showed that the addition of 1% Triton X-100, 1% Brij-58, or 1% Tween 80 to the extract and column buffers did not change the DNA-binding protein pattern [Alberts and Amodio, unpublished results, cited in (14)]. Similarly, neither 1% Triton X-100 nor 1 M urea added to the column buffers affected the single-strand specific DNA-binding protein pattern obtained from calf thymus tissue extracts (21).

C. Chromatography

An appropriate amount of DNA–cellulose is diluted in 10 volumes of Tris–EDTA, poured into a chromatography column, and allowed to settle at room temperature. The excess buffer and cellulose fines remaining above the settled cellulose are removed with a pipette. DNA–cellulose columns do not seem to be adversely affected by running dry under gravity. All subsequent column operations are carried out at 4°C. After the DNA–cellulose column is washed for several hours with rinse buffer (NMET containing 10% glycerol), it is ready for use. In some cases, the recovery of proteins applied to the column is improved if bovine serum albumin (100 μg/ml) is included in the column buffers or prerun on the column (14, 22).

There is no fixed rule for the amount of DNA needed to isolate the DNA-binding proteins from a particular kind of cell extract. For each protein, retention by the DNA–cellulose depends on the frequency of its binding sites and on its DNA affinity under the conditions employed. However, one can determine the amount of DNA that is sufficient to bind all of a particular protein or all of the major DNA-binding proteins from a cell extract. This is done by rechromatographing the breakthrough from one DNA–cellulose column on a second identical column to determine whether any additional DNA-binding proteins were present in the breakthrough. Two milligrams of DNA per milligram of protein is a generous ratio to use for analytical studies. It is sufficient to bind $> 99\%$ of the DNA-binding proteins present in a nonhistone chromosomal protein fraction where about half the proteins bind to DNA (23).

The protein extract, dialyzed against NMET buffer, is clarified by centrifugation at 20,000 g for 20 minutes at $4°C$, made 10% (v/v) in glycerol, and loaded on the DNA–cellulose column at a flow rate of 1–2 column volumes per hour. After the sample is loaded, the column is rinsed with 4–6 column volumes of rinse buffer until the amount of protein in the rinse fractions is $\leq 0.1\%$. A convenient way to load the entire sample without concern for the column running dry is to overlay the sample with rinse buffer containing only 5% glycerol (21). The DNA-binding proteins are eluted by increasing the NaCl concentration in the rinse buffer step by step or in a gradient. Approximately 6–10% of the proteins in whole cell soluble protein extracts from mouse, hamster, rat, or human cells are DNA-binding proteins (14, 23, 24 and G. Stein, unpublished results). Almost all the DNA-binding proteins can be eluted by NaCl at concentrations $\leq 2.0\,M$. However, a small fraction of DNA-binding proteins is resistant to elution with 2.0 M NaCl and can be released by digesting the DNA with DNase. When rinse buffer supplemented with 10 mM MgCl$_2$ and 40 μg/ml DNase I is run into the column and incubated overnight at $4°C$, 95% of the DNA is digested and the residual proteins can be eluted with rinse buffer (22).

Control experiments using plain washed cellulose columns show that $\leq 0.2\%$ of the proteins in a whole cell soluble extract are retained by cellulose alone. This small cellulose-binding fraction appears to be nonspecific or else involves very minor protein species. When we examined the cellulose-binding fraction from a HeLa cell extract in order to determine if any of the recognizable protein species that bind to DNA–cellulose also bind to plain cellulose, none were observed. The complete removal of particulate matter from the extract is crucial for low blank values (14). To lower blank values further, for the detection of minor DNA-binding proteins, samples should be prerun through a plain cellulose column or passed through a pad of plain cellulose layered over the DNA–cellulose and separated from it by a glass fiber disk. After an extract has been loaded and briefly rinsed through the DNA–cellulose, the plain cellulose pad is removed with a pipette.

III.   DNA–Agarose and DNA–Acrylamide Chromatography

Single-stranded DNA can be effectively trapped in an agarose gel (5,25) and single- or double-stranded DNA can be trapped in an acrylamide gel (17,26). DNA–agarose is particularly useful for preparative purposes because it is easy to prepare in large quantities.

A method for preparing very concentrated DNA–agarose has been described by Schaller et al. (5). Denatured DNA, at a concentration of 15 mg/ml in 0.02 M NaOH at 50°C, is mixed with an equal volume of 4% Sigma agarose which has been cooled to 50°C. The mixture is swirled vigorously and then poured into a thin layer in an ice-cooled glass dish where it gels rapidly. Finally, the gel is cut into small pieces, forced through a large syringe needle, and further fragmented by forcing it through a 60–75 mesh stainless-steel screen twice. The wire mesh of a hypodermic Millipore filter holder is convenient for this purpose. The fragmented DNA–agarose, suspended in Tris–EDTA containing 0.1 M NaCl, is packed in a column and washed until no more DNA appears in the runoff, or washed by several cycles of low-speed centrifugation and resuspension of the pellet in fresh buffer. Approximately 60–80% of the input DNA is retained in the agarose fragments, which can be stored for at least 1 year at 4°C in Tris–EDTA containing 1 M NaCl with no loss of DNA.

Affinity chromatography using DNA–agarose follows the same basic procedure as DNA–cellulose chromatography, including column operation at a flow rate of 1–2 column volumes per hour (not to exceed 12 ml/hour/cm²). However, the blank values for DNA–agarose are not as low as for DNA–cellulose. As much as 0.5–1.0% of the proteins from whole cell extracts of E. coli were retained on plain agarose columns composed of 4% Serva agarose gel beads (5). Because Sigma agarose gels, having the same mechanical stability as 4% Serva agarose gels, can be prepared with half as much agarose, they may have lower blank values. The DNA-specific proteins can be enriched by prerunning the extract through plain agarose or by readsorbing and re-eluting the binding proteins on a second DNA–agarose column. There is no problem with retardation of large proteins during DNA–agarose chromatography.

DNA–acrylamide gels are similar to DNA–agarose gels in their preparation, but have the advantage that double- as well as single-stranded DNA can be trapped in the gel matrix. The original procedure of Cavalieri and Carroll (17) described the preparation of a relatively dilute DNA–acrylamide gel containing 56 μg DNA/ml bead volume. We have modified their procedure slightly to prepare more concentrated DNA–acrylamide. A 50% (w/v) acrylamide stock solution is prepared from a mixture of 97% highly purified acrylamide and 3% N,N'-methylenebisacrylamide and filtered before use. To prepare 100 ml of DNA–acrylamide gel, 10 ml of the 50%

acrylamide stock are mixed with 65 ml of DNA solution, e.g., 1–3 mg/ml in Tris–EDTA, and 0.05 ,ml N,N,N',N'-tetramethylethylenediamine. Next, 0.7 ml of a freshly prepared 10% ammonium persulfate solution is added to the acrylamide mixture, which is then immediately mixed into 25 ml of 2% (w/w) agarose which has been cooled to 50°C. Although the solution gels quickly as the agarose component cools, at least 30 minutes should be allowed for polymerization of the acrylamide. The DNA–acrylamide gel is fragmented and washed in the same manner as DNA–agarose. Approximately half of the input DNA is bound. There is no significant nicking or denaturing effect of the polymerizing agents on the DNA (17).

Careful packing of the DNA–polyacrylamide columns is needed to avoid the formation of channels. Cavalieri and Carrol (17) recommend that the gel beads be packed by mixing them with an equal volume of buffer, pouring them into a column, and allowing them to pack under a constant pressure of about 60 cm until the column volume ceases to contract. The packed volume will be about 85% of that obtained by settling under gravity.

DNA–acrylamide chromatography differs from DNA–cellulose and DNA–agarose chromatography because the pore size of the matrix is small enough to retard the movement of even medium-sized proteins such as bovine serum albumin (68,000 daltons) (17). Consequently, the DNA-binding proteins will probably be eluted in broader peaks than from comparable DNA–cellulose of DNA–agarose columns, and very large DNA-binding proteins will not penetrate into the gel. On the other hand, the blank values for DNA-free acrylamide are very good, e.g., less than 0.03% of Chinese hamster cytoplasmic proteins bound to a plain acrylamide column (26).

IV. Discussion

The techniques for isolating DNA-binding proteins are relatively simple and straightforward. However, it can sometimes be difficult to compare the results from different investigations because the set of DNA-binding proteins isolated depends on the particular isolation procedure used. The type of DNA used for the chromatography is an important variable. In general, the largest set of DNA-binding proteins can be isolated using homologous DNA for the affinity chromatography, and the more heterologous the DNA used, the smaller the set of proteins bound (22,23,26–28). However, the subset of DNA-binding proteins which bind only to homologous DNA is small. For example, Johnson et al. (22) found that > 90% of the total DNA-binding proteins from Novikoff hepatoma cells bind to a mixture of calf thymus and *E. coli* DNA, and van den Broek *et al.* (*23*) found that 87% of rat liver nonhistone chromosomal DNA-binding proteins bind to *E. coli* DNA. The species-specific DNA-binding proteins are not necessarily those involved in gene regulation since regulatory proteins, such as bacterial *lac* repressor, estradiol-receptor protein complex, and glucocorticoid–receptor protein complex, bind to a variety of heterologous DNAs (*29–31*). Lin and Riggs (*32*) have suggested that a modest affinity for nonoperator DNA may be an important property of gene regulatory proteins in general.

Although most DNA-binding proteins bind to both double-and single-stranded DNA, proteins binding specifically to single-stranded DNA have been isolated from a wide variety of organisms (*1,4,21,33–37*). Beginning with the T4 gene 32 protein, a number of these single-strand specific DNA-binding proteins have been characterized as DNA-unwinding proteins. Evidence primarily from prokaryotic systems suggests that DNA-unwinding proteins have important roles in the replication, recombination, and perhaps repair of DNA [*3,33,* Mackay and Linn cited in Grossman *et al.* (*38*)]. On the other hand, some gene regulatory proteins appear to bind better to double-stranded DNA (non-locus-specific binding) (*29,31*). In addition, it is important to remember that some proteins which bind to DNA *in vitro* may not interact with DNA *in vivo*. For example, Tsai and Green (*39*) found that collagen α-chains and their protocollagen precursors bind to DNA–cellulose. It is generally assumed that this is adventitious binding.

The method used to immobilize the DNA could also have an effect on the set of proteins which bind to the DNA. As mentioned previously, very large proteins are excluded from 5% acrylamide gel beads and consequently cannot interact with DNA trapped inside the beads. However, Cavalieri and Carroll (*17*) suggest that there is sufficient immobilized DNA protruding from the beads to interact with proteins in the liquid phase. DNA–cellulose prepared by irradiation with ultraviolet light is a different substrate than dried DNA–cellulose because it contains ultraviolet photoproducts. At the level of resolution obtained by one-dimensional sodium dodecyl sulfate (SDS) polyacrylamide gel electrophoresis, the patterns of DNA-binding proteins isolated using irradiated and unirradiated DNA–cellulose are essentially identical (Fig. 1). It appears that the presence of ultraviolet photoproducts in DNA does not interfere with the binding of the major DNA-binding protein species. On the other hand, a small number of proteins have been found which bind specifically to ultraviolet irradiated DNA, and these are generally involved in DNA repair processes (*38,41*).

The manner of preparation of the cell extract is obviously critical in determining the set of DNA-binding proteins isolated. Since DNA-binding proteins are present in the cytoplasm, nucleoplasm, and chromatin portions of a cell, none of these subcellular fractions can be discarded without dis-

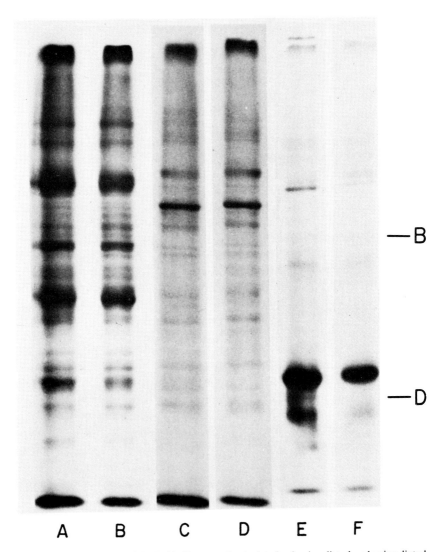

FIG. 1. Comparison of DNA-binding proteins isolated using irradiated and unirradiated DNA–cellulose. Hela cells in logarithmic growth phase in monolayer culture were incubated with [¹⁴C]leucine for 45 hours. The cells were harvested and a soluble cell extract was prepared as described in the text. Half the extract was applied to a column of irradiated double-stranded (d.-s.) calf thymus DNA–cellulose, followed by a column of irradiated single-stranded (s.-s.) calf thymus DNA–cellulose. Simultaneously, the other half of the extract was applied to comparable columns of unirradiated DNA–cellulose. The DNA-binding proteins were eluted with 0.2 M NaCl and 2.0 M NaCl steps, and analyzed by SDS gel electrophoresis on a 9% acrylamide gel using the method of Laemmli (40). To avoid overloading the gel, only half of the

carding some DNA-binding proteins. Furthermore, there is always the possibility that the DNA-binding activity of some proteins may be destroyed during the preparation of the cell extract.

The total DNA-binding protein fraction isolated from mammalian cells is quite large, approximately 6–10% of the protein in a mammalian cell extract. In general, these DNA-binding proteins have been fractionated into

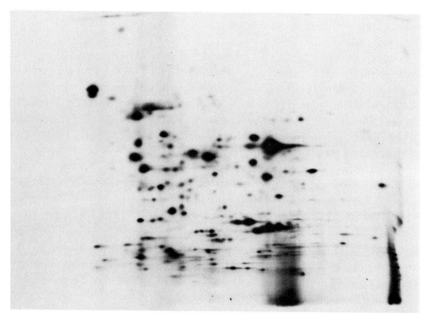

FIG. 2. DNA-binding proteins fractionated by two-dimensional gel isoelectric focusing and SDS gel electrophoresis. HeLa cells in logarithmic growth in monolayer culture were incubated with [³H]leucine for 45 hours and then harvested. A soluble cell extract was prepared and chromatographed on unirradiated double-stranded calf thymus DNA–cellulose. The DNA-binding proteins eluted with 0.2 M NaCl were fractionated in an isoelectric focusing gel (1.6% pH 5–8 ampholines + 0.4% pH 3–10 ampholines) and then in a 9% arcylamide gel according to the procedure of O'Farrell (42). The pH gradient of the isoelectric focusing gel extended from approximately pH 7.5–5 from left to right. The molecular weight distribution in the SDS gel is the same as in Fig. 1. The gel was fixed and a fluorograph was made by impregnating the gel with 2,5-diphenyloxazole according to the procedure of Bonner and Laskey (43).

0.2 M NaCl-d.-s. DNA samples and $\frac{1}{10}$ of the 0.2 M NaCl-s.-s. DNA samples were used. The gel was fixed and an autoradiograph was made. (A) 0.2 M NaCl eluate from d.-s. irradiated DNA–cellulose; (B) 0.2 M NaCl eluate from d.-s. unirradiated DNA–cellulose; (C) 2.0 M NaCl eluate from d.-s. irradiated DNA–cellulose; (D) 2.0 M NaCl eluate from d.-s. unirradiated DNA–cellulose; (E) 0.2 M NaCl eluate from s.-s. irradiated DNA–cellulose; (F) 0.2 M NaCl eluate from s.-s. unirradiated DNA–cellulose. Molecular weight marker proteins: (B) bovine serum albumin, 68,000 daltons, and (D) DNase, 31,000 daltons.

several groups, such as low and high salt eluting proteins, and then analyzed further by one-dimensional SDS polyacrylamide gel electrophoresis, where fewer than 50 bands are typically resolved. This small number of bands represent only the major species of DNA-binding proteins or groups of proteins having the same mobility in an SDS gel. When we analyzed one subset of DNA-binding proteins, the low salt eluting proteins isolated from HeLa cells on double-stranded calf thymus DNA–cellulose, by a two-dimensional gel isoelectric focusing and SDS electrophoresis procedure (*42*), we observed 250 protein spots (Fig. 2; G. Stein, unpublished results). This may be compared with 43 protein bands observed by one-dimensional SDS gel electrophoresis of a comparable DNA-binding protein fraction (Fig. 1 B). The number of DNA-binding proteins counted in the two-dimensional gel should be considered a lower limit for two reasons. First, the basic DNA-binding proteins (estimated as about one-fifth of the total) did not band in the acidic isoelectric focusing gel used. Second, we estimate that only proteins present in at least 10^4 copies per cell would have been detected by the labeling and autoradiographic procedures used in this experiment. Thus, the DNA-binding protein fraction from human cells is very complex, even at a level of analysis which may still overlook a large class of specific gene regulatory DNA-binding proteins.

ACKNOWLEDGMENTS

The author wishes to thank Dr. Glenn Herrick, Dr. David Prescott, and Dr. Rosalind Yanishevsky for their helpful comments on this manuscript.
This work was aided by grant NP-156 from the American Cancer Society.

REFERENCES

1. Alberts, B. M., Amodio, F. J., Jenkins, M., Gutmann, E. D., and Ferris, F. L., *Cold Spring Harbor Symp. Quant. Biol.* **33**, 305 (1968).
2. Litman, R. M., *J. Biol. Chem.* **243**, 6222 (1968).
3. Alberts, B. M., and Frey, L., *Nature (London)* **227**, 1313 (1970).
4. Hotta, Y., and Stern, H., *Nature (London) New Biol.* **234**, 83 (1971).
5. Schaller, H., Nüsslein, C., Bonhoeffer, F. J., Kurz, C., and Nietzchmann, I., *Eur. J. Biochem.* **26**, 474 (1972).
6. Sugden, B., and Keller, W., *J. Biol. Chem.* **248**, 3777 (1973).
7. Gilbert, W., and Muller-Hill, B., *Proc. Natl. Acad. Sci. U.S.A.* **58**, 2415 (1967).
8. Ptashne, M., *Nature (London)* **214**, 232 (1967).
9. Riggs, A. D., Reiness, G., and Zubay, G., *Proc. Natl. Acad. Sci. U.S.A.* **68**, 1222 (1971).
10. Yamamoto, K. R., and Alberts, B. M., *Proc. Natl. Acad. Sci. U.S.A.* **69**, 2105 (1972).
11. Clemens, L. E., and Kleinsmith, L. J., *Nature (London) New Biol.* **237**, 204 (1972).
12. Baxter, J. D., Rousseau, G. G., Benson, M. C., Garcea, R. L., Ito, J., and Tomkins, G. M., *Proc. Natl. Acad. Sci. U.S.A.* **69**, 1892 (1972).
13. Yamamoto, K. R., Stampfer, M. R., and Tomkins, G. M., *Proc. Natl. Acad. Sci. U.S.A.* **71**, 3901 (1974).

14. Alberts, B., and Herrick, G., *in* "Methods in Enzymology," Vol. 21: Nucleic Acids, Part D (L. Grossman and K. Moldave, eds.), p. 198. Academic Press, New York, 1971.

15. Gilham, P. T., *in* "Methods in Enzymology," Vol. 21: Nucleic Acids, Part D (L. Grossman and K. Moldave, eds.), p. 191. Academic Press, New York, 1971.

16. Shih, T. Y., and Martin, M. A., *Biochemistry* **13**, 3411 (1974).

17. Cavalieri, L. F., and Carroll, E., *Proc. Natl. Acad. Sci. U.S.A.* **67**, 807 (1970).

18. Rickwood, D., *Biochim. Biophys. Acta* **269**, 47 (1972).

19. Poonian, M. S., Schlaback, A. J., and Weissback, A., *Biochemistry* **10**, 424 (1971).

20. Fox, T. O., and Pardee, A. B., *J. Biol. Chem.* **246**, 6159 (1971).

21. Herrick, G., and Alberts, B., *J. Biol. Chem.* **251**, 2124 (1976).

22. Johnson, J. D., St. John, T., and Bonner, J., *Biochim. Biophys. Acta* **378**, 424 (1975).

23. van den Broek, H. W. J., Nooden, L. D., Sevall, J. S., and Bonner, J., *Biochemistry* **12**, 229 (1973).

24. Salas, J., and Green, H., *Nature (London) New Biol.* **229**, 165 (1971).

25. Naber, J. E., Schepman, A. M. J., and Rörsch, A., *Biochim. Biophys. Acta* **114**, 326 (1966).

26. Vaughan, S. T., and Comings, D. E., *Exp. Cell Res.* **80**, 265 (1973).

27. Patel, G. L., and Thomas, T. L., *Proc. Natl. Acad. Sci. U.S.A.* **70**, 2524 (1973).

28. Teng, C. S., Teng, C. T;, and Allfrey, V. G., *J. Biol. Chem.* **246**, 3597 (1971).

29. Lin, S-Y., and Riggs, A. D., *J. Mol. Biol.* **72**, 671 (1972).

30. Yamamoto, K. R., and Alberts, B., *J. Biol. Chem.* **249**, 7076 (1974).

31. Rousseau, G. G., Higgins, S. J., Baxter, J. D., Gelfand, D., and Tomkins, G. M., *J. Biol. Chem.* **250**, 6015 (1975).

32. Lin, S.-Y., and Riggs, A. D., *Cell* **4**, 107 (1975).

33. Sigal, M., Delius, H., Kornberg, T., Gefter, M. L., and Alberts, B., *Proc. Natl. Acad. Sci. U.S.A.* **69**, 3537 (1972).

34. Reuben, R. C., and Gefter, M. L., *Proc. Natl. Acad. Sci. U.S.A.* **70**, 1846 (1973).

35. Tsai, R. L., and Green, H., *J. Mol. Biol.* **73**, 307 (1973).

36. Banks, G. R., and Spanos, A., *J. Mol. Biol.* **93**, 63 (1975).

37. Donnelly, T. E., Jr., Westergaard, O., and Klenow, H., *Biochim. Biophys. Acta* **402**, 150 (1975).

38. Grossman, L., Braun, A., Feldberg, R., and Mahler, I., *Annu. Rev. Biochem.* **44**, 19 (1975).

39. Tsai, R. L., and Green, H., *Nature (London) New Biol.* **237**, 171 (1972).

40. Laemmli, U.K., *Nature (London)* **227**, 680 (1970).

41. Huang, A. T.-F., Riddle, M. M., and Koons, L. S., *Cancer Res.* **35**, 981 (1975).

42. O'Farrell, P. H., *J. Biol. Chem.* **250**, 4007 (1975).

43. Bonner, W. M., and Laskey, R. A., *Eur. J. Biochem.* **46**, 83 (1974).

Chapter 20

Fractionation of Nonhistone Proteins by Histone-Affinity Chromatography

ALAN McCLEARY, LARRY NOODEN, AND
LEWIS J. KLEINSMITH

*Division of Biological Sciences,
The University of Michigan,
Ann Arbor, Michigan*

I. Introduction

DNA, histones, and nonhistone proteins presumably all interact with each other in generating the structural and functional properties of chromatin. DNA–histone, DNA–nonhistone, and histone–histone interactions have all been subject to extensive investigation (*1–7*). The interaction of nonhistones with histones, on the other hand, has received little attention, in part because of the tendency of these components to aggregate and form precipitates when mixed together (*8,9*). This chapter describes a new approach which overcomes this obstacle by fractionating nonhistone proteins on affinity columns composed of purified histones linked to Sepharose.

II. Methods

A. Isolation of Histones and Nonhistones

During isolation of chromosomal proteins, the protease inhibitor phenylmethylsulfonyl fluoride (PMSF) is present throughout the procedure in a concentration of 0.1 mM (*10*). Appropriate dilutions are made from a 20 mM stock solution of PMSF in isopropanol.

Fresh beef liver is finely minced and divided into 21 aliquots of 15 gm each. To each aliquot is added 45 ml of 0.32 M sucrose–3 mM MgCl$_2$–0.1 mM PMSF, followed by homogenization with three passes of a Potter–

Elvehjem homogenizer. Ninety milliliters of homogenization medium are added to each aliquot, which is then filtered through double-napped flannelette or six layers of cheesecloth. The total homogenate is centrifuged for 7 minutes at 1000 g. The resulting crude nuclear pellet is resuspended in 2.6 M sucrose–1 mM $MgCl_2$–0.1 mM PMSF using a loose-fitting Potter–Elvehjem homogenizer at slow speed. Additional 2.6 M sucrose solution is added until the sucrose concentration reaches 2.15 M, as determined by refractive index. Additional 2.15 M sucrose is added until the volume is 640 ml. The nuclear suspension is centrifuged for 65 minutes at 76,000 g. After pouring off the supernatant, the nuclear pellets are collected and washed twice by resuspending in 10 mM Tris-HCl (pH 7.5)–0.25 M sucrose–4 mM $MgCl_2$–0.1 mM PMSF and centrifuging for 7 minutes at 1000 g.

The nuclear pellet is washed twice in 75 mM NaCl–25 mM EDTA–0.1 mM PMSF (pH 8.0) by resuspending with a Potter–Elvehjem homogenizer and centrifuging for 7 minutes at 1000 g. The resultant pellets are resuspended in 10 mM Tris-HCl (pH 8.0)–0.1 mM PMSF with a Potter–Elvehjem homogenizer and centrifuged twice for 10 minutes at 10,000 g and twice for 10 minutes at 15,000 g. The chromatin pellet is resuspended in 3 M NaCl–10 mM Tris-HCl (pH 8)–0.1 mM PMSF using a Potter–Elvehjem homogenizer. Additional 3 M NaCl solution is added to give an A_{260} of 21–23, and the preparation is stirred for 20 minutes. The dissociated chromatin is centrifuged in a Beckman 60 Ti rotor for 13 hours at 140,000 g. The resultant supernatant is separated by volume into the upper $\frac{2}{3}$ and the lower $\frac{1}{3}$, which contains some DNA. DNA is separated from protein in this fraction by gel filtration chromatography [Bio-Gel A-15m, 100–200 mesh, equilibrated with 3 M NaCl–10 mM Tris-HCl (pH 8)–0.1 mM PMSF]. The protein peak from the column is combined with the upper $\frac{2}{3}$ fraction and dialyzed against 10 volumes of 10 mM Tris-HCl (pH 7)–0.1 mM PMSF for 2 hours, against 10 volumes of 65% saturated $(NH_4)_2SO_4$–30 mM Tris-HCl (pH 7)–0.1 mM PMSF for 8 hours, and finally against 10 volumes of 94% saturated $(NH_4)_2SO_4$ buffer overnight. The precipitated proteins are collected by centrifugation, dissolved in 3 M NaCl–10 mM Tris-HCl (pH 7)–0.1 mM PMSF and dialyzed against 3 M NaCl buffer. After centrifugation to remove a fine precipitate, the protein solution is applied to a Bio-Rex 70 column (100–200 mesh), precycled with 3 M NaCl–10 mM Tris-HCl (pH 7)–0.1 mM PMSF, and equilibrated with the same buffer containing 0.4 M NaCl. The nonhistone proteins are not retained by the Bio-Rex column and elute with the 0.4 M NaCl buffer, while the histones are bound and are eluted with the 3 M NaCl buffer. The nonhistones are quick-frozen in Dry Ice–ethanol and stored at $-70°C$.

The histone fraction from the Bio-Rex 70 column is dialyzed against 25 volumes of 0.1 mM PMSF, with one change. Concentrated sulfuric acid is then added with stirring to 0.4 N. The mixture is centrifuged for 10 minutes at 15,000 g, and 4 volumes of cold ethanol are added to the supernatant. The resultant solution is stored at $-20°C$ for 24 hours. The precipitated histones are collected by centrifugation and evaporated to dryness in a vacuum desiccator at 4°.

For fractionation of histones, gel filtration chromatography is employed (11). Histone powder is dissolved in 8 M urea–1% β-mercaptoethanol–0.1 mM PMSF to give a protein concentration of 10 mg/ml. The solution is stored overnight at 4°C and applied to a 4 × 140 cm Bio-Gel P-60 column equilibrated with 0.1 M NaCl–0.02 N HCl (pH 1.7) − 0.02% NaN₃–0.1 mM PMSF. The column is run at 20°C with a 40-cm pressure head at 1–2 ml/ cm²/hour. The elution profile as determined by A_{230} indicates four peaks. Panyim–Chalkley histone gels (12) indicate that three of these peaks are pure fractions: H1, H2B, and H4. The other peak contains a mixture of H3 and H2A. The pure fractions, representing H1, H2B, and H4, are pooled, dialyzed against 0.1 mM PMSF overnight, and lyophilized. The resulting powder is dissolved in distilled water, quick-frozen in Dry Ice–ethanol, and stored at $-70°C$.

B. Labeling of Nonhistones with ³²P

The endogenous protein kinase activity present in the nonhistone protein fraction is employed to label these proteins with ³²P. Frozen nonhistone proteins are thawed and dialyzed against one change of 30 volumes of 0.1 M NaCl–10 mM Tris-HCl (pH 7.5) overnight. A small precipitate is centrifuged out, and the supernatant is added to 0.6 ml of incubation mixture containing: 5 μmol Tris-HCl (pH 7.5), 15 μmol MgCl₂, 60–80 μg nonhistone protein, and 0.5–2 nmol [γ-³²P]adenosine 5′-triphosphate (ATP) (sp act 10–25 μCi/nmol). After incubation at 37°C for 30 minutes, the reaction is stopped by adding solid urea to 4 M. Any unincorporated label is removed by exhaustive dialysis against 0.1 M NaCl–10 mM Tris-HCl (pH 7.5)–0.1 mM PMSF. The labeled nonhistone proteins are dialyzed against 20 volumes of 4 M urea–2 M NaCl–10 mM sodium phosphate (pH 6.7)–0.1 mM PMSF overnight, with one change of buffer.

C. Affinity Chromatography

Four hundred mg of CNBr-activated Sepharose-4B is washed with 0.001 N HCl on a sintered glass filter and incubated with 6–8 mg of protein in 4–6 ml of coupling buffer [0.5 M NaCl–10 mM sodium phosphate (pH 8.3)– 0.1 mM PMSF] at 4°C for 16 hours with constant mixing. Excess protein is washed away with coupling buffer and the coupled gel is incubated with 10 ml of 1.5 M ethanolamine (pH 9) at room temperature in a shaker for 3 hours. Simultaneously, 800 mg of 0.001 N HCl-washed gel is also incubated with 12–15 ml of 1.5 M ethanolamine (pH 9) at room temperature for 3 hours. The two gels are combined and washed with five alternating cycles of pH 5 and pH 9 0.5 M NaCl–10 mM sodium phosphate–0.1 mM PMSF. Finally, the affinity gel is washed several times with 4 M urea–2 M NaCl–10 mM sodium phosphate (pH 6.7)–0.1 mM PMSF.

Chromatography of ^{32}P-labeled nonhistone proteins on histone affinity gels has been performed both by conventional top loading of a column and by a specially adapted "gradient dialysis" procedure. In this latter technique, an aliquot of phosphorylated nonhistone protein, usually about 1–3 mg, is combined with 1200 mg of histone-coupled Sepharose (in about 4–5 ml), placed in a dialysis bag, and mounted on a disc rotating at 9 rpm. The whole apparatus is immersed in a beaker of dialysis medium, which is stirred magnetically. The initial dialysis medium is 4 M urea–2 M NaCl–10 mM sodium phosphate (pH 6.7)–0.1 mM PMSF. Successive dialysis buffers are obtained by diluting $\frac{1}{2}$ volume of the previous buffer with $\frac{1}{2}$ volume of 4 M urea–10 mM sodium phosphate (pH 6.7)–0.1 mM PMSF. The NaCl concentration is halved every 4 hours, from the initial 2 M to a final value of 0.25 M. After a 6-hour dialysis against 0.1 M NaCl–urea buffer, the mixture is dialyzed against 0.1 M NaCl–10 mM sodium phosphate (pH 6.7)–0.1 mM PMSF for a period of 20 hours, during which the medium is changed every 5 hours. The histone–Sepharose–nonhistone complex is poured into small columns, washed with 0.1 M NaCl buffer, and eluted via steps of increasing ionic strength: 0.15 M, 0.3 M, 0.5 M, and 2.0 M NaCl, all containing 10 mM sodium phosphate (pH 6.7)–0.1 mM PMSF. Elution profiles are determined by measuring radioactivity and/or A_{230} of each fraction.

III.   Results and Discussion

When nonhistone proteins are top loaded on histone–Sepharose columns and eluted via increasing ionic strength, the pattern depicted in Fig. 1A is obtained. Most of the protein passes directly through the column, although a significant portion of the material binds to the histone column and is not eluted until the ionic strength is raised to 0.15 M NaCl. However, when a control column of Sepharose containing no histone is employed as the matrix, virtually the same profile is obtained (Fig. 1B). Hence, the binding of nonhistone proteins to histone–Sepharose observed under these conditions

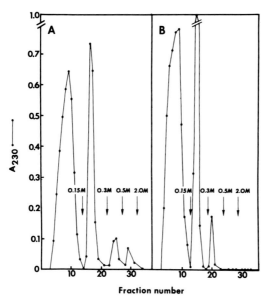

Fraction number

FIG. 1.   Affinity chromatography of nonhistone proteins on CNBr-activated Sepharose 4B columns with either (A) histone H1 as bound ligand or (B) ethanolamine (blank) as bound ligand. About 1 mg of nonhistones in 10 mM sodium phosphate buffer (pH 6.7) were top loaded onto each column and allowed to percolate through, following which the column was rinsed with phosphate buffer. Subsequently, bound nonhistones were eluted with steps of increasing NaCl concentration, as indicated by the vertical arrows on the graphs. Fractions of 1.3 ml were collected and analyzed for protein content using $A_{230\,nm}$. Note that the histone-containing and blank Sepharose columns yield similar results.

is apparently due to binding to the Sepharose rather than to the histones. Similar results are obtained in the presence or absence of 4 M urea.

If radioactive nonhistone proteins are bound to histone–Sepharose via gradient dialysis rather than by conventional top loading of the column, the results are dramatically different. As is shown in Fig. 2, no radioactivity binds to plain Sepharose under these conditions. When histone–Sepharose is employed, however, most of the radioactive nonhistone binds to the column and is not eluted until the 0.3, 0.5, and 2.0 M NaCl steps. That this binding of nonhistone proteins to the columns represents a specific attraction to histones is demonstrated by the finding that nonhistone proteins do not bind to columns of cytochrome c (Fig. 2C) or bovine serum albumin (data not shown). The control employing cytochrome c is especially noteworthy because this protein is similar in size and isoelectric point to histones.

Several other lines of evidence indicate the specificity of the nonhistone–histone interaction observed in this system. First of all, acrylamide gel electrophoresis of the individual eluted peaks reveals certain differences in the

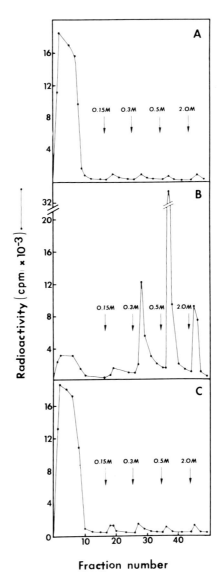

FIG. 2. Affinity chromatography of radioactively labeled nonhistone proteins on CNBr-activated Sepharose 4B columns with either (A) ethanolamine (blank), (B) histone H2B, or (C) cytochrome c as the bound ligand. One mg of nonhistone protein, labeled with ^{32}P as described in the methods, was bound to each Sepharose–ligand complex via the "gradient dialysis" procedure. After the gradient dialysis the Sepharose was poured into small columns, washed with 0.1 *M* NaCl buffer, and eluted with steps of increasing NaCl concentration as noted by the vertical arrows in the graphs. Note that the elution profile from the column containing histone H2B is quite different from the two controls.

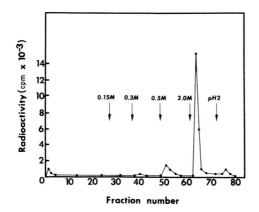

Fraction number

FIG. 3. Rechromatography of the 2.0 M NaCl-eluted peak from the Sepharose–histone H2B column illustrated in Fig. 2B on a second column of Sepharose–histone H2B, again using the gradient dialysis procedure. Note that this peak rechromatographs in the same position.

patterns of polypeptides present in each eluted peak. Second, a given eluted peak of protein will re-elute at the same ionic strength if run through the gradient dialysis procedure a second time (Fig. 3). Finally, somewhat different elution patterns are obtained when different histone fractions are used in making the histone–Sepharose columns.

These results therefore suggest the presence of a certain degree of specificity in the binding of nonhistones to histones, a finding of considerable significance for our views of chromatin structure and function. In addition to the theoretical implications of this conclusion, the technique described also provides a useful new tool for fractionating nonhistone proteins.

ACKNOWLEDGMENT

Studies on this subject in our laboratory have been supported by a grant from the National Science Foundation (BMS-23418) and the United States Public Health Service (GM 1355).

REFERENCES

1. Fasman, G. D., Schaffhausen, B., Goldsmith, L., and Adler, A., *Biochemistry* **9**, 2814 (1970).
2. DeLange, R. J., and Smith, E. L., *Annu. Rev. Biochem.* **40**, 279 (1971).
3. van den Broek, H. W., Nooden, L. D., Sevall, J. S., and Bonner, J., *Biochemistry* **12**, 229 (1973).
4. Kleinsmith, L. J., *J. Biol. Chem.* **248**, 5648 (1975).
5. Kleinsmith, L. J., Heidema, J., and Carroll, A., *Nature (London)* **226**, 1025 (1970).
6. D'Anna, J. A., and Isenberg, I., *Biochemistry* **13**, 2098 (1974).
7. Martinson, H., and McCarthy, B. J., *Biochemistry* **14**, 1073 (1975).
8. Wang, T. Y., and Johns, E. W., *Arch. Biochem. Biophys.* **124**, 176 (1968).

9. Lin, P. C., Wilson, R. F., and Bonner, J., *Mol. Cell. Biochem.* **1**, 197 (1973).

10. Nooden, L. D., van den Broek, H. W. J., and Sevall, J. S., *FEBS Lett.* **29**, 326 (1973).

11. Bohm, E. L., Strickland, W. N., Strickland, M., Thwaits, B. H., van der Westhuizen, D. R., and von Holt, C., *FEBS Lett.* **34**, 217 (1973).

12. Panyim, S., and Chalkley, R., *Arch. Biochem. Biophys.* **130**, 337 (1969).

Chapter 21

Fractionation of Nonhistone Chromosomal Proteins

GARY S. STEIN AND WILLIAM D. PARK

Department of Biochemistry and Molecular Biology,
University of Florida, Gainesville, Florida

JANET L. STEIN

Department of Immunology and Medical Microbiology,
University of Florida, Gainesville, Florida

I.   Introduction

Fractionation of the nonhistone chromosomal proteins represents a formidable task. In addition to the heterogeneity of these proteins—there are at least 450 classes of polypeptides (*1*)—major components of the nonhistone chromosomal proteins are insoluble under conditions employed for routine protein fractionation. Yet, in spite of these obstacles, some progress has been made toward separating the nonhistone chromosomal proteins.

Analytical-scale fractionation of the nonhistone chromosomal proteins has been successfully achieved by SDS–polyacrylamide gel electrophoresis. Recently a two-dimensional fractionation approach employing isoelectric focusing in one dimension and SDS–polyacrylamide gel electrophoresis in the second dimension has amplified the degree of resolution of this complex class of proteins (*1*). A limitation of such procedures is that proteins prepared by these methods are not suitable for many biological assays; this is due in part to difficulties encountered in achieving complete renaturation and in completely removing SDS. However, nonhistone chromosomal protein fractions separated in SDS–polyacrylamide gels have been used to generate antibodies which exhibit specificity for native nonhistone chromosomal proteins (*2*). The analytical procedures for fractionation of nonhistone chromosomal proteins described above are the subject of several chapters in these volumes.

Fractionation of the nonhistone chromosomal proteins on a preparative scale has been achieved to a limited extent. Solubilization of these proteins in reagents such as 5 M urea has permitted ion-exchange chromatography as well as molecular-weight sieving. Such methods yield nonhistone chromosomal protein fractions renaturable to the extent that when assayed in reconstituted chromatin systems, these proteins are capable of rendering specific genetic sequences transcribable (3–8).

This chapter will focus on preparative-scale methods for the fractionation of nonhistone chromosomal proteins. Specifically, we shall discuss general considerations important for fractionation, briefly survey some of the fractionation approaches which have been pursued, and then focus on QAE–Sephadex (Pharmacia) chromatography of nonhistone chromosomal proteins.

II.   General Considerations

In evaluating procedures for fractionation of the nonhistone chromosomal proteins, the same considerations which apply for isolation of nuclei and chromatin are operative, e.g., absence of cytoplasmic material and minimization of protease and nuclease activities. Since fractionation of chromosomal proteins frequently requires numerous steps and extensive periods of time, precautions to avoid degradation and irreversible denaturation are particularly important (9). It should be emphasized that approaches to chromosomal protein fractionation must often be modified to accommodate differences inherent in various tissues and cell types.

A problem sometimes encountered in fractionating nonhistone chromosomal proteins is that fractions may contain nucleic acid–RNA as well as DNA. Such nucleic acids can complicate interpretation of experiments in which nonhistone chromosomal protein fractions are assayed in chromatin reconstitution systems for ability to render specific genes transcribable. If nucleic acids are covalently bound to the protein fraction, separation of the nucleic acids and chromosomal proteins is extremely difficult. However, separation of noncovalently associated nucleic acids and chromosomal proteins can be executed by buoyant-density centrifugation in solutions of cesium salts and urea.

Buoyant-density centrifugation in CsCl–urea has been used by Gilmour and Paul (10) to prepare chromosomal proteins under conditions that exclude endogenous RNA. Chromatin is dissolved in 55% CsCl–4 M urea–10 mM Tris-HCl (pH 8.0)–10 mM dithiothreitol–10 mM ethylenediaminetetraacetic acid (EDTA) at a concentration of 300 μg/ml. Four–five milliliter aliquots of the dissociated chromatin are then centrifuged at 40,000 rpm for 40 hours at 8°C in an MSE 10 × 10 titanium rotor or the equivalent (the remainder of the tube is filled with paraffin oil). The chromosomal proteins are found in the top 1.5 ml of the gradient, while the DNA and RNA form a pellet.

A procedure which has been used successfully by Modak and co-workers (11,12) to separate DNA, RNA, and protein involves centrifugation in Cs_2SO_4 and urea. This method is particularly useful when the isolation of the nucleic acid components is desired. The sample is adjusted to 5 M urea–0.54 mg/ml Cs_2SO_4–30 mM NaCl–20 mM Tris-HCl (pH 7.4)–5 mM EDTA. Fifteen-milliliter aliquots overlaid with paraffin oil are centrifuged for 50–62 hours at 170,000 g in a Beckman 60 Ti rotor at 22°. RNA forms an unprecipitated band at density 1.71 gm/cm³, DNA bands at density 1.51 gm/cm³, and protein floats.

III. Selective Extraction of Nonhistone Chromosomal Proteins

One approach to fractionation of nonhistone chromosomal proteins has been to extract classes of these molecules with various concentrations of salt and urea—the urea being used to facilitate extraction of tenaciously bound molecules and to maintain extracted proteins in a soluble state. Serial extraction of rabbit liver chromatin with 5 M urea–10 mM Tris-HCl (pH 8.0) and 5 M urea–10 mM Tris-HCl (pH 8.0) containing increasing concentrations of NaCl, followed by high-speed centrifugation after each extraction, has been reported by Bekhor et al. (13). This procedure yields fractions which are extremely heterogeneous and which exhibit significant differences in their electrophoretic molecular-weight profiles. Selective extraction of chromatin with salt and urea followed by high-speed centrifugation has also been used by Hnilica and co-workers in a procedure that utilizes differences in pH in the extraction steps (14). In this procedure the majority of the nonhistone chromosomal proteins from rat liver are extracted by 5 M urea–50 mM sodium phosphate (pH 7.5), the histones are extracted by 2.5 M NaCl–5 M urea–50 mM sodium succinate (pH 5.0), and the tightly bound nonhistone chromosomal proteins are extracted with 2.5 M NaCl–5 M urea–50 mM Tris-HCl (pH 8.0). Because both of these procedures involve repeated extraction of the chromatin followed by extensive centrifugation, they are time-consuming and precautions must be taken to prevent proteolytic degradation (9).

Nonhistone chromosomal protein fractions also can be obtained by extraction with NaCl in the absence of urea. One method that has been used by several workers is to extract nuclei or chromatin with 1.0 *M* NaCl (*15–17*). This method has the advantage that when the salt concentration of the extract is lowered to 0.4 *M*, most of the DNA and histone as well as some of the nonhistone chromosomal proteins precipitate and can be removed by centrifugation. Another frequently used procedure is to extract chromatin with 0.25 *M* or 0.35 *M* NaCl (*18–21*). The proteins removed are electrophoretically similar to those that remain on the DNA (*18*), show specificity of binding to DNA (*19*), and have been implicated by Baserga and co-workers as having an important role in determining the functional and structural changes that occur in chromatin of WI-38 cells which have been stimulated to proliferate (*20*).

Extraction of nonhistone chromosomal proteins with phenol has proved to be an effective method for obtaining protein fractions, and details of the procedure have been reported (*22,23*). By performing the phenol extraction at several pHs, additional fractionation can be achieved (*24*). A note of caution is that extraction of nonhistone proteins with phenol is generally preceded by extraction of histones with dilute mineral acid; both of these procedures may result in a certain amount of irreversible denaturation of the nonhistone proteins.

IV. Nonhistone Chromosomal Protein Fractionation Procedures

Nonhistone chromosomal proteins can be isolated as a total class of macromolecules and then fractionated. Such procedures usually require separation of DNA from the total chromosomal proteins as an initial step, followed by fractionation of the total chromosomal proteins into histones and nonhistone chromosomal proteins. Dehistonized chromatin [nonhistone chromosomal protein–DNA complexes prepared by removal of histones under mild conditions as described by Hnilica and co-workers (*25,26*)] can also be used as a source of total nonhistone chromosomal proteins, and for immunological studies this material serves as an effective source of antigen (*27,28*).

The method generally used in our laboratory for separation of DNA and chromosomal proteins involves dissociation of chromatin in 3 *M* NaCl–5 *M* urea–10 m*M* Tris (pH 8.0) and centrifugation at 180,000 *g* for 36 hours. The chromosomal proteins are recovered from the supernatant fraction and should contain less than 0.2% nucleic acid. DNA and chromosomal proteins can also be separated by dissociation of chromatin in 3 M NaCl–5 M urea–10 mM Tris (pH 8.3) followed by molecular-weight sieving on agarose (29). The column is equilibrated in 3 M NaCl–5 M urea–10 mM Tris (pH 8.3), and chromatography is carried out in the same buffer. Other techniques for separation of DNA from chromosomal proteins in dissociated chromatin include precipitation of nucleic acids with lanthanum chloride (30) and chromatography on hydroxylapatite (31).

Ion-exchange chromatography is one of the methods which has been effectively employed for fractionation of nonhistone chromosomal proteins. The resins which have been used include O-(diethylaminoethyl) cellulose (DEAE-cellulose) (32,33), QAE-Sephadex (Pharmacia) (34–36), SP-Sephadex (Pharmacia) (37,38), SE-Sephadex (Pharmacia) (39), and CM-Sephadex (Pharmacia) (40,41). Chromatography is usually carried out in the presence of 5 M urea to maintain solubility of the proteins, and elution from these resins is generally achieved with a salt gradient.

DNA-affinity chromatography is an approach which has been extensively used for fractionation of the nonhistone chromosomal proteins. DNA can be physically "trapped" in cellulose (42), polyacrylamide (43), or agar (44), or covalently bound to Sephadex (19,45) or agarose (19,46,47). The immobilized DNA can then be used to select particular classes of nonhistone chromosomal proteins (19,46,48–52). Predicated on the assumption that components of the nonhistone chromosomal proteins have the ability to recognize and bind to defined genetic sequences, nonhistone chromosomal protein fractionation protocols often entail successive chromatography runs over heterologous and homologous DNA columns (48–50) and/or chromatography over DNA columns containing DNA of various $C_{r_0}t$ values (19,46). DNA binding has also been carried out in solution with DNA–protein complexes being isolated by sucrose gradient fractionation (22,51,52) or collection on nitrocellulose filters (53). Although it is difficult to establish that the interactions observed between DNA and nonhistone chromosomal proteins on these columns reflect the *in vivo* situation, nonetheless these affinity chromatography methods provide an effective method for fractionation of a complex and heterogenous group of macromolecules. It should also be noted that as defined DNA sequences become available these affinity chromatography procedures may be capable of yielding an increased resolution in the order of several magnitudes.

Other methods for nonhistone chromosomal protein fractionation on a preparative scale include molecular-weight sieving using Sepharose (40), Sephadex (39,54), or the agarose (29) and polyacrylamide (54) Bio-Gels, as well as chromatography on hydroxylapatite (31), phosphocellulose (55), or calcium phosphate gel (56).

V. QAE-Sephadex Fractionation of Nonhistone Chromosomal Proteins

One method which we have been using for fractionation of nonhistone chromosomal proteins has been ion-exchange chromatography on QAE-Sephadex. In this procedure, chromosomal proteins, which have been dissociated from chromatin by high concentrations of salt and urea and from which nucleic acid has been removed by ultracentrifugation, are dialyzed against four changes of 10 volumes of 5 M urea–10 mM Tris-HCl (pH 8.3) at 4°. The proteins are loaded on a column of QAE-Sephadex A-25 which has been equilibrated with this same buffer at a flow rate of 15–30 ml/cm²/ hour, using approximately 1 gm of Sephadex/10 mg of protein. The column is then eluted with 2.5 column volumes each of 5 M urea–10 mM Tris-HCl (pH 8.3) containing 0 M, 0.1 M, 0.25 M, 0.5 M, or 3 M NaCl at the same flow rate (Fig. 1). The total recovery of protein is approximately 85%. The SDS–polyacrylamide gel electrophoretic profiles of the column fractions are shown in Fig. 2.

VI. Assessment of Protein Purification

One of the principal objectives of nonhistone chromosomal protein fractionation is to isolate classes, and eventually individual species, which are directly involved in the regulation of specific genetic sequences. As fractionation proceeds and these complex and heterogeneous proteins are being sorted out, it is important to monitor the extent of purification.

In order to purify the molecules responsible for the regulation of specific genes, one must determine not only whether a given fraction has activity but also how much activity is present. If activity is found only in one fraction, without the ability to quantitate one cannot tell whether the activity has been destroyed in the other fractions, whether all the components of the activity are present only in that fraction, or whether the activity has actually been purified. We have recently used the techniques of chromatin reconstitution and *in vitro* transcription to assay and quantitate the activity of a nonhistone protein fraction which has been implicated in the control of histone gene transcription from chromatin.

In HeLa S3 cells histone genes can be transcribed *in vitro* from chromatin isolated from S-phase cells, but not from G1 cells (*6,57*). Using the technique of chromatin reconstitution we have recently shown that this difference can be accounted for by a component or components of the S-phase nonhistone

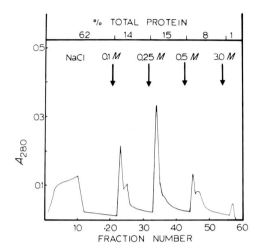

FIG. 1. Elution profile of chromosomal proteins from QAE-Sephadex. Proteins were loaded in 5 *M* urea–10 m*M* Tris-HCl (pH 8.3) and were eluted with this buffer containing 0.10 *M*, 0.25 *M*, 0.50 *M*, and 3.0 *M* NaCl. The percentage of protein eluted in each peak is shown in the upper panel.

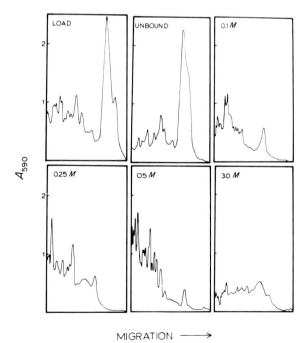

FIG 2. SDS–polyacrylamide gel electrophoretic profiles of chromosomal proteins fractionated with QAE-Sephadex.

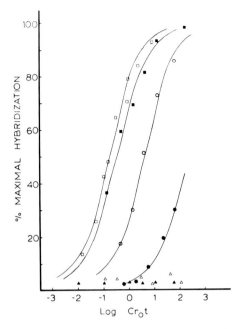

FIG. 3. Kinetics of annealing of histone cDNA to *in vitro* transcripts from G1 chromatin dissociated and then reconstituted in the presence of various amounts of S-phase chromosomal proteins. [^{3}H]cDNA (0.04 ng) was annealed at 52° in 50% formamide–0.5 M NaCl–25 mM N-2-hydroxyethyl-piperazine-N'-2-ethanesulfonic acid (HEPES) (pH 7.0)–1 mM EDTA to RNA transcripts from G1 chromatin reconstituted in the presence of 0 ($\triangle$), 0.01 ($\bullet$), 0.10 ($\bigcirc$), or 1.00 ($\blacksquare$) mg of S-phase total chromosomal protein/mg of G1 chromatin (DNA as chromatin). Histone cDNA was also hybridized to RNA transcripts from G1 chromatin reconstituted in the presence of 1.00 mg of additional G1 chromosomal protein/mg G1 chromatin ($\blacktriangle$) and to RNA transcripts from native S-phase chromatin ($\square$). *E. coli* RNA was included in each reaction mixture so that the total amount of RNA was 3.75 μg. Preparation of histone cDNA and properties of this probe have been reported (*6*). Cr_0t = (moles of ribonucleotides/liter) × time.

chromosomal proteins which has the ability to activate histone gene transcription from G1 chromatin in a dose-dependent fashion (*6,7*). As shown in Fig. 3, when chromatin from G1-phase cells is dissociated and then immediately reconstituted, transcripts from this chromatin do not show significant hybrid formation with histone cDNA even at a Cr_0t of 100. In contrast, if G1 chromatin is dissociated and then reconstituted in the presence of 10, 100, or 1000 μg of S-phase chromosomal proteins per mg of G1 DNA (as chromatin), a progressive increase in the representation of histone mRNA sequences in RNA transcripts from these chromatin preparations is observed. This increase is indicated by progressively lower $Cr_0t_{\frac{1}{2}}$ values for the hybridization reactions of chromatin transcripts and histone cDNA, suggesting that histone genes are being made available for transcription. Such "activation" of histone gene transcription is not seen when G1 chromatin is reconstituted in the presence of additional G1 chromosomal proteins.

When the S-phase chromosomal proteins are fractionated on QAE-Sephadex as described in Section V, only the fraction eluted with $0.5\,M$ NaCl is found to "activate" the transcription of histone genes from G1 chromatin when assayed in the reconstitution system. Since the $0.5\,M$ fraction has only approximately 8% of the protein, if the histone gene "activator" has actually been purified, one should be able to get the same degree of activation of histone gene transcription with 80 μg of the $0.5\,M$ fraction as with 1000 μg of the total chromosomal proteins. A similar representation of histone mRNA sequences in RNA transcripts from G1 chromatin reconstituted with either 80 μg of the $0.5\,M$ QAE fraction or 1000 μg of the total S-phase chromosomal proteins is observed. This assay procedure should be useful for monitoring further fractionation of S-phase nonhistone chromosomal proteins which are involved in the transcription of histone genes. Additionally, the procedure may be applicable for assessing the fractionation of chromosomal proteins responsible for the transcription of other genetic sequences.

ACKNOWLEDGMENT

These studies were supported in part by research grants GM 20535 from the National Institutes of Health, and BMS 7518583 from the National Science Foundation.

REFERENCES

1. Peterson, J. L., and McConkey, E. H., *J. Biol. Chem.* **251**, 548 (1976).
2. Silver, L. M., and Elgin, S. C. R., *Proc. Natl. Acad. Sci. U.S.A.* **73**, 423 (1976).
3. Paul, J., Gilmour, R. S., Affara, N., Birnie, G., Harrison, P., Hell, A., Humphries, S., Windass, J., and Young, B., *Cold Spring Harbor Symp. Quant. Biol.* **38**, 885 (1973).
4. Barret, R., Maryanka, D., Hamyln, P., and Gould, H., *Proc. Natl. Acad. Sci. U.S.A.* **71**, 5057 (1974).
5. Chiu, J.-F., Tsai, Y.-H., Sakuma, K., and Hnilica, L. S., *J. Biol. Chem.* **250**, 9431 (1975).
6. Stein, G., Park, W., Thrall, C., Mans, R., and Stein, J., *Nature (London)* **257**, 764 (1975).
7. Park, W. D., Stein, J. L., and Stein, G. S., *Biochemistry* **15**, 3296 (1976).
8. Jansing, R. L., Stein, J. L., and Stein, G. S., *Proc. Natl. Acad. Sci. U.S.A.* **74**, 173 (1977).
9. Carter, D. B., and Chae C. B., *Biochemsitry* **15**, 180 (1976).
10. Gilmour, R. S., and Paul, J., *in* "Chromosomal Proteins and Their Role in the Regulation of Gene Expression" (G. S. Stein, and L. J. Kleinsmith, eds.), p. 19. Academic Press, New York, 1975.
11. Modak, S. P., and Imaizumi, M. -T., *J. Cell Biol.* **63**, 231A (1974).
12. Modak, S., Commelin, D., Grosset, L., Imaizumi, M. -T., Monnat, M., and Scherrer, K., *Eur. J. Biochem.* **60**, 407 (1975).
13. Bekhor, I., Lapeyre, J.-N., and Kim, J., *Arch. Biochem. Biophys.* **161**, 1 (1974).
14. Chiu, J. F., Hunt, M., and Hnilica, L. S., *Cancer Res.* **35**, 913 (1975).

15. Langan, T. A., in "Regulation of Nucleic Acid and Protein Biosynthesis" (V. V. Konings-berger, and L. Bosch, eds.), p. 233. Elsevier, Amsterdam, 1967.
16. Kleinsmith, L. J., and Allfrey, V. G., Biochim. Biophys. Acta. 175, 123 (1969).
17. Kostraba, N. C., and Wang, T. Y., Cancer Res. 32, 2348 (1972).
18. Comings, D. E., and Harris, D. C., J. Cell Biol. 70, 440 (1976).
19. Allfrey, V. G., Inoue, A., Johnson, E. M., in "Chromosomal Proteins and Their Role in the Regulation of Gene Expression" (G. S. Stein, and L. J. Kleinsmith, eds.), p. 265. Academic Press, New York, 1975.
20. Nicolini, C., Ng, S., and Baserga, R., Proc. Natl. Acad. Sci. U.S.A. 72, 2361 (1969).
21. Johns, E. W., and Forrester, S., Eur. J. Biochem., 8, 547 (1969).
22. Teng, C. S., Teng, C. T., and Allfrey, V. G., J. Biol. Chem., 246, 3597 (1971).
23. Shelton, K. R., and Allfrey, V. G., Nature (London) 228, 132 (1970).
24. Shelton, K. R., and Neelin, J. M., Biochemistry 10, 2342 (1971).
25. Spelsberg, T. C., and Hnilica, L. S., Biochem. J. 120, 435 (1970).
26. Spelsberg, T. C., Hnilica, L. S., and Ansevin, A. T., Biochem. Biophys. Acta 228, 550 (1971).
27. Chytil, F., and Spelsberg, T. C., Nature (London) New Biol. 233, 215 (1971).
28. Wakabayashi, K., and Hnilica, L. S., Nature (London) New Biol. 242, 153 (1973).
29. Shaw, L. M. J., and Huang, R. C. C., Biochemistry 9, 4530 (1970).
30. Yoshida, M., and Shimura, K., Biochim. Biophys. Acta 263, 690 (1972).
31. McGillivray, A. J., Carroll, D., and Paul, J., FEBS Lett. 13, 204 (1971).
32. Wang, T. Y., and Johns, E. W., Arch. Biochem. Biophys. 124, 176 (1967).
33. Levy, S., Simpson, R. T., and Sober, H. A., Biochemistry 11, 1547 (1972).
34. Gilmour, R. S., and Paul, J., FEBS Lett. 9, 242 (1970).
35. Richter, K. H., and Sekoris, C. E., Arch Biochem. Biophys. 148, 44 (1972).
36. Augenlicht, L. H., and Baserga, R., Arch. Biochem. Biophys. 158, 89 (1973).
37. Graziano, S. L., and Huang, R. C. C., Biochemistry 10, 4770 (1971).
38. Yoshida, M., Kikuchi, A., and Shimura, K., Biochemistry J. 77, 1007 (1975).
39. Elgin, S. C. R., and Bonner, J., Biochemistry 11, 772 (1972).
40. Hill, R. J., Poccia, D. L., and Doty, P., J. Mol. Biol. 61, 445 (1971).
41. Goodwin, G. H., Nicolas, R. H., and Johns, E. W., Biochim. Biophys. Acta. 405, 280 (1975).
42. Alberts, B., and Herrick, G., in "Methods in Enzymology," Vol. 21: Nucleic Acids, Part D (L. Grossman and K. Moldave, eds.), p. 198. Academic Press New York, 1971.
43. Cavalieri, L. F., and Carroll, E., Proc. Natl. Acad. Sci. U.S.A. 67, 807 (1970).
44. Bolton, E. T., and McCarthy, B. J., Proc. Natl. Acad. Sci. U.S.A. 48, 1390 (1962).
45. Rickwood, D., Biochim. Biophys. Acta. 269, 47 (1972).
46. Allfrey, V. G., Inoue, A., Karn, J., Johnson, E. M., and Vidali, G., Cold Spring Harbor Symp. Quat. Biol. 38, 785 (1973).
47. Poonian, M. S., Schlabach, A. J., and Weissbach, A., Biochemistry 10, 424 (1971).
48. Kleinsmith, L. J., Heidema, J., and Carroll, A., Nature (London) 226, 1025 (1970).
49. Van Den Broek, H. W. J., Nooden, L. D., Sevall, J. S., and Bonner, J., Biochemistry 12, 229 (1973).
50. Kleinsmith, L. J., J. Biol. Chem. 248, 5648 (1973).
51. Teng, C. T., Teng, C. S., and Allfrey, V. G., Biochem. Biophys. Res. Commun. 41, 690 (1970).
52. Patel, G. L., and Thomas, T. L., Proc. Natl. Acad. Sci. U.S.A. 70, 2524 (1973).
53. Thomas, T. L., and Patel, G. L., Biochemistry 15, 1481 (1976).
54. Unanskii, S. R., Tokarshaya, V. I., Zotova, R. N., and Migushina, V. L., Mol. Biol. USSR 5, 270 (1971).
55. McCarty, K., personal communication.
56. Gershey, E. L., and Kleinsmith, L. J., Biochim. Biophys. Acta. 194, 331 (1969).
57. Stein, G. S., Park, W. D., Thrall, C. L., Mans, R. J., and Stein, J. L., Biochem. Biophys. Res. Commun. 63, 825 (1975).

Chapter 22

Fractionation of Chromosomal Nonhistone Proteins Using Chromatin–Cellulose Resins: Purification of Steroid Hormone Acceptor Proteins

THOMAS C. SPELSBERG

Department of Molecular Medicine, Mayo Clinic,
Rochester, Minnesota

ERIC STAKE AND DAVID WITZKE

Mayo Medical School, Mayo Clinic,
Rochester, Minnesota

I. Introduction

Research in various areas of chromosome biology has been hampered by the lack of methodology in fractionating the chromosomal proteins. In the past decade, interest in DNA affinity chromatography has increased. This approach involves the attachment of DNA to some insoluble resin followed by the exposure of the pure DNA to isolated chromosomal proteins under various conditions. Generally, proteins with high affinity for the DNA were pursued. The insoluble matrices to which the DNA has been attached have been celluloses (*1–4*), agaroses (*5*), and acrylamides (*6*). In some instances, a specificity for select DNA sequences by the chromosomal proteins has been detected. A related but unique alternative to the above-described methods is the attachment of whole (intact) chromatin to an insoluble matrix primarily through the DNA component of chromatin. This resin then can be subjected to various conditions of high ionic strength and chaotropic agents to remove various fractions of the chromosomal proteins from the DNA. One such technique has been developed in this laboratory for application in the purification of the nuclear "acceptors" which bind the progesterone receptor complex. The preparation and handling of the resin as well as the fractionation of the chromosomal proteins will be emphasized here; only a brief description of the resulting purification of the nuclear "acceptors" will be given at the end of this report.

II. Materials

A. Preparation of Cellulose

This method is essentially that of Alberts and Herrick (7). High-grade cellulose such as cellex-N-1 (BioRad Laboratories, Richmond, California) is placed in a large beaker. Boiling 95% ethanol is added at a level of 10 ml of ethanol/gm dry cellulose. The mixture is stirred with a glass rod, allowed to stand 20 minutes, and then filtered on a large glass-sintered or Teflon filter under vacuum. For relatively large volumes of cellulose, we have found that the large polyethylene filtering apparatus (Table top Büchner funnel Cat. No. H-14713 from Bel-Art Products, Pequannock, New Jersey) is superior to most other apparatuses tested. The alcohol-washed cellulose, filtered as a wet cake, is then resuspended in a similar volume of 0.1 N sodium hydroxide at room temperature. The cake resuspension is performed using a glass rod or a rubber spatula with gentle stirring in order to avoid generation of "fines" of the cellulose. After 5 minutes of incubation, the sodium hydroxide is removed by filtration. This process is repeated twice more. The cellulose is then washed several times in a similar manner with 0.001 M ethylenediaminetetraacetic acid (EDTA) and finally several times with 0.01 M HCl. The cellulose is then resuspended for 5 minutes followed by filtration in deionized water at room temperature for 6 to 10 times until the washings approach a pH of neutrality. Each of these latter washings consist of about 50 ml of water/gm dry weight of the original cellulose preparation. On occasion after the acid washes, the cellulose is washed with 1 mM of Tris (un-pH'd) and filtered before the washings with water in order to expedite the attainment of neutrality. The cellulose is filtered thoroughly on the last wash with water and lyophilized to dryness. It is then stored in an air-tight vessel until needed.

B. Isolation of Chromatin, Nucleoacidic Protein (NAP), and DNA

The chromatin or NAP from chick oviduct as well as purified hen DNA was isolated from purified nuclei as described elsewhere (8,9).

C.   Source of Ultraviolet (UV) Light

Any high-intensity short-wave light source will do. We use a mineral light R-52 short-wave light source which emits abour 10,000 ergs/sec/cm² at a 10-cm distance at a primary wavelength of 2,540 Å. The light intensity from the mineral light R-52 is routinely checked by a Blak-Ray ultraviolet intensity meter. The meter itself was calibrated with the standard traceable to the National Bureau of Standards. The R-52 light gives a reading with the Blak-Ray meter of 1000 μW/cm² at 10 cm distance or 10,000 ergs/sec/cm² (1 μW equals 10 ergs/sec). The accuracy of the meter is ±15%.

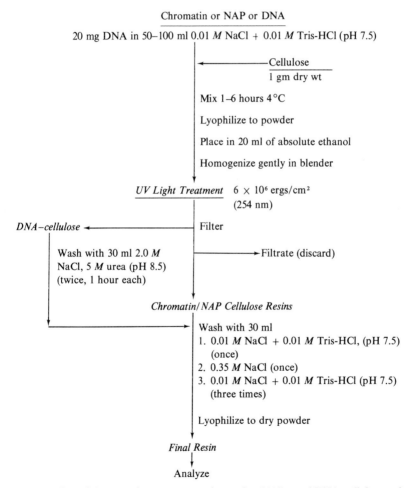

Chromatin or NAP or DNA

20 mg DNA in 50–100 ml 0.01 M NaCl + 0.01 M Tris-HCl (pH 7.5)

Cellulose
1 gm dry wt

Mix 1–6 hours 4 °C

Lyophilize to powder

Place in 20 ml of absolute ethanol

Homogenize gently in blender

UV Light Treatment   6 × 10⁶ ergs/cm²
(254 nm)

DNA–cellulose ◄—————— Filter

Wash with 30 ml 2.0 M
NaCl, 5 M urea (pH 8.5)
(twice, 1 hour each)

——————► Filtrate (discard)

Chromatin/NAP Cellulose Resins

Wash with 30 ml
1. 0.01 M NaCl + 0.01 M Tris-HCl, (pH 7.5)
   (once)
2. 0.35 M NaCl (once)
3. 0.01 M NaCl + 0.01 M Tris-HCl (pH 7.5)
   (three times)

Lyophilize to dry powder

Final Resin

Analyze

FIG. 1.   Outline of the procedure to prepare chromatin–, NAP–, and DNA–cellulose resins.

306   THOMAS C. SPELSBERG et al.

III. Method for Preparation of Chromatin–Cellulose, NAP–Cellulose, and DNA–Cellulose Resins

The quantitative removal of chromosomal proteins from the DNA requires more than high salt solutions (e.g., 2 M NaCl). In past years, the application of chaotropic agents such as urea (*10–19*), guanidine hydrochloride (GuHCl) (*16,20,21*), and formic acid (*11*) or detergents (*22*) has been used in addition to the high salt for complete dissociation of the proteins from the DNA. Consequently, for chromosomal protein fractionation studies, it is required that the DNA does not dissociate from the resin in the solvents used to remove the proteins from the DNA. In these studies, a solution consisting of 2 M sodium chloride and 5 M urea (pH 8.5) is used to assess weakly bound DNA as a test for the level of tightly bound DNA.

The basic procedure for preparing chromatin cellulose is as follows. To 1 gm of dry acid-washed cellulose, various amounts of chromatin in solution A [0.01 M NaCl–0.01 M Tris-HCl (pH 7.5)] are added and the mixture allowed to mix by gentle shaking for 1 hour at 4°C. The solution is then frozen and lyophilized. The material dries as a light-weight cake material which is then broken up into fine pieces using a glass rod. The powdered resin is then placed in absolute alcohol (20 ml/gm of starting cellulose). The solution is homogenized in a Waring blender at a rheostat setting of 30 for 10 seconds to further disperse the resin. The material is placed under a UV lamp at a distance of approximately 10 cm and mixed with a magnetic stirrer at room temperature. The resin is transferred to a sintered glass or Teflon funnel attached to a vacuum source and the alcohol filtered off. The resin is resuspended in approximately 30 ml of solution A for 20 minutes and refiltered. The resin is washed similarly with solution B (0.35 M NaCl) followed by three more washes with solution A, each for 10 minutes. The wet resin is then lyophilized and stored in air-tight vessels until needed.

As an alternative to freezing and long periods of lyophilization, the chromatin–cellulose mixture can be washed with cold ethanol or acetone to remove bound water at the beginning and end of the above procedure. The resin is still dried in the lyophilizer but only for a couple of hours. To perform this organic dehydration method, the wet cellulose or chromatin–cellulose resin from the filter is resuspended in 40 ml of absolute ethanol or acetone and incubated for 10 minutes. The solvents are removed by filtration. This is repeated several times after which the resin is allowed to dry in the filter by passing air through the resin for several minutes. The resin is then placed in the lyophilizer for a short period to evaporate the residual organic solvents. Either of these two methods (freezing and lyophilizing or organic dehydration) give a dry resin with similar properties. The NAP–cellulose and

DNA–cellulose resins are prepared as described for the chromatin–cellulose resin. The only exception is that in the preparation of DNA cellulose resin, the resin is extracted with solution C [2.0 M NaCl–5 M urea–0.01 M Tris-HCl (pH 8.5)] twice, for at least 1 hour each, after the UV light treatment.

IV. Analysis of Resins

To assess the stability of the chromatin–cellulose or NAP–cellulose resins and to discriminate the "tight" from the "weak or dissociable" binding of DNA to the cellulose, we arbitrarily selected washes of small samples of the chromatin– or NAP–cellulose resins with 30 ml of solution C. The resins were resuspended and allowed to set for at least 1 hour or longer at 4°C. Solution C is then removed by filtration and a fresh 30-ml aliquot of solution C added for 1 hour. The resins are then subjected to the solution A washes as described above.

For DNA analysis, 10 to 20 mg amounts of the dry resins are incubated at 90°C for 30 minutes in 1.0 ml of 0.3 N HClO$_4$ in 5-ml conical centrifuge tubes. Equivalent amounts of pure cellulose are used as blanks and, when combined with 20 μg of standard DNA, as the standards. After the incubation the samples are cooled, centrifuged at 1000 g for 2 minutes, and 300 λ aliquots removed for analysis of DNA by the diphenylamine method (8, 23). For histone levels, 4 to 6 mg samples of the dry-weight resin are extracted in 1.0 ml of 0.4 N H$_2$SO$_4$ for 30 minutes at 4°C. The resin is pelleted by centrifugation at 1500 rpm for 2 minutes. The supernatant is removed and analyzed for histone protein. For nonhistone protein analysis, the pellet is resuspended in 1 ml of 0.1 N NaOH and extracted as above. Total protein (histone and nonhistone protein) of whole chromatin–cellulose is quantitated in the same way as the nonhistone (acidic) proteins. The protein was quantitated by Lowry (24) as described elsewhere (8, 15). RNA contents in whole chromatin or the protein fractions are analyzed from 0.3 N KOH hydrolysates as described elsewhere (8, 15, 25).

V. Selection of Conditions for Maximal Binding of Chromatin, NAP (or Pure DNA) to Cellulose

We found that the treatment with UV light was essential for binding of DNA or chromatin to cellulose which was resistant to the high salt–urea

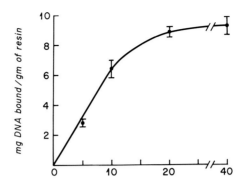

mg DNA/gm cellulose before light treatment

FIG. 2. Extent of tightly bound DNA to cellulose versus the amount of DNA added. Various amounts of purified chick spleen DNA (2 mg/ml) in solution A were added to beakers containing 0.5 mg cellulose. The mixture was frozen, lyophilized, and subjected to 10 minutes of the R-52 UV light in alcohol as described in the text. The DNA–cellulose resin was subjected to the 2 M NaCl–5 M Urea (pH 8.5) washes (solution C) in addition to the standard washes to assess only tightly bound DNA. The dried DNA–cellulose resin was then analyzed for bound DNA as described in the text and plotted as mg DNA bound/gm dry resin. The average and range of duplicate analysis are presented.

(solution C) treatments. It was also discovered that the starting ratio of either DNA or chromatin to cellulose on a weight basis was an important factor in determining the final levels of tightly bound[1] DNA. As shown in Fig. 2 when a 10-minute treatment of UV light is used, increases in the amount of DNA added to a set amount of cellulose result in a resin with an increased level of "tightly" bound DNA. This increased binding of DNA to cellulose is observed up to the ratio of 20 mg DNA/gm cellulose where it begins to level off. Similar results are observed using whole chromatin. Consequently, all subsequent preparations of the DNA or chromatin celluloses involve the addition of 20 mg of pure DNA or chromatine/gm of cellulose before light treatment. As shown in Fig. 3A and B, both the dose of light and wavelength of the light are important factors in obtaining tightly bound DNA. In Fig. 2A, application of long wavelengths of UV light (3500 Å) using a weak light source results in minimal binding of DNA to the resin. However, using a shorter wavelength from the same weak light source results in a much greater level of tightly bound DNA. Figure 3B shows results obtained using the stronger light source (the R-52 lamp). Application of this light source results in marked binding of the DNA in a much shorter period

[1]Tightly bound DNA refers to that DNA which is not dissociated from the cellulose using solution C.

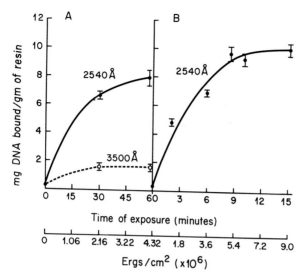

Fig. 3. Dose of UV light versus the extent of tightly bound DNA. The DNA–cellulose was prepared using 20 mg DNA added/gm cellulose as described in the text. The extent of UV light treatment was varied. (A) A low-intensity light source, UVSL:25 (Ultraviolets Products Inc., San Gabriel, California), was used which has both long wavelength (~3500 Å) and short wavelength (~2540 Å) capacity. (B) The high-intensity R-52 light source (Ultraviolets Products Inc., San Gabriel, California) was used. The energy emitted from the low wavelength light was assessed using the Blak Ray meter described in the text. The treated resins were washed with solution C [2 M NaCl–5 M urea (pH 8.5)] in addition to the regular washes to assess the levels of only the tightly bound DNA. The final dried preparations were analyzed for DNA as described in the text, and the mg DNA bound/gm of dry resin was calculated. The data are plotted as the average and range of duplicate analysis of the resin.

of time. After 10-minutes' exposure representing 6 million ergs/cm², optimal binding of the DNA is achieved. The effect of the light treatment on extent of chromatin binding varies somewhat from that of DNA binding as shown in Fig. 4. An exposure of 2 minutes results in an abrupt rise in the amount of tightly bound chromatin followed by a slower rise in DNA binding upon longer exposures (up to the 10-minute exposure time). In any case, enhanced "tightly" bound pure DNA or chromatin occurs when the resin is exposed to the 6 million ergs/cm² of short-wave UV light.

These UV treated DNA–cellulose resins are much more resistant to dissociation by high salt–urea conditions compared to untreated preparations. Figure 5 compares the stability of DNA–cellulose resins prepared with no light treatment with those prepared with a 10-minute treatment of

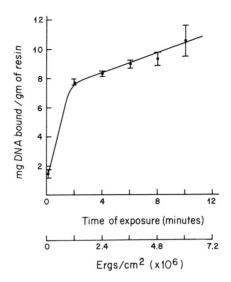

FIG. 4. Dose of UV light versus the extent of tightly bound chromatin. This experiment was performed essentially as described in the legend of Fig. 3. The chromatin–cellulose was extracted with solution C [2 *M* NaCl–5 *M* urea (pH 8.5)] after the light treatment in order to assess the tightly bound DNA.

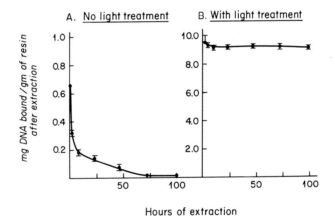

FIG. 5. Effect of light treatment on the stability of DNA–cellulose resins to 2.0 *M* NaCl–5. 0 *M* urea solutions. One DNA cellulose resin was prepared essentially as described in the legend of Fig. 2 while another received no light treatment. These resins were washed with the dilute buffer (solution A) and with solution C (2 *M* NaCl–5 *M* urea) as described in the text. The final dried resins were divided into 0.5-gm fractions and incubated with 40 ml of solution C [2 *M* NaCl–5 *M* urea (pH 8.5)] with occasional mixing and stored at 4 °C. At various periods, fractions were centrifuged at 1200 *g* for 2 minutes and the pelleted resin was washed several times with solution A and dried. The dried resins were then analyzed for DNA. The average and range of duplicate analysis of the resins are plotted.

light at short wavelength (6 million ergs/cm²) (Fig. 3). Aliquots of the DNA–cellulose resins were placed in conical centrifuges and washed with 40 ml/gm resin of the high salt–urea solution C. At various periods after incubation at 4 °C, the resins were sedimented in the conical tubes by low-speed centrifugation. The resins were then washed with solutions A and analyzed for DNA content. The resin prepared with no light treatment (Fig. 5A) shows an immediate loss of DNA when exposed to solution C; after 75 hours of extraction almost no DNA remains. It is interesting to note that the starting resin which received no light treatment shows very low level of DNA. As shown in Fig. 5B, the light treatment not only gives a much higher level of tightly bound DNA, but also the bound DNA is stable under high salt–urea conditions for over 100 hours. It appears that by using the proper light treatment and ratios of chromatin (or DNA) to cellulose, one can obtain high-capacity (high content of DNA) resins which are stable to high salt–urea and other types of chaotropic salts which dissociate proteins from the DNA.

VI. Integrity of Chromatin and Nucleoproteins Bound to Cellulose

Scanning electron micrographs of these resins are shown in Fig. 6A and B. The high-capacity resins prepared by this technique show much of the nucleoprotein fibers attached to the cellulose (Fig. 6B) compared to the lack of such fibers in resins which were treated similarly but without added DNA (Fig. 6A). The chromatins and nucleoacidic proteins (NAP) show little loss of protein during the preparation. Note that in preparing chromatin–cellulose resins for protein fractionations, the high salt–urea washes must not be used after the light treatment since this would destroy the integrity of the chromatin. When preparations of chromatin or NAP are washed only with solution A after the light treatment in alcohol, very little protein is lost, as shown in Table I. The ratios of histone to DNA in chromatin and the acidic protein to DNA in both chromatin and NAP show little change after binding to the resin when compared to the original material. Sodium dedecyl sulfate (SDS)–polyacrylamide gel electrophoretic patterns of both histones and acidic proteins from the cellulose-bound material are similar to those from the original material.

Some damage from the UV light acting on the DNA is obvious. Two approaches were taken to assess the damage due to the light treatment on the DNA. The first approach was to assess the release of DNA from the

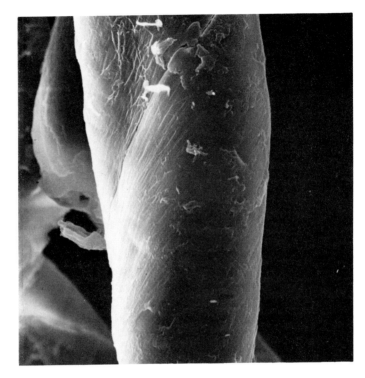

FIG. 6A.

FIG. 6. Scanning electron micrographs of cellulose resins. Preparations of pure cellulose (A) and chromatin–cellulose resins (B) washed with solution B [2 M NaCl–5 M urea (pH 8.5)] were dried and scanned at × 3000 using a scanning electron microscope (ETEC Corporation, Haywood, Calif.).

resin after heating the resin in solution D [0.002 M Tris-HCl–0.0001 M EDTA (pH 7.5)] at 90°C for 20 minutes. It is known that strong doses of UV irradiation create small DNA fragments including single strands and a lower melting temperature for the DNA (26, 28). As shown in Fig. 7, the effects of UV light up to the 15-minute period of exposure have little effect on the release of DNA from the resin in the 90°C incubator under these conditions; however, longer periods of light treatment (30–60 minutes) result in a marked loss of DNA.

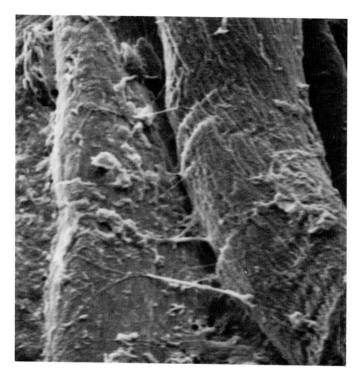

FIG. 6B.

The second approach was to assess the ability of the DNA to serve as a template for DNA-dependent RNA synthesis using isolated bacterial RNA polymerase enzyme. It is known that UV irradiation damage of DNA lowers its ability to serve as a template for DNA-dependent RNA synthesis (29). In Table II, the effect of UV light treatment on the DNA template capacity for DNA-dependent RNA synthesis is seen immediately. After 10 minutes of light exposure, one sees an immediate inhibition of the DNA to serve as template. This capacity continues to decrease up to 60 minutes of exposure. The 10-minute exposure which we have found results in maximal binding of DNA to cellulose in a stable complex also results in a DNA with a template capacity only 10% that of the original DNA not treated with UV light. Consequently, the DNA in the cellulose resins is damaged (the extent is briefly discussed later).

TABLE I

BINDING OF CHROMATIN AND DEHISTONIZED CHROMATIN (NAP)

TO CELLULOSE WITH ULTRAVIOLET LIGHT[a]

	mg DNA/ gm resin	Histone/ DNA	AP/ DNA
1. Chromatin before binding	—	1.20	1.06
2. Chromatin after binding	16.86	1.31	.99
3. Chromatin bound after high salt–urea washes	16.52	.04	.89
4. NAP before binding	—	—	.46
5. NAP after binding	11.72	—	.44
6. NAP bound after high salt–urea washes	11.50	—	.28

[a] Hen oviduct chromatin was isolated, analyzed, and added to cellulose at a level of 20 mg DNA (chromatin)/gm cellulose. Dehistonized chromatin was prepared by extracting whole chromatin with 2.0 M NaCl–5.0 M urea–0.01 M phosphate buffer (pH 6.0) followed by centrifuging 24 hours at 80,000 g_{av}. The pelleted dehistonized chromatin was resuspended and dialyzed against solution D, and the retentate was centrifuged 5 minutes at 2000 g_{av} to give the soluble preparation of NAP which was added to cellulose. The chromatin and dehistonized chromatin–cellulose preparations were lyophilized, treated with UV light, and washed as described in the text except that the high salt–urea washes were excluded. The average of duplicate analysis of DNA and protein is presented.

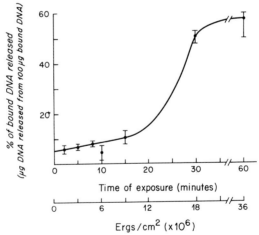

FIG. 7.   Dose of light versus the release of cellulose-bound DNA via thermal denaturation. DNA–cellulose resin was prepared as described in the text and legend of Fig. 3 with varying light treatments. The resin was washed with solution C[2 M NaCl–5 M urea (pH 8.5)] in addition to the standard washes. Two-mg fractions of the dry resins were placed in solution D [0.002 M Tris-HCl–0.0001 M EDTA (pH 7.5)] and heated for 20 minutes at 90°C. The mixtures were quickly cooled at 4°C, centrifuged at 1500 g for 2 minutes, and the DNA content in the pelleted resin as well as that in the respective unheated resins analyzed. The average and range of the three replicate analysis are plotted.

TABLE II

EFFECT OF LIGHT TREATMENT ON THE TEMPLATE
CAPACITY OF DNA–CELLULOSE[a]

Time of exposure (minutes)	Ergs/sec/cm²	Template capacity of DNA–cellulose (cpm/mg DNA)	Control (%)
0	0	2150 ± 200	100
2	1.2 × 10⁶	450 ± 50	21
5	3.0 × 10⁶	370 ± 50	17
10	6.0 × 10⁶	200 ± 33	9
15	9.0 × 10⁶	180 ± 29	8
60	36.0 × 10⁶	98 ± 15	5

[a]The DNA–cellulose preparations were prepared as described in the text with varying light treatments. Aliquots of the dry resins were then taken to give the desired amount of DNA and solubilized 12 hours in solution D $[0.002\,M$ Tris-HCl (pH 7.5)–0.0001 M EDTA$]$. They were then assayed for their ability to serve as templates for DNA-dependent RNA synthesis in an *in vitro* assay using isolated bacterial RNA polymerases as described elsewhere (*15*). The epm/mg DNA represents the incorporation of $[^3H]$uridine 5-triphosphate UTP into acid-insoluble material (RNA)/mg DNA/' 0-minute incubation.

VII. Fractionation of Chromosomal Protein from Chromatin–Cellulose Resins

As expected, these chromatin–cellulose resins can be subjected to a variety of different salt solutions, salt–urea solutions, and other types of reagents in order to dissociate the proteins from the DNA. Data relating to this salt–urea dissociation of proteins from chromatin–cellulose columns has recently been published (*30*). Table I also shows the complete removal of histones and partial removal of the nonhistone protein from chromatin using the high salt–urea washes at pH 6.0. The application of a gradient of guanidine hydrochloride from 0 to 8 M results in the elution of multiple fractions of chromosomal proteins from the chromatin–cellulose resins. As shown in Fig. 8 the elution of up to 10 fractions of chromosomal material (proteins and RNA) could be achieved using this type of gradient. A solvent volume of at least 10 times the volume of the packed resin in a column should be used in the elution of the proteins in order to achieve a decent fractionation. An even greater ratio of solvent to packed resin volumes is recommended for more refined fractions. It is of interest that one could even observe the removal of groups or fractions of proteins as the level of guanidine in the solvent was increased. Figure 8 also shows the quantitative distribution of protein which elutes from such a column when the fractions are collected according to the molarity of the guanidine. The protein values given in Fig. 8 are calculated using the protein recovered as 100%. The actual

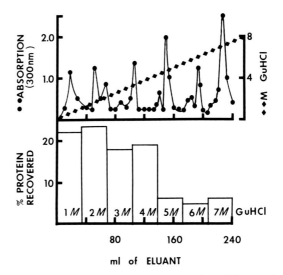

Fig. 8. Fractionation of the proteins from chromatin–cellulose resins. Chromatin–cellulose (20 gm dry weight) was prepared to give 60 mg of chromatin DNA (240 mg protein). The resin was resuspended in 100 ml of solution E [0.1 M HSETOH, 0.05 M Na$_2$S$_2$O$_5$ 0.01 M Tris-HCl (pH 8.5)] and allowed to hydrate for 6 hours at 4 °C. The resin was then collected on a column and a gradient of 0–8 M guanidine hydrochloride was passed through this column in a 4-hour period using constant levels of the buffered solution. The tubes were monitored by absorption at 300 nm (260–280 nm gave too high an absorption). The fractions were also monitored for conductivity as well as refractive index, and the gradient level of guanidine hydrochloride was plotted. The fractions were collected according to their elution with each unit of concentration of guanidine hydrochloride (e.g., 1 M, 2 M, 3 M, etc). Pooled samples were then dialyzed thoroughly versus water and lyophilized. The lyophilized materials were resuspended in a small volume of water, homogenized in a Teflon pestle–glass homogenizer, and assayed for protein (24). Total recoverable protein from these steps was estimated to be 50% of the total protein on the column as chromatin–cellulose. The relative distribution of the total recoverable protein in each fraction is shown in the lower graph. From Spelsberg *et al.* (30a). Reproduced with permission of New York Academy of Sciences.

protein recovered from these columns ranges between a yield of 50–70%. This loss includes that which occurs during the dialysis of the pooled fractions and the transfer to and from lyophilizing flasks. The actual yields from the column itself are not known, but they must be higher than those listed above.

In order to assess whether or not we are fractionating proteins from the chromatin–cellulose resins according to their affinity for DNA (instead of some other phenomenona), we isolated total chromosomal acidic proteins by a hydroxylapatite method (30b) and added these proteins (with no DNA) to the cellulose. This acidic protein–cellulose resin was subjected to the same light treatment and washes as performed for the chromatin–cellulose. The treatment of the acidic protein–cellulose resin with a gradient of

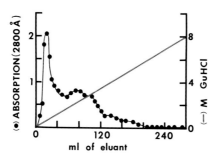

FIG. 9.   Fractionation of proteins from an acidic protein–cellulose resin. Total chromatin proteins were prepared from 100 mg chromatin–DNA by attaching the chromatin to hydroxylapatite and extracting the proteins (but not the DNA) with 5.0 M GuHCl in solution E [but substituting the Tris buffer with 0.5 M Na phosphate buffer (pH 6.0)] (*30b*). These proteins were dialyzed thoroughly and lyophilized with 20 gm of washed cellulose. This AP-cellulose was then subjected to the same UV light treatment and washes as described in the text for chromatin–cellulose, but the 2.0 M NaCl–5.0 M urea, (pH 8.5) washes were omitted. The dried AP–cellulose was then placed in solution E, collected on a column, and subjected to the same GuHCl fractionation as described in the legend of Fig. 8.

guanidine hydrochloride is displayed in Fig. 9. Most of the protein is eluted from the resin at low GuHCl molarities, lower than the eluted proteins from chromatin–cellulose resins. Also the proteins are eluted in only a couple of peaks. Consequently, we believe that most of the sharp peaks obtained with the chromatin–cellulose resins are primarily due to an affinity of these proteins to DNA and not to other interactions.

The fractions of chromosomal protein eluted from the chromatin–cellulose resins pooled according to the unit molarities of guanidine hydrochloride were dialyzed versus SDS solutions and subjected to polyacrylamide–SDS gel electrophoresis (*31*). As shown in Fig. 10, the spectrum of proteins in each of these fractions of chromosomal proteins eluted from the chromatin–cellulose resin between 1 and 7 M guanidine displays a complex pattern of protein species. Most of the histones migrate at the front (bottom of the gels) and are eluted from the chromatin–cellulose resin by the 2 M guanidine hydrochloride extraction. Even those fractions which represent a small percent of the total chromosomal proteins display a heterogeneous pattern. Table III shows a typical composition of the fractions of chromosomal proteins obtained from the chromatin–cellulose columns using unit molarities of guanidine extract as described above. These fractions were placed in a small volume of water for the analysis. Whereas most of the protein is extracted by the 4 M guanidine extract, the amount of RNA which appeared in these fractions is evenly distributed even in the final eluting fraction. The amount of DNA found in each fraction begins to increase after the 5 M to 7 M guanidine extracts. Based on these analyses of the composition of the isolated fractions, very little DNA is lost from the resin. However, analysis of the residual

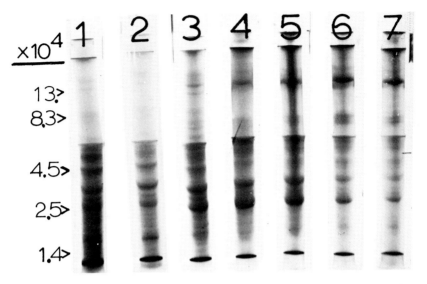

FIG. 10. Polyacrylamide–SDS gel electrophoresis of the GuHCl extracted proteins from the chromatin–cellulose resins. The proteins were extracted as described in the legend of Table III and subjected to SDS–polyacrylamide gel electrophoresis using a 0.5-cm-long, 3.5% acrylamide spacer gel (top), a 3.5-cm-long 5% acrylamide gel next, and on the bottom a 5.5-cm-long 7.0% acrylamide gel. The electrophoresis conditions and gel compositions are described elsewhere (*31*). Gels 1–7 represent proteins extracted from the chromatin–cellulose resins with 1 through 7 M GuHCl respectively. The molecular weight distributions (obtained using protein standards) are given.

resin after these extracts shows that much greater levels of DNA are sometimes dissociated from the resin when one uses 5 M and higher concentrations of guanidine hydrochloride (Table IV). This dissociation of DNA from the resin is time-dependent, and when relatively fast extractions of the resins are performed with these higher concentrations of guanidine, lower levels of DNA are subsequently removed. Generally, exposure of the chromatin–cellulose resins for 3 hours results in less than 10% loss of DNA, whereas those involving 24-hour exposure result in the loss of 30 to 60% of the DNA (Table IV).

VIII.   Isolation of the Nuclear Acceptors Which Bind Steroid Hormones

Work in this and in other laboratories is concerned with the identity of the nuclear binding site for steroid receptor complexes. Briefly, steroids enter

TABLE III

CHEMICAL COMPOSITION OF THE GuHCl EXTRACTS
OF CHROMATIN-CELLULOSE[a]

Fraction GuHCl (M)	Protein (mg)	RNA (mg)	DNA (mg)
1.	38.7	1.4	0.42
2.	40.1	1.0	0.16
3.	31.3	0.7	0.14
4.	32.8	1.1	0.23
5.	10.6	1.0	0.49
6.	9.7	0.8	0.90
7.	10.4	1.5	1.25
Total recovery	173.7	7.5	3.59
Starting chromatin	238.4	—	50.2
Yield	72.6%	—	7.2%

[a]Chromatin–cellulose resin (representing 50 mg of chromatin–DNA) was placed in a column and washed thoroughly (1 hour each) with increasing concentrations of GuHCl on a unit molarity basis. Each extract was dialyzed versus 20 volumes of H_2O for 36 hours with several changes of dialysate. Refractive indices and conductivity show that the dialysis was complete. The solutions were then lyophilized. Each fraction was resuspended in 10 ml of H_2O and subjected to analysis of protein (24), RNA (25), and DNA (23). The averages of analysis are presented. The percent yield of protein and DNA from the amount of starting material (chromatin–cellulose) is also given.

target cells and bind to receptor proteins in the cytoplasm. These complexes then undergo some form of activation (modification) and migrate into the nucleus where they interact with the chromosomal material. These events occur within 5 minutes after injecting labeled hormone into whole animals. Within 10 to 15 minutes after administration of the steroid to whole animals, changes in levels of activity of RNA polymerases occur which are followed shortly thereafter by the appearance of new messenger RNA. Formation of steroid receptor complex can be achieved *in vitro*; the complexes have been purified to homogeneity (32, 33). In addition, preparations of steroid receptor complexes have been incubated with isolated nuclei or other chromosomal material to demonstrate a receptor-dependent binding of the steroid to the nuclei. Recently, a high-affinity nuclear binding has been shown to be

TABLE IV

EFFECTS OF GuHCl ON LEVELS OF DNA IN RESIN[a]

M GuHCL	Hours of exposure	Original DNA bound to cellulose, (%)
0	24	117
2	24	103
4	24	103
6	3	90
6	24	78
7	3	104
7	24	47

[a] Fractions (0.5 gm dry wt) of chromatin–cellulose (9.5 mg DNA gm/resin) were placed in conical glass-centrifuge tubes and washed with solutions of varying GuHCl concentrations in solution E [0.1 *M* HSETOH–0.05 *M* Na$_2$S$_2$0$_5$–0.01 *M* Tris-HCl (pH 8.5)] at 4°C for either a 3-hour or 24-hour period. The solutions were centrifuged for 2 minutes at 1500 *g* and then resuspended in the same solvents as above twice more for 10 minutes each. The resins were then washed with solution A several times and then lyophilized. The resins were then analyzed for DNA as described in the text. The percent of the DNA levels in the original material were calculated and the average of duplicate analysis listed.

partially tissue specific (*34*) i.e., only the nuclei of target tissues appear to have more high-affinity nuclear acceptor sites which bind the steroid receptor complex. Using conditions which specify only high-affinity binding, isolated complexes of the progesterone-receptor from the chick oviduct have been bound to chromatin–cellulose resins using pure cellulose as a control. In addition, binding of the isolated progesterone–receptor complex has been achieved with various fractionated chromatin preparations using the cellulose technique (see Fig. 11). A chromatin–cellulose preparation was sub-divided into eight fractions and each fraction was extracted with increasing molarity concentrations of guanidine hydrochloride from 0 to 8 *M*. The residual material was then washed free of the guanidine and the resin placed in test tubes for binding of the progesterone receptor complex *in vitro*, similar to techniques described elsewhere (*30*).

Part of the resin was also subjected to chemical analysis for histone and nonhistone proteins. As shown in Fig. 11, again the histones and acidic proteins were rapidly dissociated from the chromatin–cellulose resin using solutions of increasing molarities of guanidine hydrochloride. As expected,

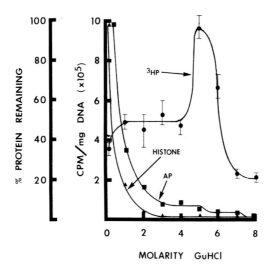

FIG. 11. Binding of [³H] progesterone-receptor complex to oviduct chromatin–cellulose extracted with GuHCl. The binding of [³H]progesterone-receptor complex to hen oviduct chromatin–cellulose preparations is described elsewhere (30). Aliquots of chromatin–cellulose were washed twice in 20 volumes of solution which contains solution E [0.1 M HSETOH–0.05 M Na₂S₂O₅–0.01 M Tris-HCl (pH 8.5)] and a specified molarity of guanidine hydrochloride. These resins were then washed in dilute Tris-EDTA buffer (solution D) several times, frozen, and lyophilized. Twenty- to 50-mg amounts of the resin were then assayed for protein by resuspending the resin in solution D, allowing it to hydrate several hours with gentle mixing, and assaying for histone using a 0.4 N H₂SO₄ extract with subsequent analysis of the acidic proteins (AP) with a 0.1 N NaOH extract as described in the text. These guanidine-treated resins were tested for their capacity for binding the [³H]progesterone-receptor complex using 20–25 μg of DNA (bound to the cellulose) together with 300 ul of the hormone-receptor complex solution (30). The ranges of three replicates of analysis for each assay of the hormone binding are shown. From Spelsberg et al. (30a). Reproduced with permission of the New York Academy of Sciences.

the histones are removed more rapidly than the nonhistone proteins. The level of binding to the residual chromatin–cellulose resin displays a slight enhancement when extracted at 4 M guanidine hydrochloride. However, after extraction of the chromatin–cellulose resin with 5 M guanidine, a greatly enhanced binding occurs. Extractions with 6, 7, and 8 M guanidine hydrochloride then cause a gradual loss of this binding activity. Present evidence suggests that this extracted chromosomal protein, when reannealed back to the DNA, results in an enhanced capacity of that DNA to bind the steroid-receptor complex. These results provide evidence that these types of affinity chromatography techniques, using chromatin bound to insoluble matrices, have the potential of being very powerful tools in the purification of various entities of the chromatin complex. Similar types of fractionation of chromatin using guanidine–HCl have been applied to hydroxylapatite with good success. Chromatin binds to hydroxylapatite through the exposed phosphate groups of DNA. When this type of resin is subjected to increasing guanidine hydrochloride concentrations, a similar pattern of eluting protein fractions is observed which is similar to that achieved with chromatin–cellulose resin.

IX. Comments on the Technique

Litman (*35*) used low doses of UV light to firmly bind DNA to cellulose. The dose of UV light (6×10^6 ergs/cm^2) used in these experiments certainly damages the DNA. The alterations of DNA by UV light are extremely complex due to the reversal of some light-induced modifications and the gradual changing of the DNA structure as the chemical modifications of the bases take place (*29*). In general, our selected dose of light (6×10^6 ergs/cm^2) on pure DNA is sufficient to create thymine dimers, hydroxylation of thymine cytosine dimers, and some cytosine hydroxylations. Some unwinding of the DNA duplex to form some single-stranded regions also occurs together with some breakage of phosphodiester bonds which yields small single-strand fragments of DNA (~ 15 nucleotides long) (*26, 27, 29*). All of this results in a DNA which has a reduced thermal denaturation, a reduced absorption at 2600 Å, a reduced hyperchromicity, and finally a reduced ability to serve as a template for DNA-dependent RNA synthesis. Some of these changes we detect in our studies. Exactly what effect this has on the "nativeness" of the association of the chromosomal proteins from the DNA in this method is not known. Some photochemical linkage of protein to the DNA must occur. Although the extent of linkage of amino acids/proteins to a DNA duplex is hundreds of times lower compared to the extent measured for protein binding to either base monomers or single-strand polynucleotides (*36*), some covalent linkage is expected. The attachment of our DNA or chromatin to cellulose probably involves entrapment in the cellulose by the cross-linking of the DNA as well as chemical linkage of the bases of single-strand regions of DNA with the hydroxyl groups of the cellulose. In any case, the use of chromatin–cellulose (presented here), chromatin hydroxylapatite (Spelsberg, in preparation), or chromatin attached to any insoluble matrix via the DNA appears useful in fractionating the chromosomal proteins. Although some DNA is lost from the chromatin–cellulose resins at extreme conditions (in 5 to 7 *M* GuHCl but not in 2 *M* NaCl–5 *M* urea), much of this can be avoided if exposure to such solvents is not lengthy. The losses of DNA from chromatin–hydroxylapatite resins appears to be much less.

Another drawback to the chromatin–cellulose resin, not found with chromatin bound to the hydroxylapatite resin, is the inability to retrieve the DNA from the resin in the former. DNA can be removed from the latter resin using buffers of high phosphate concentrations. However, the chromatin–cellulose resins have the benefit of requiring much lower volumes of extracting solvents than the hydroxylapatite resins.

We calculate exceptional yield from the chromatin–cellulose resins when one includes the steps of dialysis and lyophilization. In addition, we can obtain multiple peaks with differing patterns of protein species (as determined by SDS–polyacrylamide gels). The analysis of the elution of the progesterone nuclear "acceptors" from the chromatin–cellulose resin indicates that it has a very high affinity for the DNA. This single fractionation step allows hundreds-of-fold purification of the nuclear "acceptor" for the progesterone receptor complex. This approach examplifies the excellent possibilities in the application of this technique to many other areas of investigation involving the functions of chromosomal proteins. Note that all attempts to covalently link chromatin through its DNA to cellulose, agarose, or sephadex resins with cyanogen bromide or bifunctional agents such as 1-cyclohexyl-3-2–morpholinoethyl carbodiimide metho-p-toluene sulfonate (Pierce Chemical Co., Rockford, Illinois) failed in that much of the protein along with the DNA is attached. Consequently, much of the protein cannot be extracted from the resin.

ACKNOWLEDGMENTS

The authors thank Mr. Paul Matthai, Mrs. Patti Midthun, and Mrs. Kay Rasmussen for their excellent technical help. This work was financed by grants CA 14920 and HD 9140 from the NIH and the Mayo Foundation.

REFERENCES

1. Kleinsmith, L. J., Heidema, J., and Carroll, A., *Nature (London)* **226**, 1025 (1970).
2. Kleinsmith, L. J., *J. Biol. Chem.* **248**, 5648 (1973).
3. van den Broek, H. W. J., Nooden, L. D., Sevall, S., and Bonner, J., *Biochemistry* **12**, 229 (1973).
4. Kostraba, N. C., Montagna, R. A., and Wang, T. Y., *J. Biol. Chem.* **250**, 1548 (1975).
5. Allfrey, V. G., Inoue, A., and Johnson, E. M., in "Chromosomal Proteins and Their Role in the Regulation of Gene Expression" (G. S. Stein and L. J. Kleinsmith, eds.), p. 265. Academic Press, New York, 1975.
6. Patel, G. L., and Thomas, T. L., in "Chromosomal Proteins and Their Role in the Regulation of Gene Expression" (G. S. Stein and L. J. Kleinsmith, eds.), p. 249. Academic Press, New York, 1975.
7. Alberts, B., and Herrick, G., in "Methods in Enzymology" Vol. 21: Nucleic Acids, Part D (L. Grossman and K. Moldave, eds), p. 198. Academic Press, New York, 1971.
8. Spelsberg, T. C., Steggles, A. W., and O'Malley, B. W., *J. Biol. Chem.* **246**, 4188 (1971).
9. Spelsberg, T. C., Steggles, A. W., Chytil, F., and O'Malley, B. W., *J. Biol. Chem.* **247**, 1369 (1972).

10. Bekhor, I., Kung, G. M., and Bonner, J., *J. Mol. Biol.* **39**, 351 (1969).

11. Elgin, S. C. R., and Bonner, J., *Biochemistry* **9**, 4440 (1970).

12. Gilmour, R. S., and Paul, J., *J. Mol. Biol.* **40**, 137 (1969).

13. Gilmour, R. S., and Paul, J., *FEBS Lett.* **9**, 242 (1970).

14. Huang, R. C. C., and Huang, P. C., *J. Mol. Biol.* **39**, 265 (1969).

15. Spelsberg, T. C., Hnilica, L. S., and Ansevin, T., *Biochim. Biophys. Acta* **228**, 550 (1971).

16. Levy, S., Simpson, R. T., and Sober, H. A., *Biochemistry* **11**, 1547 (1972).

17. MacGillivray, A. J., Carroll, D., and Paul, J., *FEBS Lett.* **13**, 204 (1971).

18. MacGillivray, A. J., Cameron, A., Krauze, R. J., Rickwood, D., and Paul, J., *Biochim. Biophys. Acta* **277**, 384 (1972).

19. Monahan, J. J., and Hall, R. H., *Can. J. Biochem.* **51**, 709 (1973).

20. Arnold, E. A., and Young, K. E., *Biochim. Biophys. Acta* **257**, 482 (1972).

21. Hill, R. J., Poccia, D. L., and Doty, P., *J. Mol. Biol.* **61**, 445 (1971).

22. Shirey, T., and Huang, R. C. C., *Biochemistry* **8**, 4138 (1969).

23. Burton, K., *Biochem. J.* **62**, 315 (1956).

24. Lowry, O. H., Rosenbrough, N. J., Farr, A. O., and Randall, R. J., *J. Biol. Chem.* **193**, 265 (1951).

25. Cerriotti, G., *J. Biol. Chem.* **214**, 59 (1955).

26. Cleaver, J. E., and Boyer, H. W., *Biochim. Biophys. Acta* **262**, 116 (1972).

27. Slor, H., and Lev, T., *Biochim. Biophys. Acta* **312**, 637 (1973).

28. Umanskii, S. R., Tokarskaya, V. I., Zotova, R. N., and Migushima, V. L., *Mol. Biol. (USSR)* **5**, 270 (1971).

29. Kochetkov, N. K., and Bodovskii, E. I. (eds.), "Organic Chemistry of Nucleic Acids," Part B, Chapter 12, p. 543. Plenum, New York, 1972.

30. Webster, R., Pikler, G., and Spelsberg, T. C., *Biochem. J.* **156**, 409 (1976).

30a. Spelsberg, T. C., Webster, R., Pikler, G., Thrall, C., and Wells, D., *Ann. N.Y. Acad Sci.* **286**, 43 (1976).

30b. Spelsberg, T. C., manuscript in preparation.

31. Wilson, E. M., and Spelsberg, T. C., *Biochim. Biophys. Acta* **322**, 145 (1973).

32. Kuhn, R. W., Schrader, W. T., Smith, R. G., and O'Malley, B. W., *J. Biol. Chem.* **250**, 4220 (1975).

33. Schrader, W. T., and O'Malley, B. W., *J. Biol. Chem.* **247**, 51 (1972).

34. Pikler, G. M., Webster, R. A. and Spelsberg, T. C., *Biochem. J.* **156**, 399 (1976).

35. Litman, R. M., *J. Biol. Chem.* **243**, 6222 (1968).

36. Simukova, N. A., and Budovskii, E. I., *FEBS Lett.* **38**, 299 (1974).

Chapter 23

Methods for Assessing the Binding of Steroid Hormones in Nuclei

G. C. CHAMNESS, D. T. ZAVA, AND W. L. MCGUIRE

Department of Medicine,
University of Texas Health Science Center,
San Antonio, Texas

I. Introduction

When a steroid hormone enters a target cell it binds to a receptor molecule which is then translocated into the nucleus. There it associates with nuclear sites and stimulates a sequence of events which ultimately results in a series of biological responses. The evidence for this general scheme of steroid hormone action is extensive (*1,2*). Though a number of details have recently been critically reexamined (*3*), the importance of receptor translocation to nuclei is undisputed.

The binding of receptor–steroid complexes by nuclei can be studied by several methods. First, radioactive steriod can be administered to intact animals or to cell or organ cultures, the nuclei isolated, and the content of radioactivity then measured directly. This approach is straightforward, but it may lead to complications due to steroid metabolism, especially when nuclei are isolated after a relatively long period of time. Such experiments are also expensive since saturation of receptor requires large quantities of exogenous radioactive steroids. Second, nuclei can be isolated from cells and then incubated with radioactive receptor–steroid complexes prepared independently. This approach allows considerable manipulation of experimental conditions, such as variations in temperature or receptor concentration, or heterogeneous combinations of receptors and nuclei. It is limited, however, by the difficulty of maintaining isolated nuclei without degradation for long times and by the possibility of artifacts inherent in any cell-free system. Finally, nuclei can be isolated and the receptor–steroid complexes measured by incubation with an excess of radioactive steroid at an elevated temperature permitting exchange with the unlabeled steroid. The principal difficulty of this assay has been the technical problem of very high nonspecific steroid binding in nuclei of many tissues studied. To circumvent this problem we have developed an alternative nuclear exchange assay which is discussed in detail in this chapter. In addition, we will also discuss some of the methods which we have developed and used to examine the binding of steroid receptors to isolated nuclei.

II. Receptor Binding in Isolated Nuclei

A. Buffers

Buffers are T $[0.01\ M$ Tris-HCl (pH 7.4)], TE $[$T $+$ 1.5 mM ethylenediaminetetraacetic acid (EDTA)], TE/0.15 (TE $+$ 0.15 M KCl), and TKE (TE, pH 8.5, $+$ 0.4 M KCl). All pH's are adjusted at 4°C. The addition of 0.5 mM dithiothreitol may be helpful in some cases (4). Another buffer which has been useful in some difficult cytoplasmic receptor problems $(5,6)$, but which has not yet been used in the nuclear binding studies, is PG (5 mM sodium phosphate–1 mM thioglycerol–10% glycerol).

B. Tissues

Tissues are removed from mature female rats ovariectomized at least 2 days previously. They are either placed on ice for immediate use, or frozen in liquid nitrogen and stored at $-76°$C and then thawed at 4°C at the time for use.

C. Cytosol Receptor

All operations are at 4°C unless noted otherwise. Tissues are minced with scissors and homogenized in 4 volumes of TE/0.15 buffer using a glass–glass homogenizer. The homogenate is centrifuged 30–50 minutes at 105,000 g, and the cytosol is removed from between the floating fat layer and the pelleted particulate fractions. Those cytosols which are to be charged with estradiol are incubated with 1–5 $\times$ 10^9 M [^{3}H]estradiol for at least 1 hour, after which the remaining free steroid is removed by transferring the cytosol to a pellet of dextran-coated charcoal [2.5 gm/liter of Norit A., 25 mg/liter dextran in 0.01 M Tris-HCl (pH 8.0); a volume of this suspension equal to the volume of the cytosol is centrifuged 10 minutes at 200 g and the supernatant discarded]. The charcoal is resuspended in the cytosol for 15 minutes and then centrifuged again 10 minutes at 2000 g, leaving the supernatant free of unbound steroid.

Estrogen receptor undergoes a temperature-dependent transformation on or before entering nuclei. This transformation can be accomplished by warming cytosol alone after charging the receptor with estradiol, as long as EDTA is not present. When transformed receptor is to be used, cytosol is prepared in T buffer alone and warmed to 25°C for 30 minutes after first charging at 4°C. Addition of 1 M urea has been reported to enhance the transformation process (7), but it is not required and has not been used in nuclear binding studies.

In certain cases receptor is partially purified by precipitation with 30% ammonium sulfate; this procedure is described elsewhere (8).

D. Nuclei

Most of our experiments use crude nuclei, obtained from the high-speed pellet remaining after preparation of cytosol by rehomogenizing in a glass–glass homogenizer with at least 10 volumes of buffer and centrifuging 20 minutes at 105,000 g. The washed pellet is again homogenized in buffer, divided into an appropriate number of tubes, and recentrifuged to provide aliquots suitable for incubation with receptor.

In many cases results are compared with those in which purified nuclei are used, though no significant differences have yet been found. Our methods for purification of nuclei, including a variation designed for uterine nuclei, have been published (8).

E. Incubation

Nuclei containing 50–100 μg DNA are suspended gently in 0.5 ml of receptor solution using a small glass–glass homogenizer and incubated at 25°C for 15 or 30 minutes; 4 ml of chilled TE/0.15 are then added and the suspension is vigorously shaken before centrifugation at 105,000 g for 30 minutes. Additional washing of the pellet does not further reduce bound counts significantly. The pellet is rehomogenized in 0.5 ml TE, and 0.2 ml is dissolved in 1.0 ml of NCS solubilizer (Amersham-Searle) for radioactivity determination in 10 ml of toluene-PPO-POPOP scintillation cocktail (4 gm 2,5-diphenyloxazole (PPO), 0.15 gm 1,4-bis [2-(5-phenyloxazolyl)] benzene (POPOP), 1 liter toluene). Another 0.2 ml is used for DNA assay (9). The results are calculated as pmoles of estradiol bound per mg of DNA.

In other experiments the final pellet is extracted twice with 1.0 ml ethanol rather than being homogenized and solubilized; DNA can then be determined in the remaining pellet, which no longer contains any bound counts

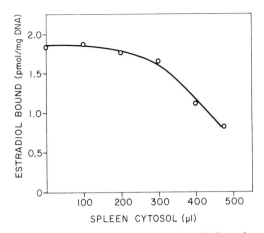

Fɪɢ. 1. Effect of cytosol concentration on the nuclear binding of rat uterine estrogen receptor. Twenty-five μl of uterine cytosol in TE/0.15, charged with [³H]estradiol, were incubated 15 minutes at 25 °C with crude uterine nuclei (30 μg DNA per pellet) and various volumes of spleen cytosol in the same buffer, in a final volume of 500 μl. From Chamness *et al.* (*8*). Reprinted with permission from *Biochemistry*. Copyright by the American Chemical Society.

as determined by solubilizing after ethanol extraction. The combined ethanol extracts are dried in a scintillation vial and then redissolved in 5 ml of toluene–PPO–POPOP for counting.

A critical point is that total cytosol protein must be kept constant in these incubations even though receptor may vary, as in experiments to show saturability of nuclear sites. This is necessary because cytosols contain an unidentified inhibitor of receptor binding, so that increasing cytosol concentrations will cause decreasing efficiency of receptor binding and an apparent but artifactual approach to saturation (*8,10*). Figure 1 shows an example of this effect; a constant amount of receptor charged with [³H] estradiol was incubated with uterine nuclei in the presence of increasing amounts of spleen cytosol, and receptor uptake by the nuclei was diminished.

F. Sucrose Density Gradients

After binding to the nuclei, receptor can be extracted by rehomogenization of the pellets in 0.4 ml TKE for 60 minutes at 4° C. The suspension is then centrifuged 20 minutes at 105,000 g, and 200 μl of supernatant are layered on 5–20% sucrose gradients in TKE. The gradients are centrifuged 15.5 hours at 308,000 g in a Beckman SW-56 rotor, fractionated, and counted. Each gradient contains about 1000 cpm of [¹⁴C]bovine serum albumin (*11*) as an internal marker.

G. Observations

These procedures applied to the estrogen receptor of rat uteri have shown that receptor binding to isolated nuclei is not limited to nuclei of target tissues (*12*) and that this binding does not appear to be saturable (*8*). Similar results have been obtained in other systems (*13,14*). We have also shown that nuclei can bind receptor at 4°C whether or not it has been transformed, but only after transformation is the sedimentation form recovered on KCl extraction identical to that found *in vivo* (*12*). These findings emphasize the importance of examining the conditions of the assay and the state of the receptor critically before establishing and interpreting a nuclear binding procedure.

III. Nuclear Exchange Assays

We now turn to a technique for studying estrogen-charged receptor in the nuclei of rat uteri. To permit study of estrogen receptor in immature rat uterine nuclei a nuclear exchange assay was first developed by Anderson *et al.* (*15*). Crude nuclei are incubated with [³H]estradiol at 30 or 37°C to permit the radioactive steroid to replace any nonradioactive ligand on the receptor. The difference in counts bound in the presence and absence of nonradioactive competitor is used to determine the amount of specific receptor present. This assay led to valuable results in immature rat uteri, but it suffers from receptor losses during incubation and from high nonspecific binding (*16*) when applied to other tissues including the mature rat uterus. We therefore developed an alternative nuclear exchange assay.

The assay was based on our previous finding that cytosol receptor could be quantitatively precipitated with protamine sulfate so that incubation of the precipitate at 30°C with [³H] estradiol permitted complete exchange of any previously bound estradiol (*17*). Incubation at 4°C permitted simultaneous evaluation of receptor unoccupied by estradiol. In order to apply these procedures to nuclear receptor, it was necessary to extract the receptor with high-salt buffers and to find conditions under which the extracted receptor could be completely precipitated with protamine. The procedure is described in detail below.

A. Buffers

The buffers are those used previously, often with the addition of 0.5 mM dithiothreitol (D). For extraction, KCl (K) is 0.6 M rather than 0.4 M as used above, but TKED like TKE is adjusted to pH 8.5.

B. Preparation of Nuclear Extract

Mature rat uteri are frozen in liquid nitrogen and stored at $-76\,°C$. The frozen uteri are pulverized with a Thermovac tissue pulverizer (Thermovac Industries) and immediately placed in a 15-ml tube in ice. All subsequent procedures unless otherwise stated are carried out at $0-4\,°C$. To 200–350 mg of tissue is added 1 ml of TED buffer, and the tissue is homogenized with a Brinkman Polytron PT–10–ST at a speed setting of 3.5 for three 10–second intervals. This is the optimum homogenization condition for virtually complete cell disruption and minimal shearing of nuclei as detected by phase microscopy. The homogenate is centrifuged at 800 *g*. The pellet is washed twice with 1 ml of TED buffer, pH 7.4, followed by centrifugation at 800 *g*. To the washed pellet is added 2 ml of TKED buffer, pH 8.5, and the suspension is mixed vigorously. We find that incubation for 1 hour with TKED buffer is sufficient to liberate the tightly bound nuclear receptor. Following incubation the suspension is centrifuged for 30 minutes at 105,000 *g* to obtain the soluble nuclear receptor. This procedure removes about 85–90% of the nuclear receptor. More exhaustive extractions fail to remove the residual amounts of receptor which remain tightly bound to the nuclear pellet. A recent report (*18*) suggests that the salt-resistant sites might represent a specific class of nuclear binding sites.

DNA in the pellet is determined according to the diphenylamine method of Burton (*9*) using calf thymus DNA as a standard. No DNA can be detected in the nuclear salt extract.

C. Protamine Exchange Assay

The nuclear extract is diluted with TED, pH 7.4, to obtain a protein concentration of about 0.25 mg/ml and a KCl concentration less than 0.1 *M*. Under these conditions, receptor precipitation by protamine is essentially quantitative (Fig. 2).

All assays are performed in triplicate. Glass test tubes 12×75 mm are incubated at room temperature for at least 10 minutes with 1 ml of TED buffer containing 0.1% bovine serum albumin. The tubes are then rinsed with 1 ml of ice-cold TED buffer. Protamine sulfate (USP injection, without phenol preservative, Eli Lilly Co.) is diluted to 1 mg/ml in TED buffer, and 0.25 ml is added to each tube. After addition of 0.5 ml of nuclear extract, the mixtures are vortexed and the precipitates sedimented by centrifugation at 2000 *g* for 10 minutes. The supernatants are removed; the precipitates firmly coat the bottoms of the tubes and can be incubated, washed, etc. without further centrifugation. Then $2-5 \times 10^{-9}\,M$ [^{3}H]estradiol is added in 0.5 ml TED to the precipitated receptor; parallel incubations

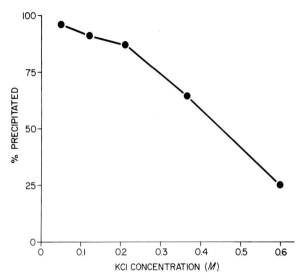

FIG. 2. Fractional precipitation of charged estrogen receptor from rat uterine nuclei by protamine sulfate at various KCl concentrations. From Zava *et al.* (*16*). Reprinted with permission from *Biochemistry*. Copyright by the American Chemical Society.

include 100-fold excess diethylstilbestrol to measure nonspecific [³H]estradiol binding. The tubes are incubated at 0°C overnight or at 37°C for 2.5 hours, then chilled to 0°C for several minutes. After three washes with 1 ml of cold TE buffer, bound radioactivity in the protamine pellets is extracted twice with 1 ml of 100% ethanol at room temperature. The extracts are dried in scintillation vials and counted in 5 ml of toluene–PPO–POPOP scintillation cocktail.

D. Comparison of Nuclear Exchange Assay Methods

The procedure just described was compared with the direct nuclear exchange method of Anderson *et al.* (*15*), each procedure using half of the same preparation of frozen mature rat uteri. The results in Fig. 3 show that the present method measures more than twice the number of nuclear receptors revealed by the older technique. In addition, standard errors are lower, and the nonspecific binding is less than 10% for the protamine assay as compared with 30–50% (70–90% using 13 nM [³H]estradiol as originally published rather than 2 nM) for the direct exchange assay.

E. Observations

Although in our hands the direct exchange assay is less applicable in frozen mature rat uteri than the protamine procedure, we do find that in

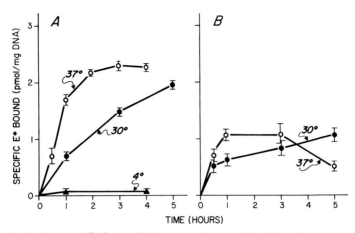

Fig. 3. Exchange of [³H] estradiol for protamine-precipitated nuclear extract (A) and crude nuclear pellet (B) as a function of time and temperature. Thirty minutes after intraperitoneal injection of 10 mg of unlabeled estradiol into mature ovariectomized-adrenalectomized rats, uteri were removed, pooled, frozen, pulverized, and then halved for each assay. From Zava *et al.* (*16*). Reprinted with permission from *Biochemistry*. Copyright by the American Chemical Society.

cell cultures where nonspecific estrogen binding is low, the direct exchange assay equals the protamine method in its ability to quantitate total charged nuclear sites. Since the protamine procedure has worked equally well in a variety of tissues studied, we believe it will have a more general use than the previous method. It has been used to study the accumulation of estrogen receptor in mature rat uterine nuclei and simultaneous depletion from the cytoplasm at various estrogen doses and time points (*16*) and is currently being applied to nuclear estrogen receptor in breast cancer tissue.

NOTE ADDED IN PROOF

We have recently found that KCl extracts of human breast cancer nuclei frequently contain proteolytic activity which interferes with the protamine exchange assay. This problem can be avoided, however, by adsorbing receptor onto hydroxylapatite prior to incubation with [³H]estradiol. This assay method and results appear in two papers by R. E. Garola and W. L. McGuire, *Cancer Res.* 1977, to be published.

ACKNOWLEDGMENT

These studies were supported in part by the NIH (CA11378 and CB-23862), the American Cancer Society (BC-23), and the Robert A. Welch Foundation.

REFERENCES

1. Jensen, E. V., and DeSombre, E. R., *Science* **182**, 126 (1973).
2. O'Malley, B. W., and Means, A. R., *Science* **183**, 610 (1974).

3. Gorski, J., and Gannon, F., *Ann. Rev. Physiol.* **38**, 425 (1976).
4. McGuire, W. L., and DeLaGarza, M., *J. Clin. Endocrinol. Metab.* **37**, 986 (1973).
5. Horwitz, K. B., and McGuire, W. L., *Steroids* **25**, 497 (1975).
6. Chamness, G. C., Costlow, M. E., and McGuire, W. L., *Steroids* **26**, 363 (1975).
7. Notides, A., and Nielsen, S., *J. Biol. Chem.* **249**, 1866 (1974).
8. Chamness, G. C., Jennings, A. W., and McGuire, W. L., *Biochemistry* **13**, 327 (1974).
9. Burton, K., *Biochem. J.* **62**, 315 (1956).
10. Milgrom, E., and Atger, M., *J. Steroid Biochem.* **6**, 487 (1975).
11. Rice, R. H., and Means, O. E., *J. Biol. Chem.* **246**, 831 (1971).
12. Chamness, G. C., Jennings, A. W., and McGuire, W. L., *Nature (London)* **241**, 458 (1973).
13. Garroway, N. W., Orth, D. N., and Harrison, R. W., *Endocrinology* **98**, 1092 (1976).
14. Higgins, S. J., Rousseau, G. G., Baxter, J. D., and Tomkins, G. M., *Proc. Natl. Acad. Sci. U.S.A.* **70**, 3776 (1973).
15. Anderson, J., Clark, J. H., and Peck, E. J., Jr. *Biochem. J.* **126**, 561 (1972).
16. Zava, D. T., Harrington, N. Y., and McGuire, W. L., *Biochemistry* **15**, 4292 (1976).
17. Chamness, G. C., Huff, K., and McGuire, W. L., *Steroids* **25**, 627 (1975).
18. Clark, J. H., and Peck, E. J., Jr., *Nature (London)* **260**, 635 (1976).

Chapter 24

Methods for Assessing the Binding of Steroid Hormones in Nuclei and Chromatin

J. H. CLARK, J. N. ANDERSON, A. J. W. HSUEH,
H. ERIKSSON, J. W. HARDIN, AND E. J. PECK. JR.

Department of Cell Biology,
Baylor College of Medicine,
Houston, Texas

I. Introduction

Steroid hormone responsive cells contain macromolecules which are called receptors. These receptors probably represent the primary mechanism by which target cells detect and respond to hormones. A general scheme for the interaction of steroids with cells is given in Fig. 1. It is generally agreed that steroids (S) enter cells by diffusion and are bound by soluble cytoplasmic receptors (R_c) to form receptor-hormone complexes (R_cS). These complexes undergo translocation to the nucleus where they bind to chromatin (R_nS). The binding interaction between the R_nS complex and chromatin is considered important in the stimulation of new RNA synthesis which ultimately changes the metabolic, developmental, or growth parameters of the cell. Subsequent to chromatin binding the R_nS complex probably undergoes some type of processing which involves the formation of intermediate or inactive forms $(R_nS)'$. This nuclear processing results in the reestablishment of activated R_c in the cytoplasm and the elimination of an inactive form of the steroid (S'). For recent reviews and more complete discussions of these points, see O'Malley and Means (*1*). King and Mainwaring (*2*), Katzenellenbogen and Gorski (*3*), Baulieu (*4*), and Clark *et al.* (*5,6*).

Early investigations of the interaction of receptor–steroid complexes with the nuclear compartment proved difficult because the complex resisted solubilization and it was not possible to quantitatively determine the amount of [³H]steroid that was specifically bound in the nuclear fraction (*7,8*).

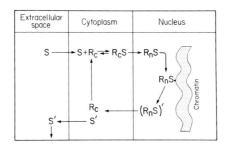

FIG. 1. Hypothetical model of receptor–steroid interactions. Steroid hormones (*S*) enter the cytoplasmic compartment and are bound by soluble receptor molecules (R_c) to form receptor–steroid complexes (R_cS). R_cS is translocated to the nucleus (R_nS) where it becomes associated with binding sites on chromatin. Dissociation of the R_nS complex from chromatin binding sites may involve processing to an inactive form (R_nS)' which dissociates into re-activated R_c and an inactive form of the steroid (*S'*) which leaves the cell.

However, Jensen *et al.* (*9*) extracted a 5 S estrogen-binding component from uterine nuclei using 0.3 *M* KCl at pH 7.4. Shortly thereafter Puca and Bresciani (*10*) described the partial solubilization and quantitation of a mac-romolecule, from rat uterine nuclei, to which estrogen was bound with high affinity and limited capacity. This procedure consists of exposing target tissues to [³H]labeled steroid *in vivo* or *in vitro*. Nuclear fractions are pre-pared and homogenized in Tris-EDTA buffer containing 0.4 *M* KCL. After centrifugation a soluble 4–5 S steroid binding component is demonstrable via sucrose gradient centrifugation or gel filtration methods. Details of these methods are adequately covered in the reports of Puca and Bresciani (*10*), Erdos (*11*), and Stancel *et al.* (*12*). The method suffers from incomplete solubilization of receptor–hormone complex and from the inadequacy of many of the methods available to assess bound steroid. Density gradient centrifugation and gel filtration methods are laborious and/or not strictly quantitative. Adsorption methods employing ion-exchange chromato-graphy are hampered by the high ionic strength extraction media. Charcoal assays should be employed with care since the 4–5 S form of some receptors is readily stripped of steroid by this absorbant. Hydroxylapatite may be employed for the quantitation of salt-extracted receptor since it is not af-fected markedly by high ionic strength (*13*). Regardless of the method used, all methods result in incomplete extraction of the receptor. Thus, the quantity of receptor that is resistant to salt extraction and remains in the nucleus is not measured.

We have developed a method for the determination of R_nS complexes that circumvents the above difficulties (*14*). The method relies on the ex-change of [³H]estradiol for any bound estrogen in the nuclear fraction. Its major advantage is the capacity to measure R_nS complexes which result from the interaction of endogenous steroids or unlabeled estrogenic compounds with receptors, thus providing a valid index of the effects of blood and tissue levels of a given hormone. In addition to strict quantitation, specificity studies are easily performed by exchange. This procedure has proven especially useful in examining the relationship between receptor-steroid interactions and steroid-induced responses (15–19).

In theory, the procedure depends on the observation that the rate of dissociation of steroid from receptors is extremely temperature-sensitive. Thus at 0–4°C the half-life of dissociation of receptor–steroid complex is on the order of days, while at 30–37° the half-life is minutes. The slow rate of dissociation at lower temperatures allows isolation and processing of subcellular fractions such as cytosols and nuclei without appreciable alternation of existing receptor–steroid complexes. The rapid rate of dissociation at elevated temperatures allows the exchange of added ^3H labeled steroid for unlabeled steroid occupying receptor sites. Thus, in general, an exchange assay consists of the preparation of subcellular fractions at low temperature, exchange at elevated temperature, and subsequent work-up and processing at low temperature. It should be noted that no single set of conditions for exchange are applicable to every tissue or receptor system. This point will be discussed in detail below.

II.   Steroid Exchange Methodology

The $[^3H]$estradiol exchange assay has proven useful for measuring nuclear receptor–estrogen complexes in the uterus (15,20,21), in the pituitary and hypothalamus (22), in lactating mammary gland (23), in the corpus luteum (24), in chick oviduct (19), and in Leydig cell tumors (Samuels et al., personal communication). Recently the method has been modified in our laboratory for the measurement of progesterone and glucocorticoid receptors in the nuclear and cytoplasmic compartments of the rat uterus (25,26), and in liver nuclei from both rat and chicken (Eriksson and Clark, unpublished observation). Other laboratories have used exchange to measure androgen receptors in the nuclei of the testes cells (27). We have also used the method to quantify the number of receptor–estrogen complexes that are bound to chromatin (28). Thus this method appears generally applicable. In this section we will present the method currently in use to measure $[^3H]$ estradiol exchange and then indicate the modifications as they apply to other steroids.

A. [³H]Estradiol Exchange Assay for Nuclear-Bound Receptor
Estrogen Complexes

Tissues are removed, weighed, and placed in ice-cold TE buffer (0.01 M
Tris–0.015 M Na$_2$ EDTA, pH 7.9, 4°C) as quickly as possible. Unless stated
otherwise all procedures should be performed at 2–4°C. Tissue/volume =
30–50 mg/ml for uterine tissue (1–2 immature uteri/ml or 4–8 mg protein/ml)
and 100–200 mg/ml for tissues which contain lower quantities of receptor
(e.g., hypothalamus or mammary gland).

Tissue is homogenized in a ground-glass homogenizer with a motor-driven
pestle. Short periods (10–30 seconds) of homogenization are used with
alternate cooling in an ice bath. For easily homogenized tissue such as hypothalamus or pituitary, a glass homogenizer with a Teflon pestle should be
used.

The homogenate is centrifuged for 20 minutes at 800 g, and the crude
nuclear pellet is washed 3 times. Washing is accomplished by adding ice-
cold TE buffer to the pellet (1 ml/50 mg equivalent for uterus), mixing with
a Vortex thoroughly, and centrifuging at 800 g for 10 minutes. The assay
may also be performed on purified nuclei or chromatin (see below). However, we usually use crude nuclear preparations in order to avoid losses due
to preparation of these components.

The washed nuclear pellet is resuspended by gentle homogenization in
TE buffer (same ratios as above) and 250 μl aliquots of the nuclear suspension are dispensed into 12 × 75 mm disposable culture tubes which contain
either [³H]estradiol or [³H]estradiol plus a 100-fold excess of diethylstilbestrol (DES) in 250 μl of TE buffer (final volume = 500 μl). 2,4,6,7-[³H]estra-
diol (sp act 90–115 Ci/mmol) should be used in most cases; however, 6,7-[³H]
estradiol (sp act 40–50 Ci/mmol) may be used in assays that require only
higher [³H]estradiol concentration ranges.

Assays to *estimate* the total number of receptors may be done using 20
nM [³H]estradiol with or without 2 μM DES. If a saturation analysis is
desired, the [³H]estradiol concentration (final) should range from 0.1 nM
to 20.0 nM and the DES concentration should be 100 times that of
[³H]estradiol. Most assays, especially at the beginning of a study, should
utilize saturation analysis. This will insure accurate determinations of appropriate binding parameters. These points are discussed in Section II,C,1.

Note: The use of DES concentrations higher than 100 times that of
[³H]estradiol may result in invalid estimates of specific binding (5).

Assay tubes are placed in a 37°C water bath for 30 minutes with mixing
on a Vortex at 5–10 minute intervals. This step can be varied for optimizing
assay conditions for other receptors, i.e., temperature can be lowered and
time of incubation can be extended. Exchange is terminated by the addition

of 1 ml of ice-cold TE buffer to each tube and centrifugation at 800 g for 10 minutes.

The nuclear pellet is washed 3 times by resuspending the pellet in 1.5 ml of ice-cold TE buffer, mixing on a Vortex, and centrifuging as described above.

Next 1.0 ml of 100% ethanol is added to the washed nuclear pellets, and assay tubes are placed in a 30°C water bath for 10–20 minutes. Tubes are vortexed and centrifuged at 800 g for 10 minutes. The total ethanolic extract is added to 5 ml of PPO-POPOP scintillation cocktail [20.84 gm Permablend (Packard, Downers Grove, Illinois) per gallon of toluene] and samples are counted.

The exchange method can be used for the determination of unoccupied receptor sites (R_n) as well as occupied complexes (R_nS). Incubation of nuclear preparations of 4°C with ^{3}H-labeled steroid will result in the binding of the steroid to R_n sites but will leave R_nS sites unaffected. Thus a comparison of assays run at 4°C with those run under exchange conditions, i.e., elevated temperatures, provides a method for the determination of both states of the receptor (5).

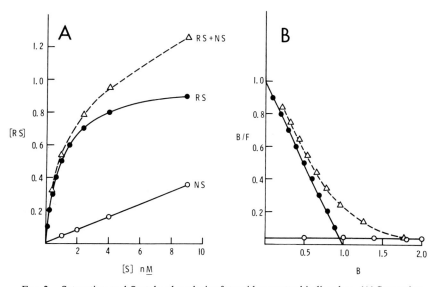

FIG. 2. Saturation and Scatchard analysis of steroid–receptor binding data. (A) Saturation analysis showing the total binding in the system ($\triangle$, $RS + NS$) which consists of contributions made by receptor–steroid binding ($\bullet$, RS) and nonspecific steroid binding ($\bigcirc$, NS). The quantity of RS on the ordinate is given in pmol/ml. (B) The data from graph A are used in a Scatchard plot to determine the number of receptor–steroid binding sites and the dissociation constant of the reaction. Bound steroid divided by the concentration of free steroid (B/F) is plotted as a function of the quantity of bound steroid (B). For details see text.

The quantities of receptor–steroid complex (RS, Fig. 2) may be determined by subtracting the nonspecifically bound steroid (tubes B, Fig. 2, NS) from the total steroid bound (tubes A, Fig. 2, $RS + NS$). These values may then be plotted according to the method of Scatchard (29) for the evaluation of the total number of RS complexes (x-intercept for RS binding, Fig. 2B), and the dissociation constant of the binding reaction (K_d = slope of the line for RS binding). It is not necessary to plot total and nonspecific binding in order to make the above determinations. These data were included only to give a complete picture of the influence of these parameters.

B. Modification of [³H]Estradiol Exchange Assay for Other Steroids

The measurement of the progesterone receptor by exchange differs slightly from the [³H]estradiol exchange assay (25,26). All differences stem from the labile nature of the mammalian progesterone receptor. Measurement of the mammalian progesterone receptor by exchange requires lower temperatures and therefore longer periods for complete exchange. The modifications described below permit exchange to occur with minimal degradation of the receptor.

Tissues are weighed and washed in cold Tris–glycerol (TG) buffer [10 mM Tris-HCl–30% glycerol (v/v) (pH 7.4)]. All subsequent steps are performed at 4° unless otherwise indicated. The crude nuclear fraction is obtained and washed as before. The washed pellet is suspended in TG buffer to a concentration equivalent to 50 mg of tissue/ml. Portions (0.5 ml) of this suspension are dispensed into two series of tubes, A and B, containing 40 μl of buffer. Series A contains 20 nM [³H]progesterone and is used to determine the total amount of [³H]progesterone exchange. Series B contains [³H]progesterone as in series A plus a 100-fold excess of nonradioactive progesterone. The nuclear fractions are incubated with shaking at 4°C for 18–24 hours. Following incubation, 1.5 ml of TG buffer is added, and the samples are centrifuged at 800 g for 10 minutes. Pellets are washed 3 times with TG buffer and extracted with 1.0 ml of ethanol. The radioactivity in the ethanol extract is determined as explained for the [³H]estradiol assay.

C. Nuclear Binding and Retention of Receptor–Hormone Complexes

The interactions between the $R_n S$ complex and nuclear binding sites are poorly understood. However, we have demonstrated that one of the important parameters involved in nuclear binding is retention of the $R_n S$ complex by the nucleus (15,30). Therefore the measurement of $R_n S$ complexes as a function of time after or during hormone exposure is of utmost importance. In our laboratory this has been done by using the ^{3}H-labeled steroid exchange assay under a variety of conditions as described below.

1. Nuclear and Chromatin Preparation

As described above the [^{3}H]estradiol exchange assay has been employed with crude nuclear preparations. However, the method can also be used to measure hormone–receptor complexes bound to purified nuclei or to chromatin. Several different methods exist for the preparation or purified nuclei from a variety of tissues. Some of these methods are very laborious and give relatively poor recovery. Recently, however, a new method has been proposed by Wray and Stubblefield (31) and modfied by Conn and O'Malley (32) and Hardin et al. (18). It uses buffered hexylene glycol in the isolation. This method has proven to be simple, reliable, and less laborious and has successfully been used in nuclear isolations from a number of cell and tissue types (18,32).

For studies on receptor–steroid binding to chromatin, this nuclear component can be prepared according to the method described by Spelsberg et al. (33). However, since the majority of receptor complexes are extracted with high-molar salt solutions, the final salt wash of the isolated chromatin has to be omitted from this procedure.

2. Nuclear Retention and Acceptor Sites

We have shown that the long-term retention of 1000–3000 receptor–estrogen complexes/cell is a requirement for uterine growth (15,30). Since uterine cells contain 15,000–20,000 receptor sites in the cytoplasm, it is apparent that only a fraction of these are required for maximal growth responses. We have suggested that the long-term retention of the receptor–estrogen complex in the nucleus is due to the binding of these complexes to a limited number of nuclear acceptor sites and that retention at these acceptor sites for greater than 4–6 hours is a requirement for the production of uterine growth (5,6,16,17).

To test this hypothesis we have used salt extraction of uterine nuclei to examine the differential extractability of the receptor–estrogen (E) complex. The rationale for the use of this technique was based on the observation by several investigators that extraction of nuclei with 0.3–0.4 M KCl does not remove all of the nuclear-bound estrogen, suggesting that some receptor–hormone complexes are bound more tightly than others. With this technique we have demonstrated that R_nE exists as two forms: (1) KCl-extractable, defined as any R_nE complex that can be extracted by 0.4 M KCl or less; and (2) KCl-resistant, defined as any R_nE which resists extraction by 0.4 M KCl. The stoichiometric and temporal relationships between these forms suggests that $R_n E$ complexes which resist KCl extraction may be bound to nuclear acceptor sites (*29*).

KCl extraction of nuclear $R_n E$ is a simple procedure which consists of exposing nuclei to various concentrations of KCl. Nuclear fractions are washed 3 times with cold TE buffer [0.01 M Tris–0.0015 M EDTA (pH 7.9), 4°C]. Salt extraction is performed by adding various concentrations of TK buffer [0.01 M Tris–0.1–0.6 M KCl (pH 8.0), 4°C] to the nuclear pellet with the mass to volume ratio maintained at 30–50 mg/ml (300–400 μg DNA/ml). The nuclear pellets are resuspended in the TK buffer and allowed to stand on ice, with mixing every 5–10 minutes, for 30 minutes. This is followed by centrifugation at 800 g for 10 minutes, and the KCl extract is carefully decanted. TE buffer is added to the nuclear pellet (30–50 mg/ml) and the pellet resuspended. If necessary, resuspension is accomplished by rehomogenization of the pellet with a glass–Teflon homogenizer. This fraction is used in the regular [³H]estradiol exchange as described in Section II,A for the determination of KCl-resistant complexes. It is very important to determine the actual salt concentration in the extract since the volume of the nuclear pellet will dilute the original salt solution. This may be done by measuring the conductivity of the extract.

The quantities of receptor–steroid complexes in the KCl extract (R_n-extractable) are determined by the hydroxylapatite exchange assay. This assay technique is a modification of the [³H]estradiol exchange procedure and the hydroxylapatite assay which was first described by Erdos (*11*) and modified by Williams and Gorski (*34*). Hydroxylapatite (HAP) is prepared from Bio-Gel HT HAP, (Bio-Rad Richmond, California) by washing extensively with TE buffer at 4°C until the pH of the wash is 7.4. The HAP is resuspended in TE buffer at a ratio of 60% HAP to 40% buffer. Portions (250 μl) of the KCl extract are dispensed in two series of tubes A and B. Series A contains various concentrations of ³H-labeled estrogen (0.10–20 nM) in 50 μl of TE buffer and series B an equal amount of labeled estrogen plus a 100-fold excess of unlabeled DES in 50 μl. The tubes are incubated at 37°C for 30 minutes. After the incubation 500 μl of the HAP suspension is added to each tube. The tubes are vortexed and kept on ice for 15 minutes with additional vortexing every 5 minutes. Then 1.5 ml of TE buffer is added to each tube, and the tubes are centrifuged at 800 g for 10 minutes. The pellets are washed 3 times with 1.5 ml of TE buffer and then extracted with 1.0 ml of absolute ethanol at room temperature for 30 minutes. The radioactivity in the ethanol extract is determined as described above.

D. Assay Validation and Optimization

Although the [³H]estradiol exchange assay has been fully validated and optimized for use in the uterus and a few other tissues, the method must be validated and the assay conditions optimized for each new tissue or hormone that is investigated. Therefore in the following sections we will outline the criteria that should be met.

1. SATURATION ANALYSIS: DETERMINATION OF TOTAL RECEPTOR CONCENTRATION AND DISSOCIATION CONSTANT

Saturation analysis should be performed routinely. From this analysis the total receptor number (R_t) and the dissociation constant (K_d) can be determined as described earlier (Fig. 2). However, the ideal circumstances presented in Fig. 2 do not always exist, and hence we felt the following suggestions should be made.

A broad range of ^{3}H-labeled steroid, ranging from 0.1 to 10 times the K_d of the reaction, should be used for saturation analysis. If receptor sites of different affinities are suspected, the [^{3}H]steroid concentrations should be spaced more closely and over a wider range (0.01–100 times the K_d). We have recently found that the $R_n E$ complex of the rat uterus contains two different estrogen-binding sites with K_ds of 0.3 and 3.0 nM. In previous investigations we had not detected the 0.3 nM site because of our failure to use a large number of steroid concentrations over a wide range.

One of the most common problems incurred while attempting to use the [^{3}H]steroid exchange assay is the occurrence of high levels of nonspecific binding. This situation, of course, decreases the probability that the quantity of receptor-bound steroid can be measured accurately. An example of this

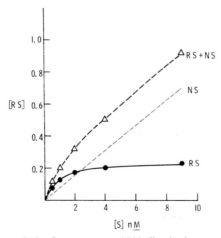

FIG. 3. Saturation analysis of receptor–steroid binding in the presence of high concentrations of nonspecific steroid binding. Nonspecific binding (NS) is great with respect to receptor–steroid binding (RS). Thus the difference between total binding (RS + NS) and NS binding is difficult to determine. The quantities shown on the ordinate and abscissa are identical to those shown in Fig. 2.

situation is shown in Fig. 3. A careful analysis of saturation is necessary to establish the existence of competitive and saturable binding under these conditions (6). One of the ways to overcome this problem is to add protamine sulfate to the nuclear solution (McGuire, personal communication). This results in the precipitation of nuclear material that contains the receptor and results in a relative decrease in nonspecific binding. The nuclear fraction is prepared as described in Sections II,A and II,C, and 3 ml of a 0.1% protamine sulfate solution are added to the tubes. The nuclear fraction is mixed thoroughly and centrifuged at 800 *g* for 10 minutes. The pellet is then washed and ^{3}H-labeled steroid exchange can be performed as described above. Other methods have been used to decrease the nonspecific binding. These include washing nuclear pellets with various detergents and/or dilute ethanol solutions. Although some success has been obtained by these methods, the protamine sulfate precipitations method appears to be superior.

2. TIME AND TEMPERATURE OF STEROID EXCHANGE

Exchange is a tool—a method for the determination of previously occupied high-affinity receptor sites. As stated above, the procedure is based on the temperature-dependence of the dissociation of receptor–estrogen complexes. Thus low temperatures can be used to "trap" receptor–steroid complexes while the extent of exchange can be manipulated by the time and temperature at which the system is exposed to excess ^{3}H-labeled steroid. Figure 4 shows the extent of exchange (RS^*) for a hypothetical receptor–ligand complex assuming that the half-life of this complex ($T_{\frac{1}{2}}$) varies dramatically with temperature. The $T_{\frac{1}{2}}$ values refer to the half-time for dis-

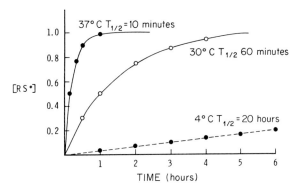

FIG. 4. Extent of steroid exchange at various temperatures. The exchange reaction is occurring in the absence of any degradation. The quantity of *RS* on the ordinate is given in pmol/ml. For details see text.

sociation of the complex in the presence of excess steroid at the given temperature. Note that in the above example about 1 hour is adequate for complete exchange at 37°C, but perhaps 1 week or longer would be required at 4°C.

Unfortunately, the system is seldom as clean as that shown in Fig. 4. Rather, the situation depicted in Fig. 5 is usually found. In this case, two processes occur during the exchange reaction. One is exchange as in Fig. 4 and the other is the temperature-dependent inactivation of receptor sites. The effect of degradation on the quantity of receptor is shown in Fig. 5. This combination of reactions—exchange tending to increase measurable RS while inactivation decreases measurable RS—was first noted in comparing 37°C and 30°C as conditions for exchange of $R_n E$ complexes in uterine tissue. However, it has been most dramatically demonstrated for the exchange of progesterone and corticosterone in our laboratory (A. J. W. Hsueh, E. J. Peck, and J. H. Clark, unpublished observation). Data for three experimental systems are shown in Fig. 6. Thus exchange of $R_n E$ occurs without substantial inactivation except at very high temperatures, while $R_n P$ and $R_n C$ exhibit complex kinetics of exchange and degradation. These studies emphasize the necessity for a series of time and temperature studies before choosing a procedure for an exchange assay. Otherwise, one is apt to measure an artificially low number of nuclear receptor–steroid complexes.

3. HORMONE SPECIFICITY

Implicit in all studies of steroid hormone binding is hormone specificity. That is, an estrogen receptor should recognize and bind estrogens and have little affinity for other steroid hormones. Therefore, estrogenic compounds

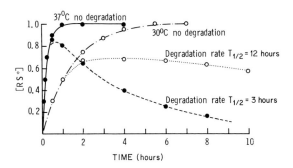

FIG. 5. Effect of receptor degradation on the quantity and apparent rate of exchange of steroid hormone for receptor binding sites. Both exchange and degradation are occurring simultaneously, hence the reduction in the quantity of RS^* with time. The quantity of RS on the ordinate is given in pmol/ml. See text for discussion.

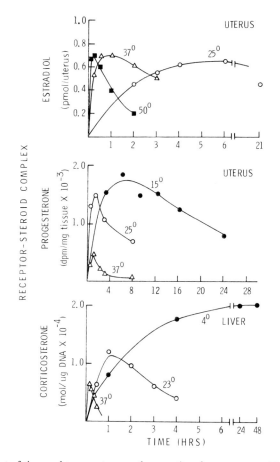

FIG. 6. Effect of time and temperature on the quantity of receptor–steroid complex in the [³H]steroid exchange assay. Nuclear fractions from the uterus and liver were prepared and the [³H]steroid exchange assay was performed at various temperatures and times of incubation.

should act as competitive inhibitors of estrogen binding. Competitive inhibition should be checked carefully. Inhibition of receptor binding may be the result of noncompetitive effects and hence make assays invalid. Noncompetitive inhibition may occur for several reasons: e.g., the inhibitor partially precipitates the receptor or the inhibitor binds to a second site and in so doing alters the receptor site. Therefore a simple demonstration of inhibition of *RS* formation by the addition of an inhibitor does not prove competitive inhibition. Competitive inhibition should be proven by the simple and classic analyses shown in Fig. 7. Figure 7A and B shows competitive inhibition of *RS* binding, and the inference can be made that *S* and *I*

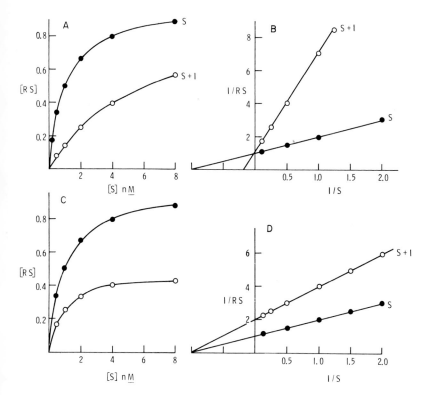

FIG. 7. A comparison of competitive inhibition of hormone binding with noncompetitive inhibition. (A and B) Saturation and double reciprocal plots of competitive inhibition of hormone binding. (C and D) Saturation and double reciprocal analyses of noncompetitive inhibition of hormone binding. The quantities shown on the ordinate and abscissa are identical to those shown in Fig. 2.

are competing for the same binding site, thus demonstrating specific (competitive) inhibition. Noncompetitive inhibition is shown in Fig. 7C and D.

4. TISSUE SPECIFICITY

Since steroid hormones appear to influence specific tissues, an exchange assay should manifest tissue specificity. This has been adequately shown in the case of estrogen receptors (35). High levels of R_nE complexes are found only in tissues which are responsive to estrogen, i.e., uterus, vagina, pituitary, mammary tumors, etc. Strict tissue specificity as shown with estrogen may not be possible, or even likely, since other steroids such as glucocorticoids may not have specific target organs. The question of tissue specificity must be evaluated but may not place the same constraints on all hormones tested.

348   J. H. CLARK *et al.*

ACKNOWLEDGMENTS

This work is supported by grants from USPHS (HD-04985 and HD-08389) and The American Cancer Society (BC-92).

REFERENCES

1. O'Malley, B. W., and Means, A. R., *Science* **183**, 610 (1974).
2. King, R. J. B., and Mainwaring, W. I. P., "Steroid-Cell Interactions." University Park Press, Baltimore, Maryland, 1974.
3. Katzenellenbogen, B. S., and Gorski, J., *in* "Biochemical Actions of Hormones" (G. Litwack, ed.), p. 187. Academic Press, New York, 1975.
4. Baulieu, E. E., *in* "Hormone and Antihormone Action at the Target Cell" (J. H. Clark, W. Klee, A. Levitzki, and J. Wolff, eds.), p. 51. Dahlem Konferenzen, Berlin, 1976.
5. Clark, J. H., Peck, E. J., Jr., and Glasser, S. R. *in* "Reproduction in Domestic Animals" (H. H. Cole and P. T. Cupps, eds.), 3rd Ed., p. 141. Academic Press, New York, in press.
6. Clark, J. H., Peck, E. J., Jr., Hardin, J. W., and Eriksson, H., *in* "Hormone Action: Steroid Hormone Receptors" (B. W. O'Malley and L. Birnbaumer, eds.). Academic Press, New York, in press.
7. Jensen, E. V., and Jacobson, H. I., *Recent Prog. Hormone Res.* **18**, 387 (1962).
8. Noteboom, W. D., and Gorski, J., *Arch. Biochem. Biophys.* **111**, 559 (1965).
9. Jensen, E. V., DeSombre, E. R., Hurst, D. J., Kawashima, T., and Jungblut, P. W., *Arch. Anat. Microsc. Morphol. Exp.* **56**, Suppl. 3–4, 547 (1967).
10. Puca, G. A., and Bresciani, F., *Nature (London)* **218**, 967 (1968).
11. Erdos, T., *Biochem. Biophys. Res. Commun.* **32**, 338 (1968).
12. Stancel, G. M., Leang, K. M. T., and Gorski, J., *Biochemistry* **12**, 2137 (1973).
13. Erdos, T., Best-Belpomme, B., and Bessada, R., *Anal. Biochem.* **37**, 244 (1970).
14. Anderson, J., Clark, J. H., and Peck, E. J., Jr., *Biochem. J.* **126**, 561 (1972).
15. Anderson, J. N., Peck, E. J., Jr., and Clark, J. H., *Endocrinology* **96**, 160 (1975).
16. Clark, J. H., Anderson, J. N., and Peck, E. J., Jr., *Adv. Exp. Med. Biol.* **36**, 15 (1973).
17. Clark, J. H., Anderson, J. N., and Peck, E. J., Jr., *Steroids* **22**, 707 (1973).
18. Hardin, J. W., Clark, J. H., Glasser, S. R., and Peck, E. J., Jr., *Biochemistry* **15**, 1370 (1976).
19. Kalimi, M., Tsai, S. Y., Tsai, M. J., Clark, J. H., and O'Malley, B. W., *J. Biol. Chem.* **251**, 516 (1976).
20. Russell, S., and Thomas, G. H., *Biochem. J.* **144**, 99 (1974).
21. Lan, N. C., and Katzenellanbogen, B. S., *Endocrinology* **98**, 220 (1976).
22. Anderson, J. N., Peck, E. J., Jr., and Clark, J. H., *Endocrinology* **93**, 711 (1973).
23. Hsueh, A. J. W., Peck, E. J., Jr., and Clark, J. H., *J. Endocrinol.* **58**, 503 (1973).
24. Richards, J. S., *Endocrinology* **95**, 1046 (1974).
25. Hsueh, A. J. W., Peck, E. J., Jr., and Clark, J. H., *Steroids* **24**, 599 (1974).
26. Walters, M. W., and Clark, J. H., *in* "Progesterone Receptors in Normal and Neoplastic Tissues" (W. L. McGuire, J. P. Raynaud, and E. E. Baulieu, eds.), p. 271. Raven Press, New York, 1977.
27. Sanborn, B. M., and Steinberger, E., *Endocrinal. Res. Commun.* **2**, 335 (1975).
28. Clark, J. H., Eriksson, H. A., and Hardin, J. W., *J. Steroid Biochem.* **7**, 1039 (1976).
29. Scatchard, G., *Ann. N. Y. Acad. Sci.* **51**, 660 (1949).
30. Clark, J. H., and Peck, E. J., Jr., *Nature (London).* **260**, 635 (1976).
31. Wray, W., and Stubblefield, E., *Exp. Cell Res.* **51**, 469 (1970).
32. Conn, P. M., and O'Malley, B. W., *Biochem. Biophys. Res. Commun.* **64**, 740 (1975).
33. Spelsberg, R. C. Steggles, A. W., and O'Malley, B. W., *J. Biol. Chem.* **246**, 4188 (1971).
34. Williams, D., and Gorski, J., *Biochemistry* **13**, 5537 (1974).
35. Clark, J. H., Peck, E. J., Jr., Schrader, W. T., and O'Malley, B. W., *Methods Cancer Res.* **12**, 367.

Chapter 25

Proteins of Nuclear Ribonucleoprotein Subcomplexes

PETER B. BILLINGS AND TERENCE E. MARTIN

Department of Biology,
University of Chicago,
Chicago, Illinois

I. Introduction

In prokaryotic cells transcription is immediately coupled with translation. Ribosomes actively engaged in protein synthesis are associated with the elongating messenger RNA while it is still bound to the DNA template. Control of gene expression in prokaryotes is therefore exercised predominantly at the transcriptional level. Several additional processing steps intervene, however, between transcription and translation in eukaryotic organisms. Rapidly synthesized RNA of DNA-like, nonribosomal base composition is transcribed as large heterogeneously sedimenting molecules (hnRNA), only a fraction of which is conserved and enters the cytoplasm as mRNA (*1–4*). At least some messenger sequences have been detected in these considerably larger nuclear molecules (*5–8*). However, selective cleavage from much larger precursors may not be involved in the processing of other messenger RNA species (*9–11*). In addition to processing by cleavage and selective transport, most mRNA sequences and some hnRNA molecules are modified to contain short, approximately 200 nucleotide segments of poly (A) at the 3'OH termini (*12–15*) added before or during nuclear processing by a post-transcriptional mechanism(*16,17*). The 5' end of hnRNA sequences destined to become mRNA is likewise modified by the addition of capping groups and the subsequent specific methylation of such "caps" resulting in 5' termini of the general structure $m^7 G^{5'}ppp^{5'} N^m pNp$—(*18,19*). In addition, hnRNA and some mRNA molecules contain internal base methylation at the N^6-position of adenosine (*18,19*).

The substrate for this processing of hnRNA by cleavage, polyadenylation, capping, methylation, and selection for transport to the cytoplasm is not the naked RNA molecule, but rather the ribonucleoprotein (RNP) fibril. The nascent hnRNA in chromatin spreads appears in shortened RNP fibrils still attached to the chromatin template (20). Various RNP complexes can be distinguished in thin sections by electron microscopy (21,22). Nascent hnRNP very likely corresponds to the perichromatin fibrils of such thin sections (22–24).

Ribonucleoprotein complexes containing rapidly labeled RNA have also been isolated from purified nuclei and characterized biochemically (25). Such RNP in general are isolated in the form of complexes sedimenting at approximately 30 S (26,27) and occasionally as much larger complexes which may subsequently be cleaved into 30–45 S RNP (28–30). RNP may also be isolated from chromatin by procedures similar to those we have employed on intact nuclei (31). Proteins characteristic of RNP have been shown to comprise a significant fraction of the total chromatin nonhistone proteins (32,33). Over the past few years, we have studied the proteins associated with 30 S RNP and the substructures of the hnRNP fibril which may easily be extracted from isolated nuclei and purified to homogeneity (34,35).

II. Expectations, Methods, and Criteria

A number of predictions can be made about the proteins which bind newly synthesized hnRNA and mRNA sequences. In analogy with the histones, a limited number of related structural proteins, serving the general function of folding nascent hnRNA, could be expected to comprise a large proportion of hnRNP protein. One might anticipate minimal tissue specificity of these structural proteins, as well as species conservation. However, the affinities of these proteins for other classes of RNA (tRNA, rRNA) would be expected to be low. Transiently associated processing enzymes may be bound to hnRNA, but would comprise only a minor fraction of the total hnRNP mass. Proteins specifically recognizing various sequences within hnRNA molecules [poly(A), oligo(A), oligo(U), and capping groups] may also bind, perhaps with greater affinities than general structural proteins. Proteins specific for unique hnRNA sequences may also exist, but individually would be below detection in RNP from cells containing a diverse population of hnRNA sequences.

A number of procedures might be expected to release RNP from purified nuclei. Nuclear rupture by sonication or lysis with detergents can be used; however, contamination by preribosomal RNP and chromatin fragments may obscure subsequent analysis. High salt lysis combined with DNase treatment would eliminate entrapment of RNP in the chromatin gel giving larger yields than by sonication; however, released DNA-binding proteins, which are maintained in solution by high salt, may interact with RNP. Gentle extraction procedures with buffered salt solutions, leaving nuclei relatively intact and avoiding exposure to high salt, should minimize rearrangements of nuclear proteins, but they suffer the disadvantage of being lengthy, resulting in considerable cleavage of hnRNA. However, the process does yield discrete reproducible substructures which can readily be purified on the basis of size, and we have taken advantage of this to characterize the protein components of these subunits.

Ideally the significance of individual proteins is assessed by determination of relevant function. "Structural proteins" are, almost by definition, notoriously difficult to evaluate by simple functional tests. The proteins of RNP complexes share such difficulties with the proteins associated with DNA in chromatin. Simple test systems are just not available; the development of functional assays is the major challenge in this field. With simple discrete substructures one can attempt dissociation and reconstruction experiments, and this approach is being followed by ourselves and others. Protein affinities for particular RNA sequences can be tested *in vitro*, but preparations of the various RNA processing enzymes will be required to determine how these, in conjunction with presently designated "structural proteins," modify nuclear RNA for intranuclear function or transport to the cytoplasm.

III.   Protein Composition of hnRNP

A.   Extraction and General Characterization of 30 S RNP

The procedure we have employed for the isolation of 30 S RNP from purified nuclei is a modification of the method originally developed by Samarina *et al.* (*26,28*). Briefly, this involves extraction of isolated nuclei at 0°C with a low salt solution at high pH. Almost any procedure for isolation and purification of nuclei at or near 0.1 *M* NaCl gives reasonable yields of 30 S RNP. The procedure of Chauveau *et al.* (*36*) has been used for most tissues; however, for the majority of experiments described here, using the Taper hepatoma ascites cells grown in Swiss–Webster mice (*37*), the procedure of Martin and McCarthy (*27*) has been utilized. Purified nuclei are extracted briefly in 0.1 *M* NaCl–0.01 *M* Tris-HCl (pH 7.0)–0.001 *M* MgCl$_2$ (STM, pH 7), washed in STM pH 9 and extracted with STM pH 9 for 4 hours at 0°C. The nuclei are removed by centrifugation, and the extract is then centrifuged 15 hours at 25,000 rpm on a 15–30% sucrose gradient in

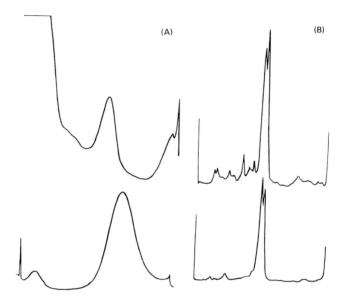

F<small>IG</small>. 1. Sucrose density gradient profiles and SDS–polyacrylamide gel tracings of crude and purified 30 S RNP. (A) STM pH 9 extracts from purified mouse ascites cell nuclei (prepared as described in Section III,A) were layered on 15–30% sucrose gradients in STM pH 8 and centrifuged for 15 hours at 25,000 rpm (SW-27 rotor). Gradients were displaced through an ISCO density gradient fractionator while recording absorption at 254 nm. Shown below is a similar gradient profile obtained when pooled 30 S RNP are pelleted overnight at approximately 100,000 g, resuspended in STM pH 8, and recentrifuged on an identical gradient. (B) 30 S RNP recovered from the first sucrose gradient (above) or second, purification gradient (below) were pelleted, dissolved in SDS-denaturing buffer [2% SDS–5% β-mercaptoethanol (BME)–0.05 M Tris-HCL (pH 7.0)], and electrophoresis was performed on 10% polyacrylamide SDS slab gels as described by Laemmli (38). Coomassie brilliant blue stained gels were scanned on a Joyce–Lobel densitometer with a red filter.

STM pH 8. The bulk of the rapidly labeled RNA in the extract is recovered as a homogeneous ribonucleoprotein complex sedimenting at approximately 30 S (Fig. 1A).

The release of hnRNP, which presumably leaches out through nuclear pores, is considerably more rapid when extracted at the elevated pH chosen than at lower values (34). Higher temperatures also hasten the release from the nuclei. However, RNP isolated at 37°C or room temperature is more highly contaminated and the nuclei appear damaged. At 0°C nuclei remain largely intact as viewed by phase-contrast microscopy even after a 4–6 hour extraction in STM (34). Nuclear lysis prior to or during the extraction results in a greatly reduced recovery of RNP which is then entrapped in the released chromatin gel. Nuclear rupture by sonication in STM pH 8, followed by high-speed pelleting of chromatin, frees 30 S RNP from the chromatin template in amounts roughly comparable to that released by extraction. However, even after purification on sucrose gradients, such RNP are more highly contaminated with high-molecular-weight proteins than extracted 30 S RNP when assayed on sodium dodecyl sulfate (SDS)-containing polyacrylamide gels. Recovery of RNP complexes larger than 30 S by this technique as reported by Pederson (29) has not been observed with mouse ascites cells.

Isolated 30 S RNP appear to have a discoid shape of about 15–25 nm across when negatively stained and viewed through the electron microscope (34). A molecular mass of 1×10^6 daltons for the 30 S RNP may be estimated based on the formula of Spirin (39). The 30 S RNP fixed in formaldehyde has a characteristic buoyant density of 1.39–1.41 gm/cm³ in CsCl [(27,28) see also Fig. 9] corresponding to 75–80% protein by the empirical formula of Spirin (40) or slightly higher values (85–90%) by the formula of Hamilton (41). Direct determinations on pelleted or ethanol-precipitated 30 S RNP have given a protein composition of 90.8 ± 2.0%; however, RNA may have been lost during the concentration of the RNP for this assay.

The amount of RNA contained in a single 30 S RNP ($1–2 \times 10^5$ daltons) is clearly insufficient to include even average-sized messenger RNA molecules, or the much larger hnRNA transcribed in the nucleus. However, since the majority of the nuclear pulse-labeled RNA (27), including sequences identical to 85–100% of the poly (A) + mRNA sequences (42), is associated with 30 S RNP, it must be assumed that extracted 30 S RNP represents subunits derived by cleavage of much larger hnRNP fibrils by endogenous RNase. The existence of larger RNP complexes comprised of 30 S subunits has been demonstrated directly (28,30) by extracting considerably larger RNP complexes in the presence of RNase inhibitor from rat liver cytoplasm supernatant (RLS), and subsequently cleaving these large RNP to 30 S subunits with low concentrations of RNase. Further evidence of the subunit nature of native hnRNP is afforded by the discovery that the poly(A) portion of hnRNA is recovered in extracts associated with a distinct 15 S RNP complex and not with the 30 S RNP subunits (35).

B. Protein Composition of 30 S RNP

There exists considerable controversy as to the protein composition of hnRNP. Numerous groups working in this area have reported a complex protein composition for hnRNP (29,43–45). On the other hand, Georgiev and co-workers have indicated a very simple protein composition of hnRNP consisting of a single polypeptide in the molecular weight range of 40,000 (28,46). There seems to be general agreement that in most systems the

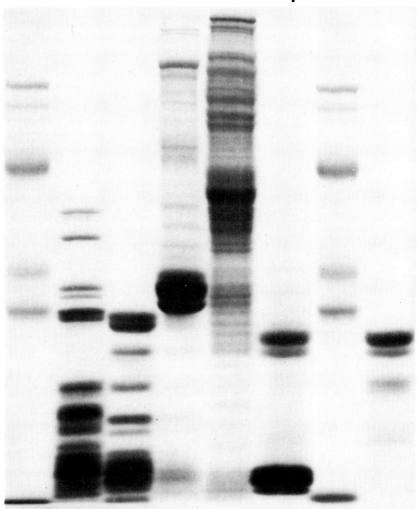

FIG. 2. Comparison of 30 S RNP proteins with ribosomal proteins and histones. Mouse ascites cell 30 S RNP and an adjacent 10–20 S region recovered from a sucrose gradient (See Fig. 1A) were pelleted overnight at 100,000 g and resuspended in SDS denaturing buffer. Electrophoresis was performed on 10% polyacrylamide, SDS slab gels as described by Laemmli (38). Residual nuclei following extraction of RNP were extracted with 0.2 $N H_2SO_4$ for total histone or 5% perchloric acid for H1. High salt–puromycin dissociated ribosomal subunits were prepared by the method of Blobel (47). Molecular weight marker proteins in slots 1 and 7 are β-glactosidase (130,000), phosphorylase a (92,500), bovine serum albumin (68,000), ovalbumin (43,000), glyceraldehyde-3-phosphate dehydrogenase (36,000), and globin (15,500). L°, S°—large and small ribosomal subunits; 30, 15–pelleted 30 S RNP proteins and proteins from the 10–20 S region; H_t, H1—total mouse ascites cell histones and isolated H1 (f_1).

"major" proteins of hnRNP are recovered in the 35,000–40,000 dalton range. The predominant proteins we find associated with 30 S RNP have apparent molecular weights of 37,500–40,000 as analyzed by SDS–polyacrylamide gel electrophoresis (*34*). These major proteins account for greater than 80% of the Coomassie blue staining protein recovered from 30 S RNP (see Figs. 1B and 2). The major polypeptides are neither ribosomal proteins nor histones as shown by their mobilities on both SDS and acid–urea polyacrylamide gels. Unlike several of the minor bands which likely arise by contamination of 30 S RNP from adjacent sucrose gradient regions, the 37,500 and 40,000 dalton proteins are not prevalent in adjacent fractions of sucrose gradients [see Fig. 2 and (*34*)].

1. ELECTROPHORETIC CHARACTERIZATION OF THE MAJOR 30 S RNP PROTEINS

At least some of the discrepancy in the protein composition of hnRNP is the result of varying degrees of purification. We have employed a second sucrose gradient to purify 30 S RNP from contaminating proteins cosedimenting with, or loosely bound to, RNP on the first gradient. Purified 30 S RNP exhibits a characteristic slightly faster sedimentation rate than originally extracted RNP; this may reflect a partial loss of free RNA tails and/or minor proteins associated with crude particles resulting in a more compact structure (see Fig. 1A). Under these conditions a simple protein composition is found for 30 S RNP consisting entirely of proteins in the molecular weight range 37,500–40,000 (see Fig. 1B). Although evidence for heterogeneity of these two size classes will be presented, the 37,500 and 40,000 dalton species appear to be present in roughly equivalent amounts suggesting that the total protein mass of the purified 30 S RNP is comprised of approximately 10 copies of each. The minor proteins lost on purification, probably including most processing enzymes, are all of greater molecular weight than the major 30 S RNP proteins and include polypeptides with molecular weights of 48,000, 53,000, 57,000, 71,000, 74,000, 150,000, and > 200,000 daltons.

Electrophoresis for longer times resolves minor components in the 37,000–40,000 dalton region (Fig. 3A). These 2–3 additional polypeptides are found only on occasions when very high resolution has been achieved on gel electrophoresis, but they suggest a degree of heterogeneity of proteins in a very narrow molecular weight range. The mobility of the major proteins is altered in the presence of 8 M urea included in Laemmli SDS gels (Fig. 3B). Under these conditions the major proteins have apparent molecular weights of 38,000–46,000, and four minor bands are more clearly resolved. However, high concentrations of urea may reduce the binding of SDS to

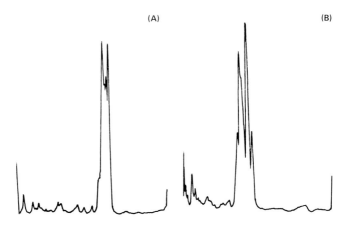

FIG. 3.   Proteins of 30 S RNP electrophoresed on SDS gels in the presence or absence of urea. Pelleted 30 S RNP was resuspended in SDS-denaturing buffer either lacking (A) or containing (B) 8 M urea and SDS slab-gel electrophoresis was performed in the absence or presence of 8 M urea on 10% polyacrylamide gels prepared otherwise as described by Laemmli (*38*). To achieve greater resolution, current was applied to the gel run in the absence of urea (A) for a time 30% longer than that required for the tracker dye to leave the bottom of the gel.

protein, possibly inducing artifactual heterogeneity and alterations in the apparent molecular weight values.[1]

Electrophoresis on acid–urea gels at pH 4.5 (*48*) resolves 30 S RNP proteins as five bands—two major and three minor (diagrams of gels are given in Fig. 4 and gel tracings in Fig. 6A). Prior reduction of the proteins with β-mercaptoethanol (BME) results in the disappearance of the slowest migrating major band (band A in Fig. 4) and the two most rapidly migrating minor bands (D and E), leaving only a single major band (B) and a slightly faster migrating minor band (C′) (see Fig. 4). Similar results are obtained with performate-oxidized RNP protein or with reduced protein electrophoresed on gels previously removed from the casting tubes and soaked in 5% mercaptoethanol. These results are consistent with the interpretation of Krichevskya and Georgiev (*46*) that several of the minor bands in this gel system result from intra- and intermolecular disulfide linkages. The total

---

[1] The SDS included in the polyacrylamide gels described in this paper was obtained from Schwarz/Mann. Recently, we have noticed an alteration in the apparent molecular weight of 30 S RNP proteins using SDS purchased from MCB or Bio-Rad (e.g., on Bio-Rad SDS gels the proteins of MW 37,500 and 40,000 reported here are resolved as 35,000 and a closely spaced doublet 36,000–37,000). These results indicate that caution must be exercised in comparing proteins by their relative mobilities even when the same gel system is employed (*47a*).

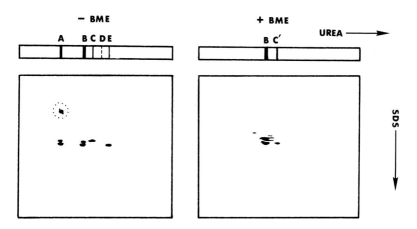

F<small>IG</small>. 4.   Acid–urea:SDS two-dimensional gel analysis of 30 S RNP proteins. Pelleted 30 S RNP was resuspended in 6 M urea–0.06 M potassium acetate (pH 6.8) and incubated 1 hour at 37°C in the presence or absence of 5% β-mercaptoethanol (BME). The first dimension employed 18 cm acid–urea gels as described by Leboy *et al.* (*48*). Longitudinal 1.5 mm-thick slices of one-dimensional gels were incubated 15 minutes in SDS-denaturing buffer containing 10% BME and inserted in denaturing buffer–10% glycerol above SDS–10% polyacrylamide slab gels (*38*) with 5% stacker gels. Incubation for longer periods or in higher concentrations of BME resulted in loss of the dimer bands (encircled with a dotted line).

complexity of RNP proteins cannot be assessed from acid–urea gels insofar as phosphorylated proteins and minor high-molecular-weight proteins with low isoelectric points do not enter such gels, which are designed to resolve basic proteins (*49*). We have attempted to determine whether the major proteins which do migrate into these gels are the major species of 30 S RNP proteins. Analysis of proteins migrating on acid–urea gels by electrophoresis in a second dimension in the presence of SDS indicates that all of the proteins are within the molecular weight range 37,500–40,000 (Fig. 4). The major band designated as B on the unreduced or reduced urea gels can be seen to be composed of both 37,500 and 40,000 dalton species when electrophoresis is performed in a second dimension (Fig. 4). Band A on the unreduced urea gel, having approximately half the mobility of band B, also contains both molecular weight species. The presence of this band on the acid–urea gels is highly variable in amount, and it is recovered as band B following reduction. We interpret this band as the result of dimerization of the two major species of proteins. Unreduced band C contains predominantly the 40,000 dalton species. Upon reduction prior to electrophoresis in the first dimension, a band is recovered in the similar position of C′; however, second-dimensional analysis of C and C′ clearly indicates that the two differ (Fig. 4). The band C of 40,000 disappears into band B upon reduction prior to the first-dimensional analysis and probably results from intramolecular disulfide linkages giving a more compact conformation. We have as yet no explanation for the appearance of the reduced protein recovered in the similar position of C′, having a molecular weight of approximately 37,500, and representing 20–24% of the total RNP protein. Unreduced band E comprises only 7% of the total fast green staining protein of the complex, which is insufficient to entirely account for band C′. A very minor band on the first-dimension unreduced acid–urea gels at position D has not been resolved on the second dimension.

Our results therefore indicate that the number of "core" proteins comprising the purified 30 S RNP is small (two to five). While heterogeneity on SDS gels is occasionally found within the range 37,000–40,000 daltons, these two major protein size classes are seen to comprise the bulk of the 30 S RNP protein moiety. The core proteins migrate on acid–urea gels indicating that they possess higher isoelectric points than the minor, loosely associated proteins, and heterogeneity can be resolved on the low pH urea gels by electrophoresis on a second dimension. The alterations in the urea gel band-

TABLE I

AMINO ACID COMPOSITION OF NUCLEAR RNP

	Mole percent recovered			
	Total protein of purified 30 S RNP		Mouse ascites cell 30 S RNP	
Amino acid	Duck liver	Mouse ascites cells	37,500 mol wt protein	40,000 mol wt protein
Lys	5.44	6.27	4.96	5.11
His	1.98	1.99	2.07	2.24
Arg	5.83	6.01	4.86	4.28
Asp	9.46	10.80	9.83	9.21
Thr	3.69	3.28	3.27	3.17
Ser	8.96	7.62	7.04	8.20
Glu	13.07	10.53	11.76	12.97
Pro	3.90	3.67	3.59	3.52
Gly	17.90	22.41	23.27	25.56
Ala	7.16	3.96	6.56	5.67
Cys	Trace	Trace	—	—
Val	4.35	4.75	5.47	4.63
Met	1.62	1.76	1.09	1.04
Ile	2.62	2.54	2.73	2.45
Leu	4.19	3.30	4.41	4.32
Tyr	3.73	5.64	4.84	3.83
Phe	4.09	5.42	4.22	3.80
X	2.01	—	Trace	Trace

ing pattern following reduction demonstrate the presence of a very small number (see Table I) of highly reactive cysteine residues. They also suggest that disulfide links may be important in assembling and/or maintaining the conformation of the protein core complex. However, such linkages are probably not entirely responsible for the stability of the core complex which is unaffected by 1% mercaptoethanol but is destroyed in high concentrations of urea or moderate concentrations of salt (see Section III,C).

2. COMPARISON OF 30 S RNP PROTEINS FROM DIFFERENT HIGHER EUKARYOTIC SPECIES

We have been interested in the question of whether the proteins associated with hnRNA represent a generally conserved class, like ribosomal proteins and histones, which also must interact with nucleic acids. The mobilities of the major 30 S RNP proteins from mammalian species, including rat, mouse, and monkey liver and the mouse tumor lines Taper ascites, Krebs, and the light chain myeloma, MOPC 41, are all indistinguishable on discontinuous SDS–polyacrylamide gels. The only exception we have found among the higher vertebrates is the slightly decreased mobility (mol wt 41,000 versus 40,000) in the larger polypeptide from duck and chicken RNP (Fig. 5). When chicken RNP proteins are analyzed on acid–urea gels four to five bands are recovered; the mobilities of major and minor components are slightly different from those of mouse Taper ascites cells. The tendency toward formation of inter- and intramolecular disulfide bands is also found in chicken RNP protein. RNP proteins from both chicken and mouse ascites cells appear moderately basic, and their mobilities are greatly affected by reduction prior to electrophoresis (Fig. 6), being resolved as only two bands in approximate ratios of 4–5:1. This high degree of similarity of the major polypeptides is also reflected in the amino acid composition of total purified 30 S RNP protein from mouse ascites and duck liver (Table 1).

3. ARE THE MAJOR RNP PROTEINS RELATED?

Evidence presented thus far suggests that the two major polypeptides comprising the core of 30 S RNP may be related. The core proteins are very similar in apparent size, differing by only approximately 20–25 amino acids. Mobilities on acid–urea polyacrylamide gels which separate basic proteins largely by charge are identical. Conventional sequencing techniques have proven unsuccessful presumably because both species contain blocked N-termini. We do not believe that the smaller protein is derived from the larger by protease activity in our extracts. Inclusion of the serine-type protease inhibitor, phenylmethane sulfonyl fluoride (PMSF), at all stages in the isolation of nuclei and purification of RNP has shown no alteration in the mobilities nor relative proportions of the major proteins.

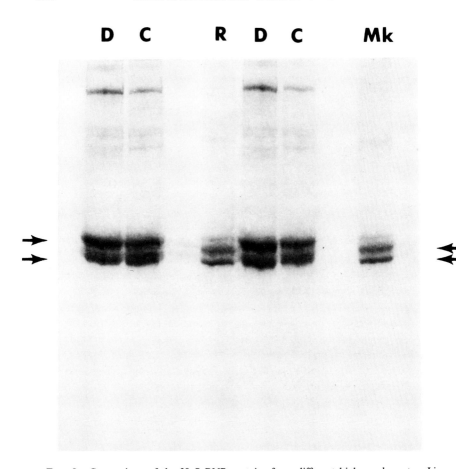

FIG. 5. Comparison of the 30 S RNP proteins from different higher eukaryotes. Liver nuclei were prepared by the procedure of Chauveau *et al.* (*36*) from duck (D), chick (C), rat (R), and monkey (Mk). Nuclear 30 S RNP isolated as described in Section III,A was pelleted and the proteins electrophoresed as described in Fig. 1. Arrows on the left mark the position of avian RNP proteins of molecular weight 41,000 and 37,500; those on the right indicate the position of the mammalian RNP proteins of molecular weight 40,000 and 37,500. From Martin *et al.* (*34*).

In order to further investigate the possible relatedness of the two major proteins, the bands separated on SDS slab gels were excised and individually eluted by electrophoresis in the presence of SDS. Amino acid analysis performed on the isolated 37,500 and 40,000 dalton proteins indicates the proteins are very similar in their amino acid composition (Table 1). The composition compares well with that of total RNP protein derived from purified mouse ascites 30 S RNP. The proteins contain only moderate

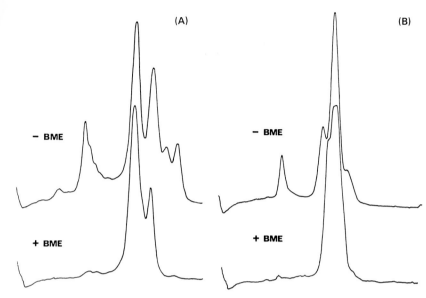

FIG. 6. Comparison of the RNP proteins of mouse ascites cells with those of chicken liver on acid–urea gels. RNP isolated as described in Section III,A, from mouse Taper ascites cells or chicken liver, was pelleted and resuspended in sample buffer for urea gels at pH 4.5 (48). To one-half of each sample, β-mercaptoethanol (BME) was added to 5%, and all four samples were incubated 1 hour at 37°C prior to electrophoresis on 0.6 × 8 cm, 7.5% polyacrylamide tube gels. Gels were stained in 1% fast green in 7% acetic acid and destained by diffusion. They were then scanned at 620 nm on a Gilford spectrophotometer. (A) Mouse ascites cell RNP; (B) chicken liver RNP.

amounts of the basic residues, but high amounts of the acidic residues, aspartic and glutamic acids, and have unusually high glycine content. Given the electrophoretic behavior of the proteins on acid–urea gels, it seems likely that a large proportion of the acidic residues could be present as Asn and Gln. The uncommon amino acid dimethylarginine coelutes from the amino acid analyzer column with a minor residue found in both 30 S RNP proteins. The authors are indebted to Drs. L. C. Boffa and G. Vidali, who identified N^G, N^G-dimethylarginine in rat liver RNP-protein (50), for the suggestion which led to the identification of this residue in the two mouse ascites RNP-proteins. This same unusual amino acid is also found as a constituent of the two major HeLa RNP-proteins (51) and in a presumably homologous nuclear protein from *Physarum* (52). In addition, both proteins contain traces of alloisoleucine and other less common residues as yet not definitively identified. Thus, the two major proteins of 30 S RNP appear to be very similar in their charge, size, and amino acid composition and may, like histones, represent a relatively conserved class of related proteins, both members of which share a common ancestry and have similar, interdependent functions in 30 S RNP. Our results, however, do not completely exclude the possibility that one polypeptide is a modified form of the other.

C. Stability of the 30 S RNP Complex

A model has been proposed (28,53) in which nascent hnRNA is bound to the surface of a number of globular protein particles termed "informofers" yielding large polysomelike structures. The 30 S RNP, isolated in the absence of inhibitors of RNase, arise by limited cleavage of such larger structures. As proposed, the model suggests that intact informofers bind RNA, rather than the free protein binding and subsequently folding the RNA into the globular structures that are isolated as 30 S RNP. In support of this model, it was reported that 30 S RNP protein cores (informofers), radioisotopically labeled by chemical iodination, could be isolated free of RNA on high salt–sucrose gradients, continuing to sediment in the approximate same position as the intact RNP (53).

In our initial studies on the salt stability of 30 S RNP, we similarly used 30 S RNP labeled in the protein moiety with ^{125}I, however, we employed the milder, lactoperoxidase-catalyzed iodination technique of Marchalonis (54). Previously, we found the major 30 S RNP proteins to be only slowly labeled in vivo with radioactive amino acids, and therefore attempts to monitor by radioactivity the fate of the major RNP-proteins on high salt–sucrose gradients were obscured by minor, more rapidly turning over

FIG. 7. Dissociation of 30 S RNP on sucrose gradients containing 0.5 M NaCl. Mouse ascites cells were incubated in minimal essential medium for 30 min in the presence of 15 μCi/ml [^{3}H]uridine as described by Quinlan et al. (35). Nuclear extracts were prepared and centrifuged on STM pH 8 sucrose gradients as shown in Fig. 1A, top. Recovered 30 S RNP were pelleted and resuspended in 4 ml STM (pH 8) on ice. Insoluble material was removed by low speed centrifugation and half of the resulting suspension was adjusted to 0.5 M NaCl and recentrifuged on 15–30% buffered sucrose gradients containing 0.5 M NaCl; the remaining fraction was run in parallel on a normal STM (pH 8) sucrose gradient. Gradients were collected in 1 ml fractions. The top panel shows the distribution of acid-insoluble radioactivity in 0.1 ml portions of each fraction from the high-salt gradient (●—●) superimposed on the A_{254} gradient profile. In the low salt control gradient >75% of the precipitable counts were in the 30 S region (fractions 16–26), <7% in fractions 1–10. The remaining portions of fractions from the high salt gradients were pooled (1–3, 4–6, etc.), precipitated at 7% TCA with 20 μg/ml carrier RNA and collected at 9000 g for 5 min. The ethanol-washed pellets were dissolved in 2% SDS and an equal volume of each analyzed for proteins on 10% polyacrylamide SDS gels. The protein gel pattern of each 2.7 ml portion is displayed in the lower panel directly beneath the respective pooled fractions from which it is derived. A smaller volume of fraction 4–6, overloaded in the gradient analysis, is shown to the left between MW marker proteins (identified in Fig. 2).

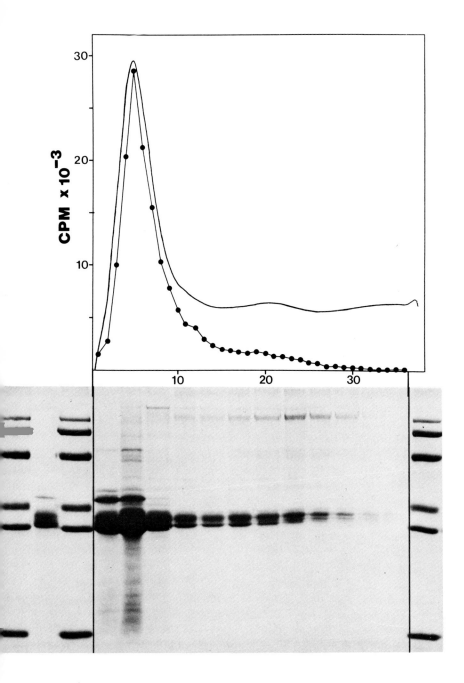

proteins having higher specific activities. Our results with iodinated 30 S RNP, like those of Lukanidin *et al.* (*53*), suggested that the protein core remained intact in the presence of 0.5–2 M NaCl and could be recovered from high-salt–sucrose gradients in about the 30 S region when account was taken of the increased density of such gradients. Separate experiments with [³H]adenosine or uridine-labeled 30 S RNP confirmed that RNA was progressively lost from the 30 S RNP region as the salt concentration was increased above 0.2–0.3 M NaCl. By 0.5 M NaCl most of the RNA is lost from the 30 S region and is largely recovered at the top of the gradient. When we attempted to correlate the labeled-RNA distribution across a high-salt–sucrose gradient (Fig. 7, top panel) with the unlabeled-protein distribution by SDS polyacrylamide gel analysis (Fig. 7, lower panel), the results obtained were completely different from those based on ¹²⁵I-labeled protein distribution. Unlabeled protein is indeed displaced from the 30 S region of the gradient in amounts approximating the loss of RNA counts toward the top of the gradient (Fig. 7). The salt dissociation of the 30 S RNP-protein core has also been confirmed using purified 30 S RNP labeled *in vivo* with tritiated amino acids. A gel analysis of iodinated 30 S RNP, as in the lower panel of Fig. 7, indicates that very little of the iodinated salt-stable 30 S RNP protein can be made to enter the gel even after extensive boiling in 10% dithiothreitol–2% SDS–8 M urea (data not shown). These and other experiments have led us to believe that the radioiodination procedures used induce covalent cross-linking of the protein moiety of 30 S RNP rendering the particles resistant to high salt. Our results are therefore consistent with those of Stevenin and Jacob (*55*) who, in studying a more complex nuclear hnRNP from rat brain, reported finding no salt-stable protein complex sedimenting at greater than 15 S. We hope our results on the artifactually induced salt stability of 30 S RNP protein core by iodination will help to clarify a minor controversy in the literature.

Purification of 30 S RNP on a second STM gradient results in a variable partial loss of RNA from the 30 S RNP region. The RNA fragments lost on purification are smaller than those remaining with the particles, and the retained RNA is enriched in mRNA sequences (*42*). Therefore, the RNA–protein interaction, although not stable under high ionic conditions, exhibits varying binding affinities for different hnRNA sequences in more moderate ionic conditions. A limited degree of specificity is shown in the binding of additional RNA to 30 S RNP. Mature rRNA and tRNA are bound to a negligible extent while hnRNA and mRNA of both poly(A) plus and minus classes do bind (*26,34*). Exposure of 30 S RNP to moderately high (0.5 M) concentrations of NaCl results in the reversible dissociation of the majority of the protein from the RNA. Current evidence suggests that the proteins associated in the nucleus with hnRNA are not found on polysomal mRNA (*56,57*). Proteins destined to be exchanged either prior to or shortly after exit from the nucleus might not be expected to be as tightly bound or as salt resistant as those bound to polysomal mRNA. Once the RNA is removed, RNP-proteins have a tendency to aggregate if the salt is removed by dialysis; similarly the protein will precipitate in low salt if 30 S RNP are subjected to ribonuclease.

The finding that the proteins of 30 S RNP are only slowly labeled *in vivo* suggests that these polypeptides are extensively reutilized in the nucleus. The important question of the validity of the "informofer" model for the formation of RNP complexes remains to be answered. The reversible dissociation of the RNA and of the protein core itself in high salt cannot attest to the state of the proteins under the physiological conditions in the nucleus in which they interact with hnRNA. However the observation that mild iodination cross-links the polypeptides into a salt-stable core suggests that intimate protein–protein interactions exist in the 30 S RNP and that these presumably contribute to the molecular organization of the native complex. The ability of 30 S RNP to bind additional messengerlike RNA *in vitro* may simply be a result of loss of RNA from binding sites during the isolation of 30 S RNP and not be an accurate model for the binding of large native transcripts *in vivo*. Reconstruction studies underway in a number of laboratories will undoubtedly improve our understanding of the nature of the interacting macromolecules in the formation of RNP complexes.

IV. Nuclear Poly(A) Containing Ribonucleoproteins

Most of the cytoplasmic mRNA and some of the nuclear hnRNA molecules contain poly(A) sequences at the 3′-OH termini (58). Experiments prompted by the recovery of most of the extracted uridine-labeled RNA in the form of 30 S RNP, were performed to determine whether the poly(A) component of hnRNA is likewise contained in 30 S RNP complexes. Extracted poly(A), however, is not recovered associated with 30 S RNP, but rather in a smaller RNP complex sedimenting in the 10–20 S region of sucrose gradients (34,35,59,60). A protein particle of similar size has been found associated with polysomal poly(A) (61). When this poly(A)–containing 10–20 S fraction of sucrose gradients is pelleted, resuspended, and recentrifuged for considerably longer periods, it is resolved into an ultraviolet (UV) absorbing peak at approximately 8 S with another biphasic peak at 15–17 S (Fig. 8). Adenosine-labeled poly(A), as monitored by Millipore filter binding at high salt (62) is seen to be coincident with the UV-absorbing 15 S material (35).

A considerably higher proportion of protein is bound to nuclear poly(A) than is associated with the bulk of nhRNA in 30 S RNP. The buoyant density in CsCl of fixed nuclear poly(A)-RNP is 1.36 gm/cm³ (59,63) as compared to 1.40 for 30 S RNP (see Fig. 9), corresponding to complexes containing 82% and 75% protein respectively, by the formula of Spirin (40).

A.   Proteins Associated with 15 S Poly(A)-RNP

Our preliminary report indicated that the protein composition of purified poly(A)-RNP is more complex than that of purified 30 S RNP (35). Each of

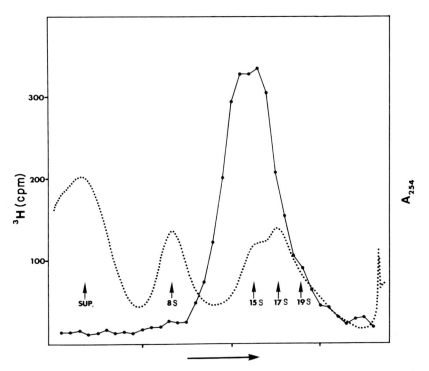

Fig. 8.   Purification of poly(A) RNP on sucrose gradients. Nuclear extracts were made from mouse ascites cells labeled 30 minutes in minimal essential medium with 10 μCi/ml [³H]adenosine as described (35) and centrifuged on gradients as shown in Fig. 1A. The crude 10–20 S region of such gradients was pooled with similar unlabeled material, pelleted for 20 hours at 100,000 g, resuspended in STM pH 8, and recentrifuged on an identical gradient for 24 hours at 27,000 rpm. Then 1-ml fractions were collected and 0.02 ml of each was diluted in high-salt buffer [700 mM KCl–10 mM Tris-HCl (pH 7.6)–1 mM MgCl₂] for binding of poly(A) (35) and passed through HAWP Millipore filters. Radioactive poly(A) was determined by counting the dried filters in 5 ml of liquifluor. Absorption at 254 nm, ·····; [³H]poly(A), ●—●.

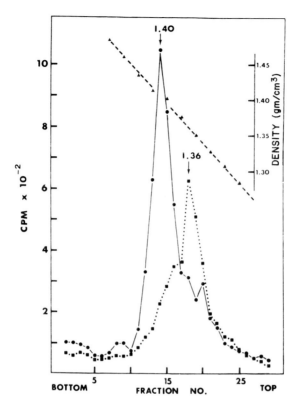

FIG. 9. Comparison of the CsCl buoyant densities of poly(A)-containing RNP and 30 S RNP. [³H]adenosine labeled 10–20 S poly(A) RNP (35) isolated on sucrose gradients containing 0.1 M NaCl–0.01 M sodium phosphate buffer (pH 7.7)–0.001 M MgCl₂ (SPM) were made 0.2 M in NaCl and digested with 20 μg/ml RNase A–10 units/ml RNase T₁ for 1 hour at 37°C. This material was mixed with ³²P-labeled 30 S RNP isolated on SPM gradients and fixed with 6% formaldehyde prior to addition of solid CsCl to an initial density of 1.38–1.39 gm/cm³. The 4.5-ml samples were overlaid with mineral oil and centrifuged at 4°C for 67 hours at 38,000 rpm in the SW-50.1 rotor. Then 15-drop fractions were collected. Buoyant densities were determined by refractometry, and radioactivity was measured by TCA precipitation counting in liquifluor. [³²P]-30 S RNP, ●—●; [³H]poly(A)-RNP, ■····■.

the UV absorbing peaks (Fig. 8) resolved on gradients through which the crude 10–20 S material has been centrifuged has a characteristic protein pattern (Fig. 10). The supernatant fraction of the gradient contains a large number of moderately sized proteins with a broad multicomponent peak of protein centering about 35,000 daltons. The homogeneous peak of material sedimenting at 8 S is composed largely of higher molecular weight polypeptides, many of which are phosphoproteins, with prominent species of mol wt 48,000 and 54,000 daltons. Poly(A) centers over the 15 S peak, which

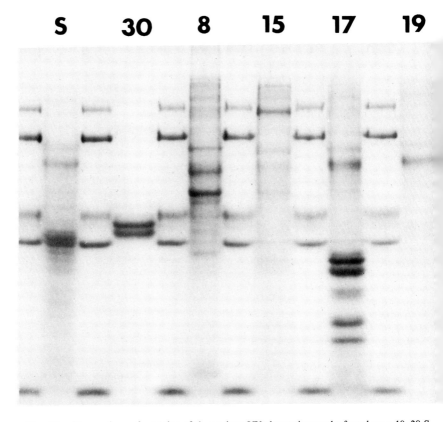

FIG. 10. Comparison of proteins of the various UV absorption peaks found on a 10–20 S poly(A)-RNP purification gradient. The 8 S, 15 S, 17 S, and 19 S regions of a gradient, as shown in Fig. 8, were pelleted for 20 hours at 100,000 g, resuspended in SDS denaturing buffer, and electrophoresed on Laemmli 10% polyacrylamide–SDS slab gels. Precipitation with TCA was used to recover the gradient supernatant proteins (S). The proteins of purified 30 S RNP were electrophoresed on the same gel for comparison. Molecular weight marker proteins in adjacent slots include phosphorylase a (92,500), bovine serum albumin (68,000), ovalbumin (43,000), glyceraldehyde-3-phosphate dehydrogenase (36,000), and globin (15,500).

consists of a major protein of 86,000 daltons with less prevalent poly-peptides of mol wt 140,000, 120,000, 63,000 and 57,500 daltons. The low-molecular-weight proteins recovered in this region appear as contaminants from adjacent 8 and 17 S fractions where such proteins are in greater abundance. The 17 S region is characterized by approximately eight poly-peptides in the molecular weight range 20,000–32,000 daltons. The pro-minent species of mol wt 30,000 and 32,000 daltons comigrate with mouse ascites cell histone Hl (f_1); however, unlike Hl, they are insoluble in 5% perchloric acid. The 63,000 and 57,500 dalton species are found in lower concentrations in the 17 S and 19 S regions of the gradient.

In the cytoplasm, a protein of mol wt 78,000 has been reported associated with the poly(A) sequences of mRNP isolated by dissociation of polysomes in the presence of 0.5 M KCl (47). To determine whether the nuclear counterpart of this protein also remained bound to poly(A) in the presence of high salt, poly(A)-RNP was recentrifuged on gradients containing 0.35–0.5 M NaCl or KCl. The apparent size of poly(A)-RNP on such gradients is considerably reduced; most of the poly(A) is recovered in the 7–10 S region, thus indicating extensive deproteinization by the high salt. Analysis of the proteins from various regions of high salt gradients indicates that the 86,000 dalton protein is among those polypeptides removed from nuclear poly(A) by high salt (data not shown). Since we recover no proteins associated with 15 S nuclear poly(A)-RNP in the molecular weight range 74,000–78,000 daltons, we presume the salt-stable 78,000 dalton protein is confined largely to cytoplasmic polysomal poly(A), replacing those proteins bound in the nucleus. Such an exchange of proteins is perhaps facilitated by the greater affinity of the 78,000 dalton protein for poly(A) as reflected in its salt stability.

B. Oligo(dT)-Cellulose Purification of Nuclear Poly(A)-RNP

Techniques for binding 15S poly(A)-RNP to oligo(dT)-cellulose are based on the method of Lindberg and Sundquist (64). Binding to oligo(dT)-cellulose T-3 (obtained from Collaborative Research, Waltham, Mass.) was performed in the cold in the presence of 0.2–0.5 M NaCl. Approximately 90% of the TCA-precipitable counts from the 15–17 S poly(A)-RNP region of purification gradients was bound to the affinity column (Table II). Contrary to the results obtained by Lindberg and Sundquist (64), for mRNP, considerable poly(A) was eluted from the column by removing the salt in the binding buffer (see Table II) as reported for deproteinized poly(A) (65). Poly(A) bound in 0.2 M NaCl and eluted in the absence of salt with increasing concentrations of formamide is largely removed in 25% formamide; however, approximately one-third is removed only upon raising the formamide to 50%. A single-step elution in 50% formamide was therefore chosen for analysis of proteins associated with poly(A) bound to oligo(dT) in 0.2 M NaCl (Fig. 11A). If the concentration of salt in the binding buffer is maintained in the elution buffer, poly(A)-RNP is bound more tenaciously, being released only in 50% formamide as indicated in Table II.

As shown earlier (Section IV,A), treatment of 15 S poly(A)-RNP with moderately high salt results in considerable loss of protein from the complex

TABLE II

PURIFICATION OF NUCLEAR POLY(A)-RNP BY OLIGO(dT)-CELLULOSE

RNP	Binding[a]		Elution					
				% Counts eluted with formamide at[b]				
	[NaCl] (M)	Counts not bound (%)	[NaCl] (M)	0%	10%	25%	50%	Additional elution
Purified 15 S	0.2	4	0.0	17	40	10	30	—
	0.2	11.2	0	—	—	55.7	33.1	—
	0.2	8.9	0	—	—	—	91.1	—
	0.2	9.0	0.2	—	—	5.8	82.6	2.6[c]
	0.5	6	0	28	38	16	10	2.3[d]
	0.5	10.8	0	—	—	81.7	7.5	—
	0.2	4.4	0.1	—	—	21.0	74.6	—
Crude 10–20 S	0.2	54.9	0.1	—	—	28.6	16.5	—

[a] Poly(A)-RNP isolated in STM pH 8 were passed through Sephadex G-100 equilibrated with 0.1 M NaCl–0.02 M Triethanolamine (pH 7.6)–0.01 M Na$_2$EDTA, adjusted to the final binding concentration of NaCl including 0.2% NP40 and cycled through oligo(dT)-cellulose 3 times before collecting the nonbound fraction. Elution was performed in the same buffer with the indicated concentrations of NaCl and formamide. Portions of recovered fractions were assayed for TCA-insoluble adenosine counts.

[b] Dashes indicate elution at the indicated formamide concentration was omitted.

[c] Additional elution with buffer containing 0.5 M NaCl, 50% formamide.

[d] Additional elution with salt-free buffer containing 90% formamide.

as reflected in the much-reduced sedimentation rate. When bound to oligo (dT)-cellulose in the presence of high salt (0.5 M NaCl), a larger proportion of the poly(A)-RNP may be eluted by removal of binding salt (Table II), and most is eluted before the 50% formamide step. This decreased affinity for oligo(dT)-cellulose upon reduction in the salt content of the buffer is characteristic of deproteinized poly(A), and presumably it reflects a considerable loss of protein by the poly(A)-RNP following exposure to high salt. Proteins bound to the column in 0.5 M NaCl and eluted with the poly(A) in 25% formamide are shown in Fig. 11B. Insufficient protein was detected in the 50% formamide elution for analysis. Proteins bound to the column in 0.5 M NaCl represent a subset of the proteins bound in 0.2 M NaCl of MW 57,500, 63,000, and a very minor amount of 67,000.

Our results on the protein composition on poly(A)-RNP differ considerably from those obtained by Kish and Pederson (66) who report finding only two proteins of MW 74,000 and 86,000 present in a 4:1 ratio associated with

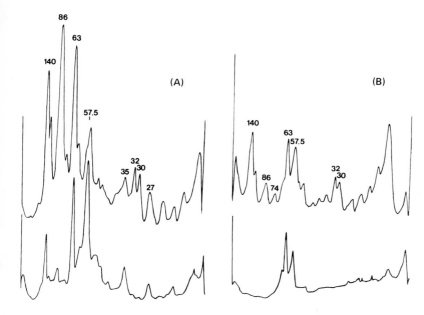

FIG. 11. Polyacrylamide gel analysis of proteins from oligo (dT)-cellulose bownd poly (A)-RNP. The purified 15–17 S poly(A)-containing regions of gradients as in Fig. 8 were desalted on Sephadex G-100 in 0.1 M NaCl–0.02 M triethanolamine (pH 7.6)–0.01 M Na$_2$EDTA and adjusted to (A) 0.2 M or (B) 0.5 M NaCl. NP-40 was added to 0.2% and samples were cycled 3 times through 0.8 gm of oligo(dT)-cellulose. The nonbound fraction was collected, and the column was extensively washed before elution in 0.02 M triethanolamine (pH 7.6)–0.01 M Na$_2$EDTA containing (A) 50% or (B) 25% formamide. Samples were taken for determination of acid-insoluble radioactivity, and the nonbound and peak fractions were dialyzed versus 10 mM ammonium acetate (pH 7), lyophilyzed twice, and electrophoresed on 10% polyacrylamide–SDS slab gels. (A) Binding at 0.2 M NaCl; elution with 50% formamide. (B) Binding at 0.5 M NaCl; elution with 25% formamide. Upper tracings—proteins of nonbound fractions; lower tracings—proteins of formamide-eluted fractions.

purified nuclear poly(A) and oligo(A) sequences. Indirect evidence suggested that the 86,000 dalton protein may be bound to oligo (U) which is associated with nuclear adenylate-rich sequences. Quinlan (63) also found high uridine 3'-monophosphate (UMP) (22%) in the base composition of oligo(dT)-cellulose purified poly(A) from the 10–20 S region of RNP gradients; however, the UMP content of 15 S RNP purified on a second gradient was found to be only 6% indicating the loss of this oligo(U) during the purification of the particle. In this regard it should be noted that greater than 50% of the crude 10–20 S adenosine-labeled poly(A)-RNP does not bind to oligo(dT) (Table II); however, this may reflect a masking of poly(A) or oligo(dT) by additional protein in the crude extracts, and not necessarily by base-paired oligo(U). The 86,000 dalton protein which we believe is poly(A)-associated in low salt is presumably not associated with oligo(U) since purified RNP, which has been shown to contain very little uridylate, was used in the protein analysis. Contrary to the results of Kish and Pederson (66), we find the 86,000 dalton protein to be easily displaced from the particle in high salt, and most is removed even at 0.2 M NaCl. We find essentially no protein of MW 74,000–78,000 in the 15 S region, but we do find a protein of MW 74,000 to be one of the most abundant of the minor 30 S proteins outside the MW range 37,000–40,000 where it may be specifically bound to oligo(A) sequences which are associated with the 30 S RNP complexes (30).

V.   Association of Ribonucleoprotein with Chromatin

A matter of concern to those studying the nonhistone protein (NHP) composition of isolated chromatin fractions from cell nuclei is the contribution made by components not involved in direct interaction with DNA and histones. Ribonucleoproteins of the nucleus can account for a significant fraction of isolated chromatin and therefore should be borne in mind when analyzing such preparations. We would expect, of course, transcriptionally active chromatin to contain attached nascent RNA chains. Electron microscope evidence suggests that nascent RNA exists as RNP fibrils (20); thus the associated proteins will contribute to the spectrum of nonhistone proteins (NHP) in active chromatin fractions. In addition there is the possibility that even after RNA chain termination some hnRNA molecules, presumably as RNP, may remain attached in some way to chromatin (67). The proteins of the 30 S subcomplex which we have described above would be expected to be prominent in these cases since the bulk of rapidly synthesized RNA is associated with these proteins, which may well bind to RNA in the nascent state.

Proteins presumed to be from RNP complexes have been detected in chromatin preparations (32,33,68), and some RNP can be extracted from isolated chromatin using the conditions we have applied to whole nuclei (31). The extent of the contribution and the tenacity of the binding of the 30 S RNP to isolated, "purified" chromatin was impressed upon us during a study of chick retina chromatin nonhistone proteins (69). We have analyzed the NHP components of chick retina nuclei by acrylamide gel electrophoresis (Fig. 12). Our fractionation procedure, a modification of Spelsberg et al. (70), gives relatively little indication of the 37,500 and 41,000 MW species of the avian 30 S RNP in the nucleoplasm fraction (some is perhaps released by a 0.35 M NaCl wash), but these polypeptides appear to make up a con-

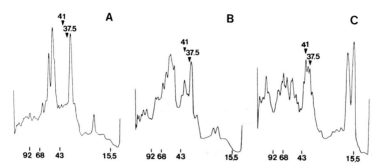

FIG. 12.   Recovery of 30 S RNP proteins in a chick chromatin NHP fraction. Embryonic chick neural retina chromatin was prepared as described (69). Chromatin was washed first in 80 mM NaCl–20 mM EDTA (pH 6.3) (Nucleoplasm fraction) and then with 0.35 M NaCl salt wash fraction). Proteins from these two fractions as well as the remaining nonhistone protein fraction (NHP) of the purified chromatin were analyzed on SDS–polyacrylamide gels (38). (A) Nucleoplasmic proteins; (B) salt wash proteins; (C) NHP. The position of protein standards indicated by the molecular weight in 10³ daltons is shown below.

TABLE III

ASSOCIATION OF RNA- OR PROTEIN-LABELED MOUSE RIBONUCLEOPROTEIN
WITH MOUSE CHROMATIN

Chromatin preparation	Labeled RNP added[a]		Cpm bound/A_{260} chromatin after			
			(1)	(2)	(3)	
	RNP	Cpm	Three low-speed washes	Centrifugation through 1.7 M sucrose	Centrifugation through 1.7 M sucrose, 0.5 M KCl	OR 0.35 M NaCl wash
Dahmus and McConnell (71)	[³H]adenosine 15 S	7,600	380	400	—	300
	[³H]adenosine 30 S	26,000	2860	1220	—	1260
	¹²⁵I-protein 30 S	4,860	680	420	200	—
Spelsberg et al. (70)	[³H]adenosine 15 S	17,230	2240	2100	—	2100
	[³H]adenosine 30 S	35,620	7480	4260	—	3900

[a] Labeled RNP fractions isolated from sucrose density gradients were mixed with lysates (specified by references indicated) of a number of ascites cell nuclei equivalent to that from which the RNP had originally been extracted. The chromatin was then purified from these lysates by the sequential steps (1), (2), and (3); the bound radioactivity was determined after each treatment.

siderable proportion of the NHP of the purified chromatin fraction (Fig. 12C).

As mentioned above some association of nuclear RNP with chromatin is expected on physiological grounds. However, we frequently observed a reduction in the yield of RNP if significant nuclear lysis occurred during the extraction of RNP from nuclei by our standard procedure. This suggested the possibility that RNP could be extensively trapped in the resulting chromatin gel. Direct evidence for the possible artifactual binding of RNP is provided by the preparation of chromatin in the presence of exogenous RNP bearing labeled RNA or protein. When purified adenosine-labeled 15 S or 30 S RNP subcomplexes are added to isolated nuclei prior to lysis and chromatin prepared by either of two rather different procedures, a significant fraction of the RNA is bound and cannot be removed by high-density sucrose and intermediate salt concentration washes (Table III). The binding of the protein components of the 30 S subcomplex to chromatin has been examined in a similar experiment. We have used iodinated 30 S RNP and, as for the RNA, find that a considerable fraction of the labeled protein is retained by the chromatin during "purification" (Table III).

Since the relative concentration of RNP in our model experiments must be several orders of magnitude lower than those naturally present in the nucleus prior to lysis, we would expect that the artifactual trapping of any free nucleoplasmic RNP during chromatin preparation to be greater than levels we have found for exogenous RNP.

ACKNOWLEDGMENTS

We thank Ms. Ljerka Urbas for expert technical assistance. The research described was supported by grants from USPHS National Institutes of Health (Grant CA-12550, and University of Chicago Cancer Research Center Grant CA-14599) and from the Frankel Medical Research Fund. P.B.B. is a trainee supported by USPHS Training Grant HD-00174.

REFERENCES

1. Shearer, R. W., and McCarthy, B. J., *Biochemistry* 6, 283 (1967).
2. Penman, S., Vesco, C., and Penman, M., *J. Mol. Biol.* 34, 49 (1968).
3. Soeiro, R., Vaughan, M. H., Warner, J. R., and Darnell, J. E., Jr., *J. Cell Biol.* 39, 112 (1968).
4. Brandhorst, B. P., and McConkey, E. H., *J. Mol. Biol.* 85, 451 (1974).
5. Wagner, E. K., and Roizman, B., *Proc. Natl. Acad. Sci. U.S.A.* 64, 626 (1969).
6. Wall, R., and Darnell, J. E., *Nature (London) New Biol.* 232, 73 (1971).
7. Williamson, R., Drewienkiewicz, C. E., and Paul, J., *Nature (London) New Biol.* 241, 66 (1973).
8. Bachenheimer, S., and Darnell, J. E., *Proc. Natl. Acad. Sci. U.S.A.* 72, 4445 (1975).
9. Daneholt, B., and Hosick, H., *Proc. Natl. Acad. Sci. U.S.A.* 70, 442 (1973).
10. McKnight, G. S., and Schimke, R. T., *Proc. Natl. Acad. Sci. U.S.A.* 71, 4327 (1974).
11. Lizardi, P. M., *Cell* 7, 239 (1976).
12. Lim, L. and Canellakis, E. S., *Nature (London)* 227, 710. (1970).

13. Edmonds, M., Vaughan, M. H., and Nakazato, H., *Proc. Natl. Acad. Sci. U.S.A.* **68**, 1336 (1971).

14. Burr, H., and Lingrel, J. B., *Nature (London) New Biol.* **233**, 41 (1971).

15. Adesnik, M., and Darnell, J. E., *J. Mol. Biol.* **67**, 397 (1972).

16. Darnell, J. E., Philipson, L., Wall, R., and Adesnik, M., *Science* **174**, 507 (1971).

17. Jelinek, W., Adesnik, M., Salditt, M., Sheiness, D., Wall, R., Molloy, G., Philipson, L., and Darnell, J. E., *J. Mol. Biol.* **75**, 515 (1973).

18. Perry, R. P., Kelley, D. E., Friderici, K. H., and Rottman, F. M., *Cell* **6**, 13 (1975).

19. Salditt-Georgieff, M., Jelinek, W., Darnell, J. E., Furuichi, Y., Morgan, M., and Shatkin, A., *Cell* **7**, 227 (1976).

20. Miller, O. L., Jr., and Bakken, A. H., *Gene Transcription Reprod. Tissue, Trans. Karolinska Symp. Res. Methods Reprod. Endocrinol. 5th, 1972*, p. 155.

21. Stevens, B. J., and Swift, H., *J. Cell Biol.* **31**, 55 (1966).

22. Monneron, A., and Bernhard, W., *J. Ultrastruct. Res.* **27**, 266 (1969).

23. Petrov, P., and Bernhard, W., *J. Ultrastruct. Res.* **35**, 386 (1971).

24. Petrov, P., and Sekeris, C. E., *Exp. Cell Res.* **69**, 393 (1971).

25. Georgiev, G. P., and Samarina, O. P., *Adv. Cell Biol.* **2**, 47 (1971).

26. Samarina, O. P., Krichevskaya, A. A., and Georgiev, G. P., *Nature (London)* **210**, 1319 (1966).

27. Martin, T. E., and McCarthy, B. J., *Biochim. Biophys. Acta* **277**, 354 (1972).

28. Samarina, O. P., Lukanidin, E. M., Molnar, J., and Georgiev, G. P., *J. Mol. Biol.* **33**, 251 (1968).

29. Pederson, T., *J. Mol. Biol.* **83**, 163 (1974).

30. Kinniburgh, A. J., Billings, P. B., Quinlan, T. J., and Martin, T. E., *Prog. Nucleic Acid, Res.* **19** 335 (1976).

31. Kimmel, C. B., Sessions, S. K., and MacLeod, M. C., *J. Mol. Biol.* **102**, 177 (1976).

32. Bhorjee, J. S., and Pederson, T., *Biochemistry* **12**, 2766 (1973).

33. Douvas, A. S., and Bonner, J., *J. Cell Biol.* **63**, 89a (1974).

34. Martin, T., Billings, P., Levey, A., Ozarslan, S., Quinlan, T., Swift, H., and Urbas, L., *Cold Spring Harbor Symp. Quant. Biol.* **38**, 921 (1973).

35. Quinlan, T. J., Billings, P. B., and Martin, T. E., *Proc. Natl. Acad. Sci. U.S.A.* **71**, 2632 (1974).

36. Chauveau, Y., Moule, Y., and Rouiller, C., *Exp. Cell. Res.* **11**, 317 (1956).

37. Taper, H. S., Woolley, G. W., Teller, M. N., and Lardis, M. P., *Cancer Res.* **26**, 143 (1966).

38. Laemmli, U. K., *Nature (London)* **227**, 680 (1970).

39. Spirin, A. S., "Macromolecular Structure of Ribonucleic Acids," p. 190. Reinhold Publ. Co., New York, 1964.

40. Spirin, A. S., *Eur. J. Biochem.* **10**, 20 (1969).

41. Hamilton, M. G., *in* "Methods in Enzymology," Vol. 20: Nucleic Acids and Protein Synthesis, Part C (K. Moldave and L. Grossman, eds.), p. 512. Academic Press, New York, 1971.

42. Kinniburgh, A. J., and Martin, T. E., *Proc. Natl. Acad. Sci. U.S.A.* **73** 2725 (1976).

43. Niessing, J., and Sekeris, C. E. *FEBS Lett.* **18**, 39 (1971).

44. Faiferman, I., Hamilton, M. G. and Pogo, A. O., *Biochim. Biophys. Acta* **232** 685 (1971).

45. Albrecht, C., and Van Zyl, I. M., *Exp. Cell Res.* **76**, 8 (1973).

46. Krichevskaya, A. A., and Georgiev, G. P., *Biochim. Biophys. Acta* **194**, 619 (1969).

47. Blobel, G., *Proc. Natl. Acad. Sci. U.S.A.* **70**, 924 (1973).

47a. Swaney, J. B., Vande Woude, G. F., and Bachrach, H. L., *Anal. Biochem.* **58**, 337 (1974).

48. Leboy, P. S., Cox, E. C., and Flaks, J. G., *Proc. Natl. Acad. Sci. U.S.A.* **52**, 1367 (1964).
49. Gallinaro-Matringe, H., and Jacob, M., *FEBS Lett.* **41**, 339 (1974).
50. Boffa, L. C., Karn, J., Vidali, G., and Allfrey, V. G., *Biochem. Biophys. Res. Commun.*, **74**, 969 (1977).
51. Beyer, A. L., Walker, B. W., Christensen, M. E., and LeStourgeon, W. M., *Cell* **11**, 127 (1977).
52. Christensen, M. E., Beyer, A. L., Walker, B., and LeStourgeon, W. M., *Biochem. Biophys. Res. Commun.* **74**, 621 (1977).
53. Lukanidin, E. M., Zalmanzon, E. S., Komaromi, L., Samarina, O. P., and Georgiev, G. P., *Nature (London) New Biol.* **238**, 193 (1972).
54. Marchalonis, J. J., *Biochem. J.* **113**, 299 (1969).
55. Stevenin, J., and Jacob, M., *Eur. J. Biochem.* **47**, 129 (1974).
56. Lukanidin, E. M., Olsnes, S., and Pihl, A., *Nature (London) New Biol.* **240**, 90 (1972).
57. Kumar, A., and Pederson, T., *J. Mol. Biol.* **96**, 353 (1975).
58. Lewin, B., *Cell* **4**, 11 (1975).
59. Samarina, O. P., Aitkhozhina, N. A., and Besson, J., *Mol. Biol. Rep.* **1**, 193 (1973).
60. Samarina, O. P., Lukanidin, E. M., and Georgiev, G. P., *Acta Endocrinol. (Copenhagen) Suppl.* **180**, 130 (1973).
61. Kwan, S.-W., and Brawerman, G., *Proc. Natl. Acad. Sci. U.S.A.* **69**, 3247 (1972).
62. Lee, S. Y., Mendecki, J., and Brawerman, G., *Proc. Natl. Acad. Sci, U.S.A.* **68**, 1331 (1971).
63. Quinlan, T. J., Ph. D. Thesis, University of Chicago, 1976.
64. Lindberg, U., and Sundquist, B., *J. Mol. Biol.* **86**, 451 (1974).
65. Aviv, H., and Leder, P., *Proc. Natl. Acad. Sci U.S.A.* **69**, 1408 (1972).
66. Kish, V. M., and Pederson, T., *J. Mol. Biol.* **95**, 227 (1975).
67. Herman, R. C., Williams, J. G., and Penman, S., *Cell* **7**, 429 (1976).
68. Douvas, A. S., Harrington, C. A., and Bonner, J., *Proc. Natl. Acad. Sci. U.S.A.* **72**, 3902 (1975).
69. Banks-Schlegel, S. P., Martin, T. E., and Moscona, A. A., *Dev. Biol.* **50**, 1 (1976).
70. Spelsberg, T. C., Steggles, A. W., and O'Malley, B. W., *J. Biol. Chem.* **246**, 4188 (1971).
71. Dahmus, M. E., and McConnell, D. J., *Biochemistry* **8**, 1524 (1969).

Chapter 26

Isolation and Characterization of Ribonucleoprotein Particles Containing Heterogeneous Nuclear RNA

VALERIE M. KISH[1] AND THORU PEDERSON

Worcester Foundation for Experimental Biology,
Shrewsbury, Massachusetts

I. Introduction

Ribonucleoprotein particles containing heterogeneous nuclear RNA (hnRNA) have been identified both cytologically and biochemically in eukaryotic cell nuclei (*1–11*). This chapter describes methods for the isolation of these ribonucleoprotein particles from eukaryotic cells and discusses some of the properties of hnRNP particles isolated from HeLa cell nuclei.

II. Isolation of Nuclear hnRNP Particles

A. HeLa Cells

1. Cell Growth and Labeling Conditions

HeLa cells (S_3 strain) are maintained in suspension culture at $2–4 \times 10^5$ cells/ml by daily dilution with fresh medium (*12*) containing 3.5% each of calf and fetal calf serum. Cells are monitored for *Mycoplasma* at monthly intervals using immunofluorescent and microbiological assays (*13*).

For labeling of RNA, cells are concentrated approximately 10-fold in fresh, prewarmed medium and incubated at 37°C for 5 or 10 minutes with actinomycin D (0.04 μg/ml) prior to labeling (20 or 40 minutes) with either $[2,8-^3H]$adenosine (20 μCi/ml; sp act 35–40 Ci/mmol) or $[5-^3H]$

[1]*Present address*: Department of Biology, Hobart and William Smith Colleges, Geneva, New York.

uridine (20 μCi/ml; sp act 27–30 Ci/mmol). For ^{32}P-labeling, cells are concentrated as above in Joklik–modified Eagle's medium containing one-tenth the usual concentration of phosphate. After a 10-minute preincubation with 0.04 μg of actinomycin D/ml, cells are labeled for 60 minutes with [^{32}P]phosphoric acid (carrier free) at a final concentration of 50–100 μCi/ml. For labeling with [^{3}H]leucine, cells are suspended at 2 $\times$ 10^5 cells/ml in medium containing one-half the standard leucine concentration and incubated for 18–24 hours with 1 μCi L-[4,5-^{3}H]leucine (sp act 47 Ci/mmol) per milliliter. For [^{35}S]methionine labeling, cells are resuspended at 2 $\times$ 10^5 cells/ml in fresh medium containing one-half the standard methionine concentration. Labeling is for 48 hours with 2 μCi/ml [^{35}S]methionine (sp act 200–400 Ci/mmol). For experiments requiring label in both RNA and protein, cells are labeled with [^{14}C]leucine (0.01–0.05 μCi/ml; sp act 312 mCi/mmol) or [^{35}S]methionine for 24–48 hours in the appropriate medium. Cells are then resuspended in fresh complete medium for pulse labeling with a tritium-labeled RNA precursor. Growth of cells under all these conditions is normal.

2. Cell Fractionation and Isolation of hnRNP Particles

The method developed in this laboratory for isolation of HeLa hnRNP particles involves the isolation of nuclei followed by their mechanical disruption and purification of hnRNP particles by sucrose gradient centrifugation. All procedures are carried out at 4°C.

a. Isolation of nuclei. In pulse-labeling experiments, cells are diluted with 5–10 volumes of ice-cold balanced salt solution (*14*) and centrifuged at 600 *g* for 5 minutes in 600-ml glass conical centrifuge bottles (Bellco Glass Company, Vineland, New Jersey). In long-term labeling situations, cells are harvested directly from culture as above, without dilution with cold balanced salt solution. Cell pellets are washed twice with 10 volumes of cold Earle's salts (3 minutes, 1000 *g*), and the volume of packed cells is measured. A packed cell volume of 1 ml is equivalent to 2.2 $\times$ 10^8 HeLa cells. Cells are suspended in 10 volumes of 10 m*M* Tris-HCl (pH 7.2)–10 m*M* NaCl–1.5 m*M* MgCl$_2$ (RSB) and after 20 minutes are broken in a Dounce homogenizer (clearance = 0.0015 in.) using 5–10 strokes. This results in breakage of >95% of the cells as monitored by phase-contrast microscopy. Nuclei are pelleted (3 minutes, 1000 *g*) and washed 3 times using 10 volumes of RSB each time.

b. Nuclear disruption by mild sonication and isolation of hnRNP particles. When nuclei in hypotonic or isotonic buffers are disrupted by sonication, the nucleoli can be selectively removed by low-speed centrifugation (*15–17*). The hnRNP particles are isolated from the remaining fraction by sucrose gradient centrifugation (*9*).

Washed nuclei are suspended in RSB at a concentration of $\leq 4 \times 10^7$/ml, and the suspension is transferred to a polyallomer tube (Beckman Spinco) packed in a beaker of ice water. Nuclei are disrupted by brief sonication (two 5–8 second bursts at 40 W; Model W185, Heat Systems Ultrasonics, Plainview, New York) using the standard microtip. Prolonged sonication is not necessary and should be avoided. Over the course of several experiments (10–15 minutes of total sonication time) the probe surface becomes pitted, which can lead to uneven energy dissipation and erratic performance. The probe tip can be reground 3 or 4 times before replacement is necessary. Nuclear breakage is >98% as determined by phase-contrast microscopy. Up to 7 ml of the sonicated suspension are layered on 30 ml of 30% (w/v) sucrose in RSB (autoclaved before use) and centrifuged for 15 minutes at 5000 rpm (4500 g) in the Beckman SW27 rotor. The pellet contains nucleoli and any unbroken nuclei. The opalescent material on top of the sucrose (postnucleolar supernatant) which contains chromatin, hnRNP particles, nuclear membrane fragments, and soluble macromolecules is collected and layered on sucrose gradients. Two procedures have been developed for sucrose gradient purification of hnRNP particles.

Method 1: Four to five ml of the postnucleolar supernatant are layered on 33-ml 15–30% linear sucrose gradients (RSB) and centrifuged for 17 hours at 15,000 rpm in the Beckman SW27 rotor. The distribution of hnRNP particles in the gradient is monitored by assay of acid-insoluble radio-activity in each gradient fraction (9). Figure 1 shows a typical sucrose gradient profile of [³H]uridine-labeled HeLa hnRNP particles.

Method 2: This method involves discontinuous sucrose gradients and was devised to facilitate the rapid preparation of large quantities of hnRNP particles (18). The gradients are composed of 25 ml of 60% sucrose, 2 ml of 45% sucrose, and 8.0 ml of 10% sucrose (all sucrose solutions made up in RSB and autoclaved). Four to five ml samples are layered, and the gradients are centrifuged in the SW27 rotor for 90 minutes at 26,000 rpm. The hnRNP particles prepared in this manner are recovered as a narrow opalescent band spanning the 45% sucrose layer. This technique offers the distinct advantage of concentrating the particles, thereby minimizing subsequent dialysis and ultrafiltration procedures, in addition to substantially reducing the time required for their isolation. The hnRNP particles isolated by these two methods are similar in that they contain high-molecular-weight hnRNA (Section III,C,1) and possess a similar distribution of polypeptides (Section III,D,1). However, we have noted a small amount of contaminating chromatin in particles isolated on discontinuous sucrose gradients.

Three procedures are used for concentrating hnRNP particles collected from preparative sucrose gradients: (1) precipitation of particles in 67% ethanol (at least 2 hours, at −20°C) followed by centrifugation (20 minutes,

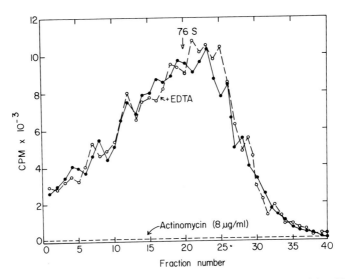

FIG. 1. Sucrose density gradient centrifugation of HeLa hnRNP particles. HeLa cells were incubated with actinomycin D (0.04 μg/ml) and two-thirds of the suspension was labeled for 10 minutes with [³H]uridine. The remaining one-third was first incubated 5 minutes with 8.0 μg of actinomycin D/ml and then labeled similarly. Nuclei were isolated and one-half of the postnucleolar supernatant (from low-actinomycin cells) was made 10 mM in Na₂EDTA before centrifugation in a 15–30% sucrose gradient [w/v in 10 mM Tris-HCl (pH 7.0)–10 mM Na₂EDTA–10 mM NaCl]. The remainder was centrifuged on a similar gradient prepared in RSB. Centrifugation was in the Beckman SW27 rotor for 17 hours at 15,000 rpm (3°C). Gradients were collected in 1.0-ml fractions and assayed for trichloroacetic acid (TCA)-precipitable radioactivity. The position of 76 S single ribosomes is indicated by the arrow (determined by centrifuging a portion of the cytoplasmic fraction, made 0.5% in sodium deoxycholate, in a fourth gradient and monitoring the absorbance at 260 nm during gradient fractionation). ●——●, control gradient; ○---○, + EDTA; ---, material from cells treated with 8 μg of actinomycin D/ml. Reprinted from Pederson (9) with permission of the publisher.

12,000 g) (11); (2) dilution of particles to a sucrose concentration of ≤10% followed by high-speed centrifugation (6 hours, 368,000 g) (9,19); (3) dialysis of particles to remove sucrose followed by ultrafiltration (Amicon PM-10 membrane, 30 lb/in.² N₂ pressure, 4°C) (18).

c. Tests for specificity. HeLa hnRNP particles exhibit a heterogenous size distribution, from 40 to 250 S in 15–30% linear sucrose gradients containing 0.01 M or 0.15 M NaCl (Fig. 1). More than 95% of the chromatin is pelleted under these conditions. The exposure of HeLa cells to a high concentration of actinomycin D (8 μg/ml), which inhibits total RNA synthesis by more than 98%, results in no detectable incorporation of [³H]uridine into the 40 to 250 S material (Fig. 1), indicating that the isotopic precursor is incorporated into RNA. In addition, the hnRNP particles are not dissociated by treatment with 0.01 M ethylenediaminetetra-acetic acid (EDTA), thereby ruling out the possibility that they represent contaminating polyribosomes (Fig. 1). Although it is difficult to unequivo-cally establish that the hnRNP particles are authentic components of the cell nucleus, several lines of evidence strongly suggest that these structures are not artifactually derived. HeLa hnRNP particles prepared under conditions of low ionic strength, which favor the nonspecific association of RNA and protein, have the same sedimentation behavior in sucrose gradients, as well as the same complement of polypeptides, as hnRNP particles prepared in isotonic buffers (9). Moreover, mixing of radio-actively labeled deproteinized HeLa hnRNA with unlabeled intact or sonicated nuclei does not result in the binding of nuclear proteins to the naked hnRNA molecules, as judged by the absence of any shift in sedimen-tation velocity of the exogenous radioactive hnRNA (9). These experiments were done at a vast excess of nuclear protein relative to the added radio-active RNA (about 500,000:1), thus favoring the binding of protein to the tracer RNA. This type of control has also been performed, with similar results, with nuclear hnRNP particles from the cellular slime mold *Dictyo-stelium discoideum* (11) and sea urchin embryos (20). These results support the conclusion that hnRNP particles do not reflect the nonspecific associa-tion of nuclear proteins with hnRNA during cell fractionation.

B. Rat Liver

To label hnRNA, adult female white rats (Charles River) are injected intraperitoneally (i.p.) with 250 μCi [5-^{3}H]orotic acid (sp act 12.6 Ci/mmol). After 45 minutes the animals are sacrificed and the livers are excised and rinsed in cold 10 mM Tris-HCl (pH 7.0)–3 mM CaCl$_2$–2.2 M sucrose. After mincing, the livers are homogenized in a loose-fitting Dounce homo-genizer and nuclei are isolated as described by Ishikawa *et al.* (21). The nuclear pellet is washed 3 times in 50 mM Tris-HCl (pH 7.0)–25 mM KCl–5 mM MgCl$_2$–0.88 M sucrose and twice in 10 mM Tris-HCl (pH 7.0)–0.15 M NaCl–1.5 mM MgCl$_2$. Nuclei are resuspended in the latter buffer at 10^7/ml and disrupted by sonication. The hnRNP particles are prepared as described for HeLa cells, except that isotonic buffers are used through-out (10).

C. *Dictyostelium discoideum*

1. Cell Culture and Labeling Conditions

Cells (axenic strain A-3, Loomis) are grown at 2×10^4–2×10^6 cells/ml in 2-(N-morpholino)ethane sulfonic acid (MES)-HL-5 medium on a rotary shaker as described by Jacobson (22). The cell generation time is 9–11 hours (22°C). Efficient and highly selective long-term labeling of RNA can be achieved by overnight incubation of log phase cultures at 22°C with [5-³H]uridine (3 μCi/ml; sp act 25–30 Ci/mmol). Under these conditions, >97% of the acid-insoluble [³H]uridine radioactivity is sensitive to ribonuclease A digestion at low ionic strength. Pulse labeling of RNA is most efficient with ³²PO₄ (HCl-free) and can be carried out in 10-fold concentrated cultures (22). Labeling of proteins with amino acid precursors is generally unsuccessful, due to the large quantity of peptides in the growth medium. We have found that *Dictyostelium* vegetative amoebae can be labeled with amino acids using the following procedure. Cells are resuspended at 5 × 10⁶ cells/ml in medium consisting of 1% (v/v) glucose–6 mM 2–(N-morpholino-ethanesulfonic) acid (MES) (pH 6.5). The desired [¹⁴C]-labeled amino acid (0.5–1.0 μCi/ml) is added for 1–2 hours, and the cells are then resuspended in regular growth medium at 2.5 × 10⁶ cells/ml for the desired period of time. For example, the specific activity of total cytoplasmic protein is about 75,000 dpm/mg when [¹⁴C]leucine, arginine, and lysine (each having a specific activity of 312 mCi/mmol) are used in combination at 0.33 μCi/ml (for each amino acid) for 1 hour, followed by 4 hours in growth medium.

2. Cell Fractionation and Isolation of hnRNP Particles

a. Isolation of nuclei. At the end of the labeling period, cells are poured into 4 volumes of cold 0.2% NaCl and centrifuged 5 minutes at 600 g. The pellet is washed once more with cold NaCl, and the washed cells are suspended at 5 × 10⁸ cells/ml in cold 50 mM N–2–hydroxyethylpiperazine–N′–2–ethanesulfonic acid (HEPES) (pH 7.2)–10 mM KCl–50 mM Mg acetate–10% sucrose–0.5% Triton X-100 (cell lysis buffer). The suspension is vortexed for 30 seconds, and the cells are then broken in a Dounce homogenizer (clearance = 0.0010 in.) using three strokes of the pestle. After centrifugation at 1000 rpm (Sorvall SS–34 rotor) for 4 minutes, only a very small pellet should be visible. (A large pellet is indicative of ineffective cell lysis and can be resuspended in the supernatant and disrupted by additional vortexing and homogenization.) The supernatant is centrifuged 5 minutes at 4000 rpm to produce a crude nuclear pellet. The pellet is washed repeatedly in cold lysis buffer lacking Triton X-100 until the supernatant is clear. The pellet of clean nuclei is suspended in 10 mM Tris-HCl (pH 7.5)–0.15 M NaCl–1.5 mM MgCl₂ (TNM) at 5 × 10⁸ cells/ml prior to nuclear disruption.

b. Purification of hnRNP particles. Nuclei are sonicated for no more than 5 seconds, as described for HeLa cells (Section II,A,2,b), and the sonicate is layered on 30% sucrose (in TNM buffer). After centrifugation for 20 minutes at 5000 rpm in the Beckman SW27 rotor, 2 ml of the resulting postnucleolar supernatant are layered on an 11–ml 15–30% linear sucrose gradient (in TNM buffer) and centrifuged in the SW41 rotor at 30,000 rpm for 2 hours. The distribution of labeled hnRNP particles is determined as described for HeLa cells. *Dictyostelium* hnRNP particles display a monodisperse gradient profile with a modal sedimentation coefficient of 55 S (*11*). Chromatin is quantitatively pelleted in these gradients.

D.   Evaluation of Alternative Methods for Obtaining hnRNP Particles from Nuclei

1.   NUCLEAR LYSIS BY HIGH-SALT BUFFER

In 1966, Penman (*23*), introduced a method of disrupting HeLa cell nuclei which permitted the high recovery of heterogeneous nuclear RNA, but in the presence of large amounts of nucleohistone. This procedure involves the lysis of nuclei in high-salt buffer (0.5–0.8 *M* NaCl) followed by DNase digestion to reduce the viscosity of the nucleohistone, which forms a gel at high ionic strength. The hnRNP particles derived from nuclei lysed in this manner are aggregated and extensively contaminated with histones and other non-hnRNP-related proteins (*9*). This method is therefore not suitable for preparation of hnRNP particles.

2.   EXTRACTION OF hnRNP PARTICLES FROM INTACT NUCLEI

The extraction of ribonucleoprotein particles containing hnRNA from intact nuclei by repeated washings in cold isotonic buffer at pH 8 (*23a*) has been employed routinely by Georgiev and co-workers, using a variety of cell types such as rat liver (*5*), mouse liver (*24*), and Ehrlich ascites cells (*23a*). This procedure does not release hnRNP particles from intact nuclei of most cultured cells unless the temperature of extraction is at least 20°C (*9,25,26*). The hnRNP particles prepared by this method are highly degraded (30–45 S) (*27*).

III.   Physicochemical Characterization of HeLa hnRNP Particles

A.   Size

Particles recovered from various regions of the preparative sucrose gradient can be visualized with the electron microscope after fixation with 2.5% (v/v) glutaraldehyde and negative staining in 1.0% (w/v) uranyl

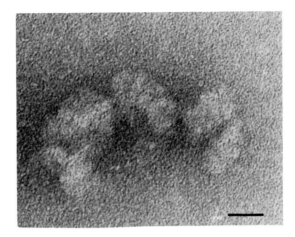

FIG. 2. Electron micrograph of a typical HeLa hnRNP particle. The hnRNP sedimenting at 250 S was recovered from the appropriate gradient fraction and applied to 400-mesh copper grids with a carbon film support. The grids were rinsed 3 times in 10 mM Tris-HCl (pH 7.0)– 10 mM NaCl–1.5 mM MgCl$_2$ (RSB) to remove sucrose, fixed for 5 minutes in 2.5% (v/v) glutaraldehyde [in 10 mM sodium phosphate (pH 7.0)], rinsed 3 times in RSB, stained for 2 minutes with 2% (w/v) uranyl acetate in water, and then rinsed 3 times in RSB and air-dried. Grids were rotary shadowed with platinum-palladium (80%:20%, #29 wire, 0.008 in. diameter, Pella Company, Tustin, California) at 60 rpm at an angle of 10°. Grids were examined in a Phillips EM 300 electron microscope and photographed at a magnification of ×49,000. The bar in the photograph represents 200 Å.

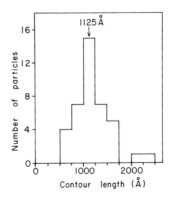

FIG. 3. Histogram of 200 S HeLa hnRNP particle contour length. The hnRNP sedimenting at 200 S was recovered from sucrose gradients and prepared for electron microscopy as described in Fig. 1. Forty individual particles were photographed at a magnification of × 49,000 and their contour lengths measured.

acetate. The negatively stained hnRNP particles appear as linear arrays of globular subunits as shown in Fig. 2. The overall length of the particles roughly correlates with their position in the sucrose gradient. The 200–250 S hnRNP particles (as in Fig. 2) have a number average contour length of about 1125 Å (Fig. 3) and contain RNA molecules of approximately 4×10^6 molecular weight. These RNA molecules have a (single-stranded) contour length of 3.16 μm,[2] so that the RNA:RNP packing ratio is about 28 (3.16 μm/1125 Å). There is a good correlation between the size of the hnRNP and the molecular weight of its deproteinized RNA [$S_{RNP} = 2.2\ S_{RNA}$, where S_{RNP} is the hnRNP sedimentation coefficient in sucrose gradients containing 0.15 M NaCl and S_{RNA} is that of the deproteinized RNA in gradients containing 0.5% sodium dodecyl sulfate (SDS)]. This suggests that the observed particles are not aggregates of smaller structures.

The subunit organization of hnRNP, first suggested by Georgiev and co-workers (5), provides a useful diagnostic measure of putative RNP isolated from novel sources. A limited digestion with ribonuclease A at low ionic strength (e.g., 0.1 μg RNase A/ml for 15 minutes at 4°C) should quantitatively convert the particles to 225 Å subunits. This property establishes that the structure of native hnRNP is based upon continuity of the RNA component, as opposed to strong intersubunit protein:protein interactions. The conversion of hnRNP to subunits can also be examined by velocity sedimentation in sucrose gradients. Samples are digested with RNase A as above, loaded on 15–30% sucrose gradients, and the centrifuge started to terminate the RNase reaction. With increasing concentration of RNase A, there is a sudden and quantitative conversion of the usual 40–250 S hnRNP particles to particles sedimenting at 45 S. At higher enzyme concentrations, there is a corresponding decrease in the amount of acid-insoluble radioactivity sedimenting at 45 S, rather than a further decrease in the sedimentation rate of the acid-insoluble material (9).

B. Equilibrium Density Gradient Analysis in Cesium Salts

A useful way of determining the ribonucleoprotein nature of a given structure is to examine its behavior by equilibrium density gradient centrifugation. The association of RNA and protein can be stabilized by fixation with 6% glutaraldehyde. Particles are first dialyzed to remove sucrose and to reduce the salt concentration to <0.15 M. One volume of 25% glutaraldehyde is neutralized to pH 7 by the addition of 0.05 volumes of 1 M NaHCO$_3$. One volume of this solution is added to 3 volumes of

---

[2] Based upon calibration of Sindbis virus RNA (glyoxal-treated, spread in formamide) where 4.3×10^6 daltons = 3.4 μm. From Hsu et al. (28).

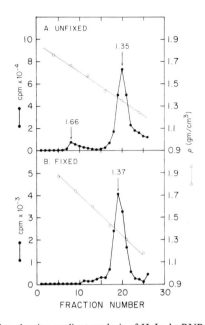

FIG. 4.   Cesium sulfate density gradient analysis of HeLa hnRNP particles with and without aldehyde stabilization. [³H]uridine pulse-labeled hnRNP particles were isolated as described in the text. (A) Unfixed: hnRNP was dialyzed 18 hours against 10 mM Tris-HCl (pH 7.2)–10 mM NaCl (4°C) and layered on preformed Cs₂SO₄ gradients having an initial density range of 1.25–1.75 gm/cm³. Centrifugation was in the Beckman SW50.1 rotor at 34,000 rpm for 64 hours (20°C). (B) High-salt-treated hnRNP plus glutaraldehyde fixation: hnRNP was isolated as above and dialyzed against 10 mM Tris-HCl (pH 7.5)–10 mM Na₂EDTA–0.5 M NaCl (4°C). The sample was fixed with glutaraldehyde (see text for details) and layered on a performed Cs₂SO₄ gradient formulated as above and centrifuged in the SW50.1 rotor at 43,000 rpm for 45 hours (20°C). In both panels, the direction of centrifugation is from right to left. The density of every fourth fraction was determined by weighing a 10-μl aliquot. Each fraction was mixed with 1 ml of water, and 500 μg of yeast transfer RNA was added as carrier. Samples were made 10% in trichloroacetic acid (TCA), and the precipitates were collected on glass-fiber filter discs (Whatman GF/C) which were rinsed with 95% ethanol, dried, and counted in a toluene-based scintillation cocktail. ●——●, cpm; ○——○, density.

hnRNP at 4°C. The hnRNP sample (0.5 ml) is layered on a preformed 4.8-ml linear gradient of Cs₂SO₄ having an initial density range of 1.25–1.75 gm/cm³. Unfixed [³H]uridine-labeled hnRNP particles (Fig. 4 A) band as a symmetric peak at a modal density of 1.35 gm/cm³, with a smaller peak at 1.66 gm/cm³, which is the density of protein-free single-stranded RNA. The 1.66 gm/cm³ peak accounts for approximately 6% of the total [³H]uridine radioactivity recovered from the gradient. Presumably this material results from the release of protein from some hnRNP particles in the presence of cesium sulfate. This conclusion is supported by the absence of a peak of free RNA in gradients of glutaraldehyde-fixed hnRNP particles. In this case, all of the radioactivity bands as RNP at a density of 1.37 gm/cm³ (Fig. 4B). The resistance of the majority of hnRNP to dissociation in Cs_2SO_4 gradients, in the absence of aldehyde fixation, is a useful diagnostic property of this material, particularly in relation to possible contamination by ribosomes, which are unstable in these gradients [see also Wilt et al. (20)]. The stability of hnRNP particles in cesium sulfate is also shared by polyribosomal messenger ribonucleoprotein particles (29).

C. Properties of hnRNP-derived hnRNA

1. Size

The hnRNP particles isolated from HeLa cells by mechanical disruption of nuclei contain approximately 95% of the total nuclear hnRNA (9). The hnRNP is deproteinized in the presence of 0.5% sodium dodecyl sulfate (SDS) using three extractions with phenol at 22°C according to the method described by Penman (23), except that the phenol is saturated with 50 mM Tris-HCl (pH 9.0) and is mixed with chloroform containing 1% isoamyl alcohol immediately before use. The final aqueous phase is made 0.1 M in NaCl, and yeast tRNA (100–250 μg/ml) is added prior to precipitation of the RNA with 2.5 volumes of 95% ethanol ($-20°C \geq 4$ hours). The hnRNA pellet recovered after centrifugation (37,000 g, 45 minutes) is dissolved in the appropriate buffer for further analysis.

For sucrose gradient analysis, the hnRNA is dissolved in 10 mM Tris-HCl (pH 7.0)–10 mM Na_2EDTA–0.1 M NaCl–1% SDS and layered on 15–30% linear sucrose gradients prepared in the same buffer containing 0.5% SDS. Centrifugation is for 16 hours at 19,000 rpm in the Beckman SW41 rotor (20°C). Using this technique, the hnRNA has a heterogeneous sedimentation ranging from 20–60 S (9).

2. Base Composition

HeLa cells are labeled with ³²P (Section II,A,1), and the isolated hnRNP particles are deproteinized by phenol–chloroform extraction (Section III, C,1). After ethanol precipitation, the RNA is hydrolyzed 18 hours at 37°C in 0.3 N NaOH or KOH or for 1 hour at 100°C in 1 N NaOH or KOH. Following adjustment of the samples to pH 2 with cold 50% perchloric acid and centrifugation to remove the precipitated potassium perchlorate, the supernatant is adjusted to pH 7.0 and concentrated by evaporation to approximately 10 μl. After chilling and low-speed centrifugation to remove precipitated salt, the ³²P-labeled sample is spotted on thin-layer cellulose plates (Eastman 13254 with fluorescent indicator No. 6065), with 15 μg of each of the four $2',3'$-ribonucleoside monophosphates as standards. Development is in two dimensions as described by Monckton and Naora (30). After drying the chromatogram overnight at $20\,°C$, ultraviolet (UV)-absorbing spots are cut out and counted in toluene scintillation fluid, or by Cerenkov radiation.

3. SPECIFIC NUCLEOTIDE SEQUENCES

a. Polyadenylate sequences. Some, but not all, HeLa cell hnRNA molecules contain a sequence of polyadenylic acid 150–200 nucleotides long, added posttranscriptionally to the $3'$-OH termini (31–33). Shorter poly (A) sequences, located internally and estimated to be 20–30 nucleotides in length ("oligo(A)") have also been detected (34). The ability to isolate and study these specific nucleotide sequences is based on their inherent resistance to digestion by ribonuclease A at high ionic strength (35). The hnRNA derived from [^{3}H]adenosine-labeled HeLa hnRNP particles is ethanol-precipitated, dialyzed against 10 mM Tris-HCl (pH 7.5)–10 mM Na$_2$ EDTA–0.5 M NaCl (TEN), and digested with RNases A (10 μg/ml) and T$_1$ [40 units (approximately 0.08 μg/ml] for 75 minutes at $37°C$. [RNase A is prepared as a 1 mg/ml stock solution in 10 mM Tris-HCl (pH 8.0)–10 mM NaCl. The solution is heated 20 minutes at $90°C$ and is stored at $-20°C$. RNase T$_1$ is diluted to approximately 1900 units/ml with TEN buffer immediately before use.] The digest is loaded on a 0.7×1.0 cm column of oligo(dT)-cellulose (Type T-3) previously equilibrated at 20–$24°C$ with TEN buffer. When the sample is enterely loaded, the column is washed with approximately 15 ml of TEN buffer to remove all unbound RNA (as monitored by acid-insoluble radioactivity in column fractions). Elution is at 10–13 ml/hour. The bound RNA is eluted from the column with 10 mM Tris-HCl (pH 7.5)–10 mM Na$_2$EDTA. The RNA is made 0.1 M in NaCl and is collected by ethanol precipitation. This ribonuclease-resistant, oligo(dT)-cellulose-purified fraction can then be analyzed by electrophoresis in polyacrylamide gels containing 99% formamide. Ten percent polyacrylamide cylindrical gels (0.6×12.0 cm) are prepared essentially as detailed by Duesberg and Vogt (36). First 100 ml of formamide is deionized by stirring with 5% (w/v) mixed bed resin [AG501-X8 (D)] and buffered to 20 mM sodium phosphate (pH 7.0) by adding 142 mg of Na$_2$HPO$_4$ and 138 mg of NaH$_2$PO$_4$. Overnight stirring ($22°C$) is generally necessary in order to dissolve the salts. The buffered formamide is stored at $-20°C$. To prepare 12 gels, 6.0 gm of acrylamide and 1.05 gm of N,N'-methylenebisacrylamide are dissolved in 59.25 ml of buffered formamide, and the solution is deionized by stirring for 20 minutes with approximately 0.5 gm of mixed bed resin. The beads are removed by low-speed centrifugation, and the supernatant is stirred with fresh resin for an additional 20 minutes. The conductivity of the resulting deionized solution should be less than

10 mμ mho. To 60 ml of the deionized acrylamide solution, add 0.12 ml of N,N,N',N'-tetramethylenediamine (TEMED) and 0.75 ml of a freshly prepared solution of 10% ammonium persulfate made up in water. The gels are poured to a height of 12 cm, and the surface is overlaid with water. Gels are generally prepared the day before use; however, they can be stored under buffered formamide at 4°C for 1–2 weeks (22). Prior to electrophoresis, gels are capped at the bottom with dialysis tubing and a 4-cm column of buffered formamide is pipetted on top of the gel bed. The RNA sample (either a lyophilized powder or an ethanol pellet) is prepared for electrophoresis using either of two methods: (1) RNA is dissolved in 25 μl of solution X and then 15 μl of solution Y is added; (2) RNA is dissolved in 5–10 μl of 40 mM sodium phosphate (pH 7.0) and then 50 μl of solution Z is added. Solution X: 99% formamide containing 2 mM sodium phosphate buffer; prepared by adding 9 volumes of deionized formamide to 1 volume of formamide buffered with 20 mM phosphate. Solution Y: 1 volume of deionized formamide plus 1 volume of glycerol plus 0.01% Bromophenol blue. Solution Z: 3 volumes of solution X plus 2 volumes of solution Y.) The sample is layered on top of the gel (through the formamide) and upper and lower reservoirs are filled with 40 mM sodium phosphate, pH 7.0. Electro-

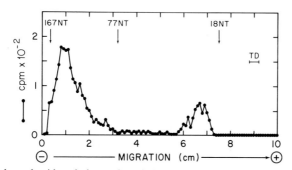

FIG. 5. Polyacrylamide gel electrophoresis in 99% formamide of ribonuclease-resistant, oligo(dT)-cellulose-purified RNA derived from HeLa hnRNP particles. [³H]adenosine pulse-labeled hnRNP particles were isolated and dialyzed 18 hours against 10 mM Tris-HCl (pH 7.5)–10 mM Na$_2$EDTA–0.5 M NaCl (TEN) (4°C). After concentration by ultrafiltration and ribonuclease digestion (Section III,C,3), the digest was extracted with phenol–chloroform, and the ribonuclease-resistant RNA was collected by ethanol precipitation. After dialysis against TEN buffer (4°C) and a second ribonuclease digestion, the sample was applied to an oligo (dT)-cellulose column as described in the text. The RNA bound to the column was eluted with 10 mM Tris-HCl (pH 7.5)–10 mM Na$_2$EDTA. Pooled fractions were made 0.1 M in NaCl and ethanol-precipitated. This RNA fraction was then analyzed in a 0.6 × 12 cm 10% polyacrylamide gel containing 99% formamide (see text for details). Electrophoresis was for 5.8 hours at 100 V. The arrows indicate the positions of RNA markers run in a parallel gel: ¹⁴C-poly(A), 167NT; yeast transfer RNA, 77NT; oligo(rA), 18NT. |—| internal Bromophenol blue dye marker. Direction of migration is from left to right.

phoresis is at 100 V for 6–7 hours, at which time the Bromophenol blue tracking dye has migrated 9–10 cm into the gel. The gel is frozen in dry ice and sliced using a device consisting of multiple, evenly spaced razor blades (Hoefler Company, San Francisco, California). Each 1 mm segment is shaken at 37°C overnight in a scintillation vial containing 5 ml of toluene-based scintillation fluid containing 3% Protosol prior to liquid scintillation counting. A typical profile of [³H]adenosine-labeled, ribonuclease-resistant, oligo (dT)-cellulose-purified RNA is shown in Fig. 5. There are two size classes of RNA present in this fraction, one with a mean length of 148 nucleotides and the second corresponding to a sequence length of 25–30 nucleotides. These adenylate-rich sequences are similar in size to poly(A) and oligo(A) sequences previously isolated from HeLa cell hnRNA (*31–34*).

   b. Double-stranded sequences. Approximately 2–5% of [³H]uridine-

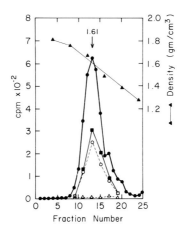

FIG. 6. Cesium sulfate density gradient analysis of double-stranded RNA isolated from HeLa hnRNP particles. [³H]uridine pulse-labeled hnRNP particles were prepared and dialyzed against TEN buffer as detailed in Fig. 4. Following digestion for 75 minutes (37°C) with RNases A and T_1 (Section III,C,3), the digest was made 12.5 mM in $MgCl_2$ and incubated with iodoacetamide-treated DNase I, as described by Zimmerman and Sandeen (*37*), in order to remove any contaminating chromatin DNA. Ribonuclease-resistant RNA was chemically deproteinized and collected by ethanol precipitation. The precipitate was dissolved in TEN buffer, layered on a preformed Cs_2SO_4 gradient, and centrifuged at 34,000 rpm for 60 hours in the SW50.1 rotor (20°C). The acid-insoluble [³H]uridine-labeled RNA appears as a monodisperse peak, with a modal density of 1.61 gm/cm³. ●——●: This density is characteristic of double-stranded RNA. In order to confirm the double-stranded character of the RNA, aliquots of alternate fractions (9–19) were digested with RNases A and T_1 for 30 minutes (37°C) in low- or high-salt buffers (see text for details). ■——■: control (no RNase treatment); O---O: RNase in high salt; ▲---▲: RNase in low salt. The loss of acid-insoluble radioactivity across the peak after RNase digestion in low salt, and the resistance of the RNA to digestion in high salt, is further evidence for the double-stranded nature of this RNA species. These data were obtained by Dr. James P. Calvet.

labeled hnRNA in hnRNP particles is resistant to hydrolysis by RNAase A (10 μg/ml) and T$_1$ (40 units/ml) after 60–75 minutes at 37 °C in 10 mM Tris-HCl (pH 7.5)–10 mM Na$_2$EDTA–0.5 M NaCl. Following phenol–chloroform deproteinization, the ribonuclease-resistant RNA is layered on Cs$_2$SO$_4$ gradients (Section III,B) and centrifuged in the Beckman SW50.1 rotor for 60–66 hours at 34,000 rpm (20°C). A single peak of acid-insoluble radioactivity is present with a modal density characteristic of double-stranded RNA (1.61 gm/cm^3; Fig. 6). The absence of acid-insoluble radioactivity at the density characteristic of single-stranded RNA (1.65–1.68 gm/cm^3), coupled with the high recovery of radioactivity from the gradient (90–95%), indicate that there is no significant fraction of protein-protected single-stranded RNA in the ribonuclease-treated hnRNP particles, since such material should appear at a density of 1.65–1.68 gm/cm^3 after deproteinization. One can verify the double-stranded nature of this RNA by analyzing its susceptibility to hydrolysis by RNases A and T$_1$ at high and low ionic strength. Alternate fractions from 9–19 (Fig. 6) were dialyzed separately against 10 mM Tris-HCl (pH 7.2)–10 mM NaCl (4°C), and each was divided into three equal aliquots. A control aliquot from each fraction was acid-precipitated after incubation at 37°C without enzyme. A second aliquot of each fraction was made 0.5 M in NaCl (high-salt conditions). The third aliquot remained in low-salt buffer. The latter two sets of samples were incubated 30 minutes (37°C) with RNases A and T$_1$ (10 μg/ml and 40 units/ml, respectively) and the amount of acid-insoluble radioactivity present at the end of the digestion was determined for each fraction. The results, presented in Fig. 6, demonstrate that the RNA in the peak is substantially resistant to digestion at high ionic strength, but is converted to acid-soluble radioactivity in the presence of RNases A and T$_1$ at low ionic strength, indicating that this RNA is double-stranded.

D.   Analysis of hnRNP Proteins

1. MOLECULAR WEIGHT

hnRNP particles are concentrated by ultrafiltration or ethanol precipitation (Section II,A,2,b) and dissolved in 10 mM sodium phosphate (pH 7.0)–1% SDS–1% 2-mercaptoethanol. After dialysis for 18 hours (20°C) against buffer containing one-tenth the concentration of detergent and mercaptoethanol, the protein concentration is determined by the method of Lowry et al. (38). Proteins are analyzed by polyacrylamide gel electrophoresis in SDS as detailed by Bhorjee and Pederson (39). Cylindrical gels (6 × 75 mm) are prepared with the following final composition: 7.5% acrylamide–0.28% N,N'-methylenebisacrylamide–0.1% SDS–0.1 M sodium phosphate (pH

7.0)–0.5 M urea–5 mM Na$_2$EDTA–0.05% N,N,N',N'-tetramethylenedia-
mine (TEMED)–0.1% ammonium persulfate. Gels are overlaid with 0.5 ml
of 0.1% SDS–0.005% TEMED–0.1% ammonium persulfate and polymerized for 1 hour (20°C). The overlay solution is decanted, and a 20-mm-
long spacer gel of the same composition, but containing 2.5% acrylamide and 0.01 M phosphate (pH 6.0), is cast on top of the 7.5% gel. The
spacer gel is overlaid with water and allowed to polymerize overnight. The
stock solution of acrylamide and bisacrylamide [30%:1.12% (w/v), respec-
tively] is deionized before use (Section III,C,3,a). Samples containing 30–45
μg of protein in a volume of 25–100 μl are made 0.4 M in sucrose and 0.1%
in Bromophenol blue before loading on the gels. The electrophoresis buffer
is 0.1 M sodium phosphate (pH 7.0)–0.1% SDS–5 mM Na$_2$EDTA. Electrophoresis is for 6.5–7.0 hours at 7.5 mA/gel. Parallel gels contain molecular
weight calibration standards (18). Gels are stained with 0.05% Coomassie
brilliant blue and destained as described by Fairbanks et al. (40). Molecular
weights of hnRNP polypeptides are estimated from a plot of the mobility
of the standards versus log molecular weight. A typical profile of Coomassie-
blue-stained polypeptides derived from HeLa hnRNP particles is shown in
Fig. 7A. For comparative purposes, the profile of mouse L-cell hnRNP
proteins is also shown (Fig. 7B). Although the range of polypeptide molecular weights is approximately the same in both cases (40,000 to 180,000),
there are obvious differences between the two profiles. Neither histones
(9,000–30,000 MW) nor ribosomal proteins (10,000–50,000 MW) are present
(9,19).

2. Two-Dimensional Analysis by Charge and Molecular Weight

[^{35}S]methionine-labeled hnRNP particles are concentrated either by
high-speed centrifugation (Section II,A,2,b) or by dialysis at 4°C over a
period of 36–48 hours against 4 × 400 volume changes of 1 mM ammonium
acetate (pH 7.0) followed by lyophilization. The hnRNP sample is dissolved
in 1 mM Tris-HCl (pH 7.5)–10 M urea–0.1% SDS, and 0.1 volume of 0.3 M
lysine (pH 4.2)–25 mM ZnSO$_4$–0.1% SDS is added. Removal of all single-
stranded nucleic acid from the hnRNP preparation is required for two-

FIG. 7.   Electrophoresis of hnRNP proteins isolated from HeLa and mouse L-cells. Nuclei
were isolated from exponentially growing HeLa or mouse L-cells using isotonic buffer [10
mM Tris-HCl (pH 7.0)–0.15 M NaCl–1.5 mM MgCl$_2$], and hnRNP particles were purified on
15–30% linear sucrose gradients in the same buffer as detailed in the text. Appropriate gradient
fractions were pooled, and the particles were pelleted by high-speed centrifugation (Section
II,A,2,b). The hnRNP pellet was prepared for electrophoresis (Section III,D,1), and samples
containing approximately 45 μg of protein were electrophoresed in 7.5% polyacrylamide gels
containing 0.1% SDS. Direction of electrophoresis is from top to bottom. (A) hnRNP proteins
from HeLa cells; (B) hnRNP proteins from mouse L-cells.

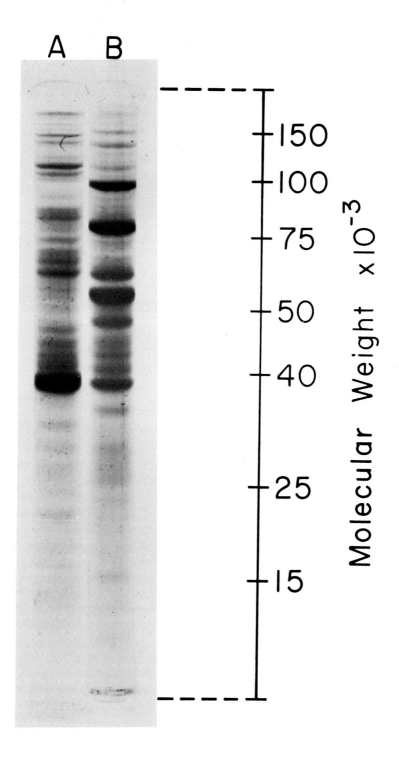

dimensional analysis of hnRNP proteins. S_1 nuclease (100,000 units/ml) is added to a final concentration of 9,000 units/ml, and incubation is carried out at 45°C for 5 minutes. The reaction mixture is then adjusted to pH 7–8 using 1 volume of 1 M Tris-HCl (pH 7.5)–10 M urea–0.1% SDS–5% 2-mercaptoethanol–4% (w/v) NP-40–4% ampholines (comprised of 3.2% pH range 5–7 and 0.8% pH range 3–10). The sample can be stored in this form at −20°C.

Isoelectric focusing gels (2.5 × 130 mm) are prepared using ampholines in the range of pH 3–10 as described by O'Farrell (41); after prerunning for 2 hours, the sample is layered on the gel, overlayered with 10 μl of a solution

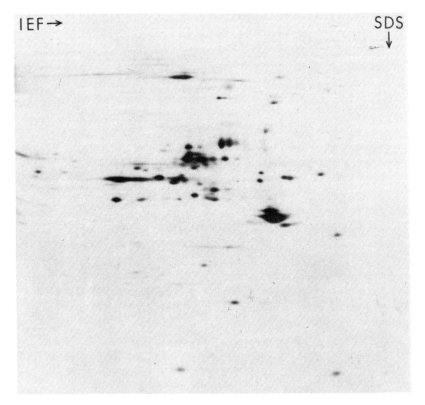

FIG. 8. Two-dimensional analysis of HeLa hnRNP proteins. [^{35}S]methionine-labeled hnRNP particles were isolated from HeLa cells, and the proteins were prepared for isoelectric focusing as described in the text. Proteins were analyzed on an isoelectric focusing gel prepared with an ampholyte pH range of 3–10, and the gels were electrophoresed 11.5 hours at 400 V and 1 hr at 800 V. Electrophoresis in the second dimension was carried out in a 5–22.5% polyacrylamide slab gel containing SDS. Electrophoresis was at 12.5 mA/gel for six hours. The gel was fixed in trichloroacetic acid and prepared for autoradiography as described in the text.

containing 9.0 M urea–1% ampholines (0.8% pH 5–7 + 0.2% pH 3–10), and the tube and gel chambers are filled with 0.02 N NaOH. Electrophoresis is at 400 V for 11.5 hours. Polyacrylamide slab gels are prepared for the second dimension using a linear gradient of 5–22.5% polyacrylamide (running gel) and a 4.75% polyacrylamide stacking gel, as detailed by O'Farrell (41). The isoelectric focusing gel is placed on top of the slab polyacrylamide gel—1% agarose is used to hold the gel in place; then the upper and lower chambers are filled with 25 mM Tris-HCl–0.192 M glycine–0.1% SDS. Electrophoresis is at 12.5 mA/gel for approximately 6 hours (22°C). The slab gel is fixed in 50% trichloroacetic acid for 20 minutes and then transferred to 7% glacial acetic acid for 16–18 hours. After drying down on filter paper under vacuum (42), radioactive proteins are visualized by contact autoradiography using Kodak NS2T Medical x-ray film. Sensitivity can be greatly increased by impregnating the slab gel with 2,5-diphenyloxazole (PPO) scintillant using dimethylsulfoxide as a vehicle, followed by fluorography (43,44). The complexity of the HeLa hnRNP proteins analyzed by this two-dimensional technique is shown in Fig. 8.

3. PHOSPHORYLATION OF hnRNP PROTEINS

HeLa hnRNP proteins are phosphorylated *in vivo* and have phosphorylation levels considerably higher than either histone or nonhistone chromosomal proteins (T. Pederson, unpublished results). Cells are labeled for one generation (19 hours for HeLa) with ^{32}P (1 μCi/ml) in medium containing one-tenth the standard phosphate concentration (Section II,A,1), and hnRNP proteins are analyzed by SDS–polyacrylamide gel electrophoresis (Section III,D,1). After staining and destaining, gels are stirred for 30 minutes in 600 ml of 5% tricholoroacetic acid at 90°C in order to hydrolyze any traces of nucleic acid in the gel (45, 46). The gels are then washed for 3 hours each in two successive 1000-ml changes of 5% trichloroacetic acid at 20°C to remove nucleic acid hydrolysis products. Gels are then stirred for 16 hours (20°C) in 1000 ml of 10% acetic acid. This hardens the gel matrix somewhat, thereby facilitating longitudinal slicing for autoradiography. Gels are sliced longitudinally for autoradiography or transversely for Cerenkov or liquid scintillation counting. Longitudinally sliced gels are dried down on filter paper (42) and applied to Kodak SB54 x-ray film for autoradiography (–20°C). Figure 9 illustrates the results of such an analysis of ^{32}P-labeled HeLa hnRNP proteins.

E. Isolation of Poly(A)-Rich Ribonucleoprotein Complexes

An understanding of the specificity of RNA-protein interactions in native hnRNP particles is impeded by the nucleotide sequence complexity of

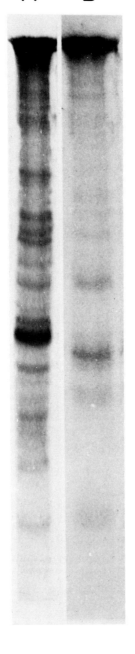

hnRNA and by the analytical complexity of the proteins. Despite this, the specific association of protein with poly (A)-rich hnRNA sequences has been demonstrated (18). This ribonucleoprotein complex is isolated on the basis of (1) its resistance to hydrolysis by ribonucleases A and T_1 in 0.5 M NaCl and (2) its affinity for poly (U)-Sepharose. The hnRNP particles are dialyzed against 10 mM Tris-HCl (pH 7.5)–10 mM Na$_2$EDTA–0.5 M NaCl (TEN) and concentrated approximately 10-fold by ultrafiltration. Particles are then digested for 75 minutes at 37°C with 10 μg of RNase A and 40 units of RNase T_1 per ml. The digest is made 0.5% in Sarkosyl, incubated for 30 minutes at 4°C, and loaded on a column of poly (U)-Sepharose prepared as follows. Polyuridylic acid is linked to cyanogen bromide-activated Sepharose 4B using the procedure described by Firtel and Lodish (47), modified to include a pH 9 wash to remove any nonspecifically adsorbed poly (U) (48). The final capacity of the settled resin is 150–165 μg poly (U)/ml. The poly (U)-Sepharose is suspended in TEN buffer containing 0.5% Sarkosyl (TENS buffer) and packed in a 0.9-cm column to a height of 10 cm (4°C). Approximately 25 bed volumes of TENS buffer are eluted through the column prior to sample application. After the sample has entered the column, approximately 20 ml of the TENS buffer are eluted through the column to remove all unbound RNA and protein. The bound ribonucleoprotein fraction can be recovered from the poly (U)-Sepharose by elution with 10 mM Tris-HCl (pH 7.5)–10 mM Na$_2$EDTA–50% formamide–0.5% Sarkosyl. [Formamide is purified by the method of Robberson et al. (49)]. Fractions representing the bound RNP are pooled, and the sample is dialyzed at 4°C for 36–48 hours against three to four 400-volume changes of 1 mM ammonium acetate (pH 7.0) prior to lyophilization. The lyophilized powder is then prepared for analysis of RNA or protein as described in earlier sections. The ribonucleoprotein complex contains poly(A) and oligo(A) sequences, as well as two polypeptides, with molecular weights of 74,000 and 86,000 (18).

FIG. 9. *In vivo* phosphorylation of HeLa hnRNP proteins. 6 × 10⁸ HeLa cells were resuspended in 2 liters of prewarmed medium containing one-fifth the usual phosphate concentration and grown for 10 hours in the presence of ³²P (phosphoric acid, carrier-free) at 2.5 μCi/ml. The hnRNP particles were purified, dissociated in SDS, and electrophoresed as described in the text and Fig. 6. After staining and destaining, the gels were heated in 5% trichloroacetic acid to hydrolyze nucleic acid and prepared for autoradiography (see text for details). After 12 days of exposure to Kodak SB54 x-ray film (−20°C), the films were developed and photographed with transmitted white light. (A) Coomassie-blue-stained gel after longitudinal slicing; (B) autoradiogram of gel slice shown in (A).

IV. Sources of Chemicals

[³H]adenosine, [³⁵S]methionine, [³²P]phosphoric acid, [³H]uridine and Protosol: New England Nuclear, Boston, Massachusetts. Actinomycin D, sucrose (ribonuclease-free), urea (Ultra-Pure grade), [¹⁴C]arginine, [¹⁴C]leucine, [³H]leucine, [¹⁴C]lysine, [³H]orotic acid, [¹⁴C]polyadenylate: Schwarz/Mann, Orangeburg, New York. Acrylamide, N,N'-methylenebisacrylamide, 2-mercaptoethanol, glutaraldehyde, and iodoacetamide: Eastman Organic Chemicals, Rochester, New York. Pancreatic deoxyribonuclease I and ribonuclease A: Worthington Biochemical Corporation, Freehold, New Jersey. Ribonuclease T₁: Sigma Chemical Company, St. Louis, Missouri. Polyuridylate: Miles Laboratories, Elkhart, Indiana. CNBr-activated Sepharose 4B: Pharmacia Fine Chemicals, Piscataway, New Jersey. Sarkosyl NL97: Geigy Chemical Corporation, Ardseley, New York. Formamide and sodium dodecyl sulfate: Matheson, Coleman and Bell, East Rutherford, New Jersey. Ion exchange resin AG501-X8 (D): Bio-Rad Laboratories, Richmond, California. Oligo(rA)$_{18}$, oligo(dT)-cellulose (Type 3): Collaborative Research, Waltham, Massachusetts. Cesium sulfate: Kawecki Berylco Industries, Inc., Revere, Pennsylvania. NP-40: Shell Chemical Company. Phenol (crystalline): Mallinkrodt Chem. Co., St. Louis, Missouri. Ampholines [supplied as 40% (w/v) solutions: LKB Instruments Inc., Rockville, Maryland. All other chemicals used are of the highest purity available.

ACKNOWLEDGMENTS

This work was supported by NIH research grant GM21595. T.P. is the recipient of a Scholar Award from the Leukemia Society of America. We gratefully acknowledge the advice of Dr. Ronald Luftig on electron microscopy and Dr. Edwin H. McConkey for details of the two-dimensional electrophoresis system. We also thank Dr. James P. Calvet for his critical review of the manuscript.

REFERENCES

1. Gall, J. G., J. Biochem. Biophys. Cytol. 2 (Suppl.), 393 (1956).
2. Callan, H. G., and Lloyd, L., Philos. Trans. R. Soc. London, Ser. B. 243, 135 (1960).
3. Gall, J. G., and Callan, H. G., Proc. Natl. Acad. Sci. U.S.A. 48, 562 (1962).
4. Stevens, B. J., and Swift, H., J. Cell Biol. 31, 55 (1966).
5. Samarina, O. P., Lukanidin, E. M., Molnar, J., and Georgiev, G. P., J. Mol. Biol. 33, 251 (1968).
6. Monneron, A., and Bernhard, W., J. Ultrastruct. Res. 27, 266 (1969).
7. Niessing, J., and Sekeris, C. E., Biochim. Biophys. Acta 247, 391 (1971).
8. Miller, O. L., and Bakken, A., Trans. Karolinska Symp. Res. Methods. Reprod. Endocrinol. 5th, 1972. Gene Transcription Reprod. Tissue p. 155 (1973).
9. Pederson, T., J. Mol. Biol. 83, 163 (1974).
10. Pederson, T., Proc. Natl. Acad. Sci. U.S.A. 71, 617 (1974).

11. Firtel, R. A., and Pederson, T., *Proc. Natl. Acad. Sci. U.S.A.* **72**, 301 (1975).
12. Eagle, H., *Science* **130**, 432 (1959).
13. Barile, M. F., and DelJudice, R., *Pathogenic Mycoplasmas Ciba Found. Symp., 1972.*
14. Earle, W. R., *J. Natl. Cancer Inst.* **4**, 165 (1943).
15. Maggio, R., Siekevitz, P., and Palade, G. E., *J. Cell. Biol.* **18**, 293 (1963).
16. Muramatsu, M., Smetana, K., and Busch, H., *Cancer Res.* **23**, 510 (1963).
17. Schildkraut, C. L., and Maio, J. J., *Biochim. Biophys. Acta* **161**, 76 (1968).
18. Kish, V. M., and Pederson, T., *J. Mol. Biol.* **95**, 227 (1975).
19. Kumar, A., and Pederson, T., *J. Mol. Biol.* **96**, 353 (1975).
20. Wilt, F. H., Anderson, M., and Ekenberg, E., *Biochemistry* **12**, 959 (1973).
21. Ishikawa, K., Kuroda, C., and Ogata, K., *Biochim. Biophys. Acta* **179**, 316 (1969).
22. Jacobson, A., *in* "Methods in Molecular Biology" (J. Last, ed.), Vol. 8, p. 161. Dekker, New York, 1976.
23. Penman, S., *J. Mol. Biol.* **17**, 117 (1966).
23a. Samarina, O. P., Asrijan, I. S., and Georgiev, G. P., *Dokl. Acad. Nauk SSSR* **163**, 1510 (1965).
24. Georgiev, G. P., and Samarina, O. P., *Adv. Cell Biol.* **2**, 47 (1971).
25. Köhler, K., and Arends, S., *Eur. J. Biochem.* **5**, 500 (1968).
26. Lukanidin, E. M., Zalmanzon, E. S., Konaromi, L., Samarina, O. P., and Georgiev, G. P., *Nature (London) New Biol.* **238**, 193 (1972).
27. Quinlan, T. J., Billings, P. B., and Martin, T. E., *Proc. Natl. Acad. Sci. U.S.A.* **71**, 2632 (1974).
28. Hsu, M.-T., Kung, H.-J., and Davidson, N., *Cold Spring Harbor Symp. Quant. Biol.* **38**, 943 (1973).
29. Greenberg, J. R., *Biochemistry* **15**, 3516 (1976).
30. Monckton, R. P., and Naora, H., *Biochim. Biophys. Acta* **335**, 139 (1974).
31. Edmonds, M., Vaughan, M. H., Jr., and Nakazato, H., *Proc. Natl. Acad. Sci. U.S.A.* **68**, 1336 (1971).
32. Darnell, J. E., Philipson, L., Wall, R., and Adesnik, M., *Science* **174**, 507 (1971).
33. Nakazato, H., Kopp, D. W., and Edmonds, M., *J. Biol. Chem.* **248**, 1472 (1973).
34. Nakazato, H., Edmonds, M., and Kopp, D. W., *Proc. Natl. Acad. Sci. U.S.A.* **71**, 200 (1974).
35. Beers, R. F., *J. Biol. Chem.* **235**, 2393 (1960).
36. Duesberg, P. H., and Vogt, P. K., *J. Virol.* **12**, 594 (1973).
37. Zimmerman, S. B., and Sandeen, G., *Anal. Biochem.* **14**, 269 (1966).
38. Lowry, O. H., Rosebrough, N. J., Farr, A. L., and Randall, R. J., *J. Biol. Chem.* **193**, 265 (1951).
39. Bhorjee, J. S., and Pederson, T., *Biochemistry* **12**, 2766 (1973).
40. Fairbanks, G., Steck, T. L., and Wallach, D. F. H., *Biochemistry* **10**, 2606 (1971).
41. O'Farrell, P., *J. Biol. Chem.* **250**, 4007 (1975).
42. Fairbanks, G., Levinthal, C., and Reeder, R. H., *Biochem. Biophys. Res. Commun.* **20**, 393 (1965).
43. Bonner, W. M., and Lasky, R. A., *Eur. J. Biochem.* **46**, 83 (1974).
44. Lasky, R. A., and Mills, A. D., *Eur. J. Biochem.* **56**, 335 (1975).
45. Auerbach, S., and Pederson, T., *Biochem. Biophys. Res. Commun.* **63**, 149 (1975).
46. Bhorjee, J. S., and Pederson, T., *Anal. Biochem.* **71**, 393 (1976).
47. Firtel, R. A., and Lodish, H. F., *J. Mol. Biol.* **79**, 295 (1973).
48. Robberson, D. L., and Davidson, N., *Biochemistry* **11**, 533 (1972).
49. Robberson, D., Aloni, Y., Attardi, G., and Davidson, N., *J. Mol. Biol.* **60**, 473 (1971).

Chapter 27

Chromosomal Association of an Epstein–Barr Virus-Associated Nuclear Antigen

INGEMAR ERNBERG

Department of Tumor Biology,
Karolinska Institutet,
Stockholm, Sweden

I. Introduction

The Epstein–Barr viral nuclear antigen (EBNA) was first detected by anticomplementary immunofluorescence (ACIF) in 1973 (*1–3*). Epstein–Barr virus (EBV) is a human DNA virus first isolated in 1964 (*4*), and it is a member of the Herpes-type virus group. Subsequently it has been shown to be the dominant etiologic agent of infectious mononucleosis, a benign lymphoproliferative disease in the human (*5–8*). It is now also suspected of having oncogenic potential in the human. In both African Burkitt lymphoma· and nasopharyngeal carcinoma there is a strong correlation between tumor presence and high titers of antibodies to EBV-associated antigens and the presence of viral genetic material in the tumor cells (*9–13*). Furthermore, EB virus has been shown to cause malignant lymphoma upon injection into marmoset monkeys (*14–16*). The *in vitro* infection of human or monkey lymphoid cells leads to establishment of continuously growing cell lines with lymphoid characteristics (*17–19*). This type of virus infection is now termed transformation or immortalization of lymphoid cells, since the cells have a limited lifespan in culture normally (*20*). There is a lack of a fully permissive system for EB virus infection, thus excluding the use of many routine virological assays, such as a plaque assay. Because of this, the characterization of EBV thus far has been quite dependent on the detection of virus-associated antigens by immunofluorescent methods. Several antigens associated with EBV have been detected by immunofluorescence: the early antigens (EA) (*21,22*), the viral capsid antigen (VCA) (*23*) the membrane antigens (MA) (*24,25*) and EBNA.

The structural virus antigens MA and VCA, together with EA, are expressed only in producer cell cultures and are not detectable in tumors. However EBNA is expressed in all EBV-DNA containing cells both in tumors and in established cell lines derived from biopsies or as a result of *in vitro* infection (*11,12*). Since EBNA is therefore ubiquitous, it is the subject of much investigation and interest.

Little is known about the structure of EBNA, and its functions are purely speculative. It is tempting to relate EBNA to the nuclear antigens of other DNA tumor viruses, notably the T antigens of the papova and adenoviruses [for review, see Tooze (*26*)]. Both EBNA and the SV40 and polyoma T antigens seem to be DNA binding, at least in part (*27,28*). One difference seems to be that EBNA is clearly localized on metaphase chromosomes, while during the same phase of the cell cycle, the T antigens are found in the cytoplasm (*1,29*). Being the only detectable virus-associated intracellular gene products regularly present in virus-transformed cells, the interest is focused on whether these antigens play an important role in transformation or maintenance of the transformed cell state. No conclusive data are as yet available, however, to substantiate these hypotheses. One important point, which has not been settled for any of the systems, is whether these nuclear antigens are coded for by the viral genome itself, or if they are host genome products induced by the virus.

MA, EA, and VCA were detected originally by direct and indirect immunofluorescence, while EBNA was not detected until the more sensitive ACIF was applied. However an antigen associated with EBV nonproducer cell lines has been detected by complement fixation (CF antigen) (*30*) and by immunodiffusion (S antigen) (*31*). Indirect evidence suggests that these antigens are the same, at least in part (*32*).

ACIF is a three layer immunofluorescence technique, which provides an increased sensitivity compared to the direct or indirect immunofluorescence techniques. ACIF is also, however, more difficult to use than the other two techniques, due to more problems with reagents, complement receptor binding, and increased background fluorescence (*2,3,33*).

II.   Method for Detection of EBNA

The principle of the anticomplementary immunofluorescence is shown in Fig. 1 and is compared with direct immunofluorescence. The higher sensitivity of ACIF is due to the increased number of fluorescent marker molecules binding per antigen molecule (*2,3*). Thus, it allows for detection of

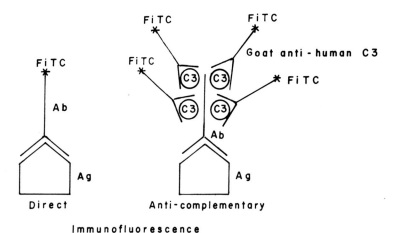

Fig. 1. Schematic representation of direct immunofluorescence and anticomplementary immunofluorescence (ACIF). FiTC, Fluorescein-isothiocyanate; Ab, antibody; Ag, antigen C3, Complement factor 3.

fewer antigen molecules per surface area than direct immunofluorescence. We have estimated the number of anti-EBNA antibodies binding to each EBNA positive cell, to be 10^3–10^4, while we find approximately 10^6 anti-EA antibodies binding to an EA positive cell. Only the latter antigen, EA, may regularly be detected by direct immunofluorescence.

A. Preparation of Cell Smear

Monolayer cells may be grown on coverslips, which can then be fixed and stained for EBNA. The lymphoblastoid cell lines that harbor EBV are cell lines growing in suspension, and they should be processed in the following way. Harvested cells are washed gently by spinning down twice in balanced salt solution (BSS) at 3000–4000 rpm in a small table centrifuge. The cell pellet is resuspended in a hypotonic solution: 0.8 mM MgCl$_2$, 1.0 mM CaCl$_2$, 30 mM glycerol in distilled water, and just before use, four drops of fetal calf serum is added per milliliter of this hypotonic solution (approximately 20%). This mild hypotonic treatment causes swelling of the cytoplasm and thereby reveals the nucleus better. The cells may be resuspended in isotonic BSS, but in the lymphoblastoid cell lines the nucleus may hardly be distinguished from the cytoplasm under these conditions. One drop of BSS or the hypotonic solution is added for each 10^6 cells in the tube; 2.5 × 10^5 cells are needed for the preparation of each cell patch ("smear"). One drop of the cell suspension is placed on an ordinary glass slide with a Pasteur

pipette and the excess cell suspension is immediately sucked back into the
pipette. On the glass a moist patch of cells will be left, but not a full drop
This procedure provides an even layer of cells, which dries quickly at room
temperature.

B. Fixation

After air drying, the smear is fixed by dipping into methanol: acetone (1:2)
at $-20\,°C$ for 2 to 5 minutes. The fixation will denature proteins and extract
hydrophobic molecules. Holes are thereby made in the cell wall, so that
the inside of the cell becomes accessible to reagents. It is not a good preservation method for the antigens, however. The cell smear can be stored for
only a few days at $4\,°C$ before staining, or EBNA reactivity disappears. It
is recommended that the staining be done as soon as possible after preparation of the smear.

C. Staining with Fluorescein

Human serum with antibodies to EBNA (titer $\geq 1:80$) is diluted $1:10$
with BSS. One drop of this serum is added to each cell smear to be stained,
thus covering the patch of cells. The cell smear is incubated for 30 minutes
at room temperature in a "moist chamber." The moist chamber consists
of a box whose bottom is covered by a moist sponge, and it should be sealed
with a tight cover. This is to prevent evaporation of the added reagent during
incubation. After incubation, the glass slide is dipped subsequently into
three beakers containing fresh BSS and the excess BSS is wiped off the glass
slide with a tissue. The BSS should be wiped off also on the "up" side of the
slide around the cell patch, but the cell patch should be left moist. Human
complement (C', i.e., a human serum with preserved complement activity,
but in this case with no antibodies to EBNA) is added to the cell patch after
dilution with BSS (1:10). Incubation is again performed in the moist chamber for 30 minutes at room temperature. After incubation, the smear is
washed in fresh BSS with a magnetic stirrer for 5 minutes with two changes
of BBS. Again excess BSS is wiped off as described above and FITC-labeled
goat antihuman-β_{I_c}(FITC; fluoresceinisothiocyanate, Hyland Laboratories,
Los Angeles) diluted 1:40 is added. Incubation is for 45 minutes at room
temperature in the moist chamber. Thereafter the glass slide is washed by
dipping into three beakers of BSS; it is left in the last beaker for 5 minutes.
Finally the slide is dipped quickly into a beaker with distilled water (to
remove salt).

D. Counterstaining

The slide is put into a solution of Evans blue (100 mg/l in distilled water), (*3,34*) for 5 to 10 minutes at room temperature, subsequently washed by dipping into two beakers of BSS, and air-dried. A small drop of BSS: glycerol (1:1) is added on top of the cell patch, and it is covered by a coverslip. The smear may be sotred at +4°C up to a month before reading in the microscope.

E. Microscopy

For details of fluorescence microscopy, see Goldman (*3*). The stained preparations are studied in a fluorescence microscope with a filter setting for fluorescein emission, at magnification 600–1000 ×. Microscopes with inverted illumination or epi-illumination may be used and are equally suitable.

III.   Visualization of EBNA Metaphase Chromosomes

If metaphase chromosomes are present in a cell smear, they are brilliantly and relatively equally stained by the method described above. The number of cells in metaphase may, of course, be increased by application of a standard method. An actively growing lymphoblastoid cell line may be treated for 4–12 hours with 40–60 ng/ml of colcemid (Gibco Bio-Cult, Paisley, Scotland). The harvesting and staining procedures are similar to those described above.

IV.   Discussion on Methods

A. Fixation

Acetone alone, methanol alone, methanol: ether (1:1), methanol: hexylene glycol: water (6:1:3), butanol, butanol: ether (1:1), and carbon tetrachloride have been used successfully as fixatives (*35,36*). Fixation times may be short, except for the methanol: hexylene glycol: water system which requires more than a 2-hour fixation. Fixation is performed at −20°, except for carbon tetrachloride, which gives the best results when used at room temperature (*36*). It is important to note that smears cannot be stored for long periods after fixation and before staining, but that the smears may be well preserved after staining.

B.   Sera

Sera from patients with nasopharyngeal carcinoma or Burkitt's lymphoma are usually used in the first step of the staining procedure, since they are a source of high titer antibodies against EBV-associated antigens (8,37). However, 90% of the adult population in Western countries have antibodies to EBNA, and if such sera are screened for antibody titer to EBNA, it should be possible to find a usable positive serum. The sera should be stored in small batches at $-20°C$. The complement of the positive reference serum of the laboratory should be inactivated by heating at 56°C for 30 minutes. All of these sera will have antibodies to EBV antigens other than EBNA. They usually have antibodies to VCA and MA, and some of the patient sera also have antibodies to EA (8,37). Staining of MA is not a problem under the fixation conditions recommended here. Staining of EA and VCA will be even more brilliant than the EBNA staining and is mostly localized to nucleus and cytoplasm, while EBNA is seen only in the nucleus. Early after synthesis EA is localized only to the nucleus (38). Presence of EA and VCA positive cells is only a problem in virus-producing cell lines and after P3HR-I EB virus superinfection or infection of cells (13,39). The frequency of such cells can be screened by direct or indirect immunofluorescence, which will not stain the EBNA-positive cells. Other types of antinuclear antibodies have to be excluded by staining of any EBNA negative human cells.

C.   Serum Titrations

As described above, the EBNA staining method is feasible for detecting EBNA in a tumor biopsy or a cell line. With a known EBNA-positive cell line it is possible to screen different sera for antibodies to EBNA and to titrate the positive sera by two-step or ten-step dilutions in BSS.

D.   Controls

To exclude nonspecific staining, a negative serum should be included as a control instead of the EBNA-positive serum. As mentioned, it is vital to exclude antinuclear antibodies other than those directed to EBNA. As a control of the anti-β_{Ic} reagent the complement in the second step can be inactivated at 56°C for 30 min before incubation. As a positive control for the complement and anti-β_{Ic} step, other known antigen–antibody reactions may be introduced, e.g., antinuclear antibodies to human nuclei. This is not necessary as a routine, when the staining of EBNA has been set up with established positive controls.

E. Complement

A human serum with no antibodies to EBNA is used as a source of complement. After drawing blood from the serum complement donor, the serum should be collected as soon as possible after clotting at 4°C, and it should be transferred into small vials in batches feasible for one staining operation (0.1–0.2 ml in each batch). This serum should then be frozen and stored at −70°C. Freezing and thawing repeatedly destroys the complement activity. The batch to be used is thawed and diluted with BSS just before addition to the cell smear, and it is kept in ice during use. Half an hour at room temperature might abolish the complement reactivity. Liabeuf *et al.* (*40*) have studied the complement reaction during the EBNA ACIF staining procedure. They found that the complement was activated through the classical pathway, and that the anti-β_{Ic} reagent reacts with activated C3. It is essential that the staining is performed in three steps, as described here, to avoid complement consumption by antigen–antibody complexes in the serum during the first incubation step (*41*).

F. The Labeled Anti-β_{Ic} Reagent

An excellent fluorescein-labeled goat antihuman-β_{Ic} reagent is obtained from Hyland laboratories (California). For special purposes commercial antihuman-β_{Ic} may be reacted with rhodamine (tetrametylrhodamin; TRITC, Becton, Dickonson & Co., Cockeysville, Maryland) after immunoglobulin purification (*42*), thus providing a reddish fluorescent reagent.

This one cannot be combined with the reddish Evans blue counterstain. The anti-β_{Ic} reagent may be stored prediluted (1:40 dilution with BSS is recommended for the Hyland reagent) at −20°C. Repeated freezing and thawing does not influence this reagent significantly.

G. Counterstain

This step is not necessary, but generally the reddish staining of the fluorescence-negative cells due to the Evans blue simplifies reading of the slides in the microscope. There is a need for counterstaining in the event that few EBNA-positive cells are expected in a preparation. It should be noted that the counterstain is by no means specific. It does mask the green background of the fluorescein staining, but at higher concentration or upon prolonged staining, it may even mask weak specific fluorescence staining. The background fluorescence will generally cause more problems with the ACIF than with the direct or indirect immunofluorescence. This background may be of both specific and nonspecific origin. The specific non-EBNA reactivity is due to binding of complement to cell surface receptors for complement on lymphoid lines of B cell origin (33). Nonspecific background reactions are due to antibody–cellular protein nonspecific bindings [see Goldman (3)].

V.   EBNA Is a Chromatin-Associated Protein

Apart from the visualization of EBNA by ACIF on metaphase chromosomes (1), Bahr et al. (43) have shown by an immunoelectronmicroscopic technique that EBNA is bound to interphase chromatin, at least in part. T. D. K. Brown (personal communication) could detect EBNA on partially purified chromatin with ^{125}I-labeled antibodies. Baron et al. (44) report a 100-fold purification to EBNA utilizing adsorption chromatography with polycytidine cellulose as a vital step. They use a sensitive complement fixation assay to detect EBNA with ^{51}Cr-labeled sheep red blood cells for detecting lysis. A similar assay system was used by Luka et al. (27), when they recently showed that EBNA binds to both single-stranded and double-stranded calf thymus DNA. The binding was stronger to double-stranded DNA. All evidence so far indicates that EBNA may be found in chromatin and that it is bound to cellular DNA.

In the studies cited above, it cannot a priori be assumed that the different types of assays used detect the same antigen, even though the same or similar sera were used for detection with ACIF, detection by iodinated antibodies, and detection with complement fixation. Evidence on a serological basis has been presented by Klein and Vonka (32) that EBNA detected by ACIF and by complement fixation are the same. This is also supported by the work of Lenoir et al. (45). Brown et al (46) showed that the Raji cell line contains one EBNA subcomponent that is lacking in another EBNA positive, EBV-carrying cell line, Rael (47). Both cell lines also have one common component. The biochemical analyses performed so far by Lenoir et al. (45), Luka et al. (27), and Baron et al (45) demonstrate one relatively homogenous antigen fraction in the Raji cell line. Luka et al. and Baron et al., however, use DNA-binding as their crucial purification step, and they might lose EBNA subcomponents which do not bind to DNA directly or bind with low affinity in vitro. The molecular weight of EBNA was approximately 180,000 in Lenoir's study and 300,000 in Baron's study. Fresen et al. (48) report the presence of a transiently induced nuclear antigen (TINA) after iododeoxyuridine treatment of an Epstein–Barr virus-converted, EBNA-containing cell line. This antigen could be distinguished from EBNA on the basis of morphological and serological evidence. It was only present in a fraction of the EBNA positive cells.

Ernberg *et al.* (*49*) have shown that EBNA is preferentially increasing during G1 in the cell cycle. This was shown by immunofluorometry and by positioning the cells in different phases of the cell cycle with measurement of DNA content (Feulgen) [^{3}H] thymidine pulse and dry mass measurement on single cells. If this reflects the synthesis pattern of EBNA, it is in line with the synthesis pattern of nonhistone proteins rather than histones, which preferentially are made during the S phase (*50,51*).

The number of EB viral genome copies per cell varies between 1 and 200 in different EBV-carrying cell lines. Using immunofluorometry it was found that the amount of EBNA per cell and the number of virus genome equivalents per cell were positively correlated (*52*). Whether this is due to a virus gene-dose effect or to the requirement of a certain number of EBNA molecules for each virus genome present is not known.

EBNA is always present where there is EBV infection or immortalization of cells. In most instances it is the only virus-associated product to be detected. There is accumulating evidence that it is a nonhistone DNA-assoicated protein. It is possible that EBNA exerts some overriding control in the cell, thus initiating or maintaining the immortalized status of the infected cell. Alternatively, EBNA may be involved in controlling and repressing late EB viral genes of the usually multiple EB viral copies present in the cell. When such late genes are expressed in the infected cell, virus production and cell death proceed. Thus the EB viral genes must be repressed in the immortalized, growing cell. The latter hypothesis is not compatible, however, with the presence of EBNA on all metaphase chromosomes in the infected cell.

ACKNOWLEDGMENTS

This work was supported by grants from the Swedish Cancer Society, King Gustav V Jubilee Fund, and by Contract NO1 CP 333 16 within the virus ċancer program of the National Cancer Institute. The author is a recipient of a fellowship in Cancer Immunology from the Cancer Research Institute in New York.

REFERENCES

1. Reedman, B. M., and Klein, G., *Int. J. Cancer* **13**, 755 (1973).
2. Hinuma, H., Ohta, R., Miyamoto, T., and Ishida, N., *J. Immunol.* **89**, 19 (1962).
3. Goldman, M., *in* "Fluorescent Antibody Methods," pp. 61, 160, 173. Academic Press, New York, 1968.
4. Epstein, M. A., Achong, B. H., and Barr, Y. M., *Lancet* **1**, 702 (1964).
5. Henle, G., Henle, W., and Diehl, V., *Proc. Natl. Acad. Sci. U.S.A.* **59**, 904 (1968).

6. Niederman, J. C., McCollum, R. W., Henle, G., and Henle, W., *J. Am. Med. Assoc.* **203**, 205 (1968).
7. Evans, A. S., Niederman, J. C., and McCollum. R. W., *New Engl. J. Med.* **279**, 1121 (1968).
8. Henle, W., Ho, H.-C., Henle, G., and Kwan, H. C., *J. Natl. Cancer Inst.* **51**, 361 (1973).
9. Zur Hausen, H., and Schulte-Holthausen, H., *Nature (London)* **227**, 245 (1970).
10. Nonoyama, M., and Pagano, J. S., *Nature (London)* New Biol. **223**, 103 (1971).
11. Lindahl, T., Klein, G., Reedman, B. M., Johansson, B., and Singh, S., *Int. J. Cancer* **13**, 764 (1974).
12. Klein, G., *Cold Spring Harbor Symp. Quant. Biol.* **39**, 783 (1974).
13. Klein, G., Giovanella, B. C., Lindahl, T., Fialkow, P. J., Singh, S., and Stehlin, J. S., *Proc. Natl. Acad. Sci. U.S.A.* **71**, 4737 (1974).
14. Shope, T., Dechairo, D., and Miller, G., *Proc. Natl. Acad. Sci. U.S.A.* **70**, 2487 (1973).
15. Miller, G., *J. Infect. Dis.* **130**, 187 (1974).
16. Werner, J., Wolf, H., Apodaca, D., and zur Hausen, H., *Int. J. Cancer* **15**, 1000 (1975).
17. Nilsson, K., Klein, G., Henle, W., and Henle, G., *Int. J. Cancer* **8**, 443 (1971).
18. Miller, G., Lisco, H., Kohn, H., and Stitt, D., *Proc. Soc. Exp. Biol. Med.* **137**, 1459 (1971).
19. Falk, L., Wolfe, L., Deinhardt, F., Paciga, J., Dombos, L., Klein, G., Henle, W., and Henle, G., *Int. J. Cancer* **13**, 363 (1974).
20. Miller, G., Robinson, J., Heston, L., and Lipman, M., *in* "Oncogenesis and Herpesviruses II" (de Thé, G., Epstein, M. A., and H. zur Hausen, eds.). IARC Tech. Publ. 11, p. 395. Lyon, 1975.
21. Henle, W., Henle, G., Zajac, B. A., Pearson, G., Waubke, R., and Scriba, M., *Science* **169**, 188 (1970).
22. Henle, G., Henele, W., and Klein, G., *Int. J. Cancer* **8**, 272 (1971).
23. Henle, W., Hummeler, K., and Henle, G., *J. Bacteriol.* **92**, 269 (1966).
24. Klein, G., Clifford, P., Klein, E., and Stjernswärd, J., *Proc. Natl. Acad. Sci. U.S.A.* **55**, 1628 (1966).
25. Ernberg, I., Klein, G., Kourilsky, F. M., and Silvestre, D., *J. Natl. Cancer Inst.* **53**, 61 (1974).
26. Tooze, J., "Molecular Biology of Tumor Viruses," Cold Spring Harbor Lab, Cold Spring Harbor, New York, 1973.
27. Luka, J., Siegert, W., and Klein, G., *J. Virol.* **22**, 1 (1977).
28. Carrol, R. B., Hager, L., and Dulbecco, R., *Proc. Natl. Acad. Sci. U.S.A.* **71**, 3754 (1974).
29. Stenman, S., Zeuthen, J., and Ringertz, N. R., *Int. J. Cancer* **15**, 547 (1975).
30. Pope, J. H., Horne, M. K., and Wetters, E. J., *Nature (London)* **222**, 186 (1969).
31. Old, L. J., Boyse, E. A., Oettgen, H. F., de Narven, E., Geering, G., Williamson, B., and Clifford, P., *Proc. Natl. Acad. Sci. U.S.A.* **56**, 1699 (1966).
32. Klein, G., and Vonka, V., *J. Natl. Cancer Inst.* **53**, 1645 (1974).
33. Theophilopoulos, A. N., Bokish, V. A., and Dixon, F. J., *J. Exp. Med.* **139**, 969 (1974).
34. Nichols, R. L., and McComb, D. E., *J. Immunol.* **89**, 545 (1962).
35. Suzuki, M., Hinuma, Y., *Int. J. Cancer* **14**, 753 (1974).
36. Ohno, S., Wiener, F., and Klein, G., *Biomedicine*, in press.
37. Henle, G., Henle, W., Clifford, P., Diehl, V., Kafuko, G. W., Kirya, B. G., Klein, G., Morrow, R. H., Manube, G. M. R., Pike, P., Tukei, P. M., and Ziegler, J. L., *J. Natl. Cancer Inst.* **43**, 1147 (1969).
38. Ernberg, I., Masucci, G., and Klein, G., *Int. J. Cancer* **17**, 197 (1976).
39. Gergely, L., Klein, G., and Ernberg, I., *Virology* **45**, 10 (1971).
40. Liabeuf, A., Nelson, R. A., Jr., and Kourilsky, F. M., *Int. J. Cancer* **15**, 533 (1975).
41. Henle, W., Duerra, A., and Henle, G., *Int. J. Cancer* **13**, 751 (1974).
42. Goldstein, G., Klein, G., and Pearson, G., *Cancer Res.* **29**, 749 (1969).
43. Bahr, G., Mihel, U., and Klein, G., *Beitr. Pathol.* **155**, 72 (1975).

44. Baron, D., Benz, W. C., Carmichael, G., Yocum, R. R., and Strominger, J. L., *in* "Epstein-Barr Virus Production, Concentration and Purification," IARC Technical Rep., 75,003, pp. 257–262. Lyon, 1975.
45. Lenoir, G., Berthelon, M.-C., Favre, M.-C., and de Thé, G., *J. Virol.* **17**, 672 (1976).
46. Brown, T. D. K., Ernberg, I., and Klein, G., *Int. J. Cancer* **17**, 197 (1975).
47. Klein, G., Dombos, L., and Gothoskar, B., *Int. J. Cancer* **10**, 44 (1972).
48. Fresen, K.-O., and zur Hausen, H., *Proc. Natl. Acad. Sci. U.S.A.* **74**, 363 (1977).
49. Ernberg, I., Killander, D., and Lundin, L., to be published.
50. Elgin, S. C. R., and Weintraub, H., *Annu. Rev. Biochem.* **44**, 725 (1975).
51. Mitchison, J. M., in "The Biology of the Cell Cyle", p. 153. Cambridge Univ. Press, London, 1971.
52. Ernberg, I., Andersson-Anvert, M., Klein, G. K., Hander, D., and Lundin, L., *Nature (London)* **266**, 269 (1977).

SUBJECT INDEX

CONTENTS OF PREVIOUS VOLUMES

Volume I

Volume IV

Volume V

Volume VI

Volume VII

Volume XIII

Volume XIV

Volume XV

A
B
C 8
D 9
E 0
F 1
G 2
H 3
I 4
J 5